INTRODUCTION TO
MATHEMATICAL BIOLOGY

INTRODUCTION TO MATHEMATICAL BIOLOGY

S. I. RUBINOW

> **Graduate School of Medical Sciences**
> **Cornell University**

A WILEY-INTERSCIENCE PUBLICATION

JOHN WILEY & SONS, New York . London . Sydney . Toronto

Library of Congress Cataloging in Publication Data:

Rubinow, S   I
    Introduction to mathematical biology.

    Bibliography: p.
    Includes index.
        1. Biomathematics. I. Title. [DNLM: 1. Bi-
ology. 2. Biometry. 3. Mathematics. QH323.5
R896i]

QH323.5.R8   574'.01'51   75-12520
ISBN 0-471-74446-8

TO MY FATHER, R. R.

Natural science must search for the laws of phenomena

H. von Helmholtz

The basic strategy of science in the analysis of phenomena is the ferreting out of invariants

J. Monod

When you can measure what you are speaking about, and can express it in numbers, you know something about it; but when you cannot measure it, when you cannot express it in numbers, your knowledge is of a meagre and unsatisfactory kind

W. Thomson, Lord Kelvin

PREFACE

This book is based on the course in mathematical biology that I have given for many years at the Graduate School of Medical Sciences of Cornell University. The course was designed to develop in the graduate biology student an awareness of and familiarity with some mathematical techniques and methods which have been applied to biology. The need for such a course has arisen because biological information has become more quantitative, and the quantitative approach to biology has concomitantly grown in importance. The mathematics required for the text is a course in elementary calculus.

Biology deals with phenomena that are intrinsically more complex and more difficult to investigate than those normally studied in the "hard" sciences such as physics. Consequently, it is more susceptible to the introduction of hypotheses whose correctness cannot be adequately tested. Furthermore, mathematical models of biological systems, with their inherent propensity for simplification and idealization, are sometimes so speculative that they have only the merest pretense of being accountable to biological observations. In the past, this lack of accountability on the part of such speculative models has been a disservice to "mathematical biology," and has caused biologists to view this subject with a healthy skepticism. Here I use the term "mathematical biology" in a restricted sense that excludes biostatistics, the latter discipline having its own accepted and distinct role in the history of biology.

Nevertheless, there is a body of mathematical ideas and models that is so closely intertwined with biological observations, it has a more or less established place in the literature of modern biology. These diverse mathematical contributions to biology compose what may perhaps be called "classical" mathematical biology. It is upon the literature of these contributions that this book is largely based.

Often, these mathematical contributions have been presented to the biology student in a fragmented fashion. Consequently students of biology learn to use mathematical formulas and techniques whose theoretical justification is hazy in the minds of many of them. Furthermore, nowhere is there to be found a presentation of this body of knowledge in a single volume. This book aims at remedying, at least partially, these defects.

I have made no attempt to be encyclopedic. Rather, the plan of the work has been to select five areas of mathematical biology and present them in a unified fashion, the unifying thread being the underlying mathematical structure. The classification scheme of the text is mixed: The first three subjects, cell growth, enzymatic reactions, and physiological tracers, are biological. The last two subjects, biological fluid dynamics and diffusion, are biophysical. The sequence in which these subjects are presented is so designed that the mathematical subject matter appears in increasing order of difficulty. Mathematically, the subject matter essentially follows a course in elementary differential equations, although linear algebra and graph theory are also touched upon. However the book is not designed to be and is very far from being a textbook in mathematics.

I trust I do not have to apologize for all the mathematical topics in biology which have been excluded. It seemed more sensible from the very beginning to proceed in depth in a few areas, rather than to treat a broader set of topics, at the cost of being more superficial or mathematically repetitive. Some topics were excluded because they did not satisfy the constraint on selection imposed by the underlying mathematical structure. And inevitably, my own particular interests (those of a mathematical physicist who came late to biology) and those of the Graduate School of Medical Sciences of Cornell University have exerted their influence in shaping this book. In any case, it is well to keep in mind that these notes are meant to be an introductory textbook only.

Every subject has its own particular jargon or vocabulary consisting of words having a technical meaning known only to the specialists in that subject. Biology and mathematics are no exceptions to this rule. It is a common experience that a lack of understanding of these special words is the greatest obstacle to reading and understanding the technical literature in a new subject. Since these notes are designed to remove the aura of mystery that surrounds the mathematical approach to biological data, I have made an effort to alert the reader to the usage of this mathematical jargon. This is accomplished, first, by bold facing all words that have a special technical meaning, and second, by providing some explanatory comment as to usage or meaning.

It is my hope that this book can also be useful to mathematically sophisticated students such as those with backgrounds in mathematics, physics, and engineering, who have little knowledge of biology but are nevertheless interested in the quantitative approach to it. Therefore, I have taken pains to treat the technical jargon of biology in the same manner that I have treated the technical jargon of mathematics. The resulting explanatory biological comments will, of course, be superfluous to the student of biology, who I hope will be as tolerant of this practice as the mathematically sophisticated student will be of the mathematical comments.

The sections marked with an asterisk are more mathematically difficult to follow than the others. It is worthwhile to read through Section 2.2*, which is concerned with the early time behavior of enzymatic reactions, even if the mathematical discussion, which is in any case sketchy, is slighted. However, Sections 3.11* and 5.12* can be conveniently skipped, as I have often done in classroom practice, without any loss of continuity. They contain a more advanced mathematical discussion of material that has been previously presented.

It has been my teaching experience that students find some refresher lectures on calculus useful at the beginning of the course. I have also given supplementary lectures on the elements of linear algebra, as needed. These lectures form the substance of the mathematical appendices.

Problems that are more difficult to solve than others are also marked with an asterisk. These problems are intended for mathematically sophisticated students only. In general, the problems tend to be difficult to solve. Partially, this is an unavoidable consequence of my having made many of the problems relate to biological research investigations reported in the literature. However, the problems are generously provided with hints, and I urge the reader to try them, even if he finds he gets "stuck" at some point and cannot proceed further. If he will then, and only then, consult the solutions provided at the end of the book, he will surely have learned something from the attempt, as the experience of my students seems to bear out.

I wish to take this opportunity to thank my colleagues and students who, wittingly or not, have aided my biological education. In particular, my colleagues M. Earl Balis, Rudy H. Haschemeyer and Merry M. Sherman deserve special mention for their assistance while this book was in preparation. I am deeply grateful to my good friend and colleague Joseph B. Keller who read the entire manuscript and made numerous useful suggestions. Finally, I am grateful to my wife Shirley for her help and encouragement.

S. I. RUBINOW

New York, New York
December 1974

CONTENTS

CONTENTS

1
CELL GROWTH

The cell is the fundamental unit of structure and function in all living things, except viruses. Cells are intensively investigated in the field of microbiology, which is the study of **microorganisms** or "**microbes**." Microbes are defined as living organisms too small to be seen by the naked eye, or less than 0.1 mm in diameter. Because of their small size, microorganisms are usually studied as aggregates in a **culture**, a colony of microorganisms grown in a medium in a laboratory.

The understanding of cell function has been greatly facilitated through quantitative study of the growth of cells. By the growth of a colony of unicellular microorganisms is usually meant (and we shall mean here) changes in their number, rather than the change in size of the individual organisms. A common method for cells to reproduce and increase their number is **binary fission**, in which a cell divides into two cells. The cell number in a culture can be determined in a variety of ways, such as the determination of the cell count per unit volume, the cell mass, or the biochemical activity of the cells.

1.1. EXPONENTIAL GROWTH OR DECAY

The essential role of a useful mathematical theory in the sciences is to predict the future. That is to say, it attempts to answer the question: Given the state of a system at some initial time, what is the state of the system at some future time? The answer is often provided as the solution to a differential equation. Let us see how such an equation might arise.

Consider the growth of a population of N organisms. We shall think of the population, for definiteness, as representing the number of bacteria in a colony, although it could equally well represent the number of people in the United States. In either case, we can assume that the population is so large that the change of a few members is "infinitesimal." Therefore, we shall consider the number N to be a continuous variable that changes with time.

1

We ask, during a small time interval Δt subsequent to t, on what does the change in N depend? It is reasonable to suppose that this change ΔN is proportional to N as well as to Δt, or

$$\Delta N \propto N \, \Delta t. \tag{1.1}$$

The proportionality of the change ΔN to N is an example of the **law of mass action**. Many reactions and processes in biology, chemistry, physics, and so on, obey such a law, at least over some range of variation. According to this law, if the population N were doubled, the change ΔN would also be doubled. The relationship (1.1) may be written as an equality by introducing the proportionality constant k:

$$\Delta N = kN \, \Delta t. \tag{1.2}$$

If we divide both sides of (1.2) by Δt and take the limit as $\Delta t \to 0$, we obtain the derivative. Thus,

$$\frac{dN}{dt} = kN. \tag{1.3}$$

This is a relation between the quantity N and its derivative, which is assumed to hold true at any time. Such an equation is called a **differential equation**. We would like to know $N(t)$, that is to say, to find an explicit function that satisfies this relation. When we have found it, we have **solved** the differential equation. The function $N(t)$ is called the **solution** or **integral** of the differential equation.

How do we solve differential equations? There is no single answer to this question. Perhaps the most fundamental method of solution is to **guess the answer**. In doing this, we attempt to take advantage of our familiarity with known functions and their derivatives. In the previous case, our familiarity with the exponential function might lead us to attempt as a solution

$$N = ce^{kt}, \tag{1.4}$$

where c is a constant. We have only to substitute this function into our equation to **verify** that it satisfies the equation, so that it is indeed a solution of our equation.

We observe that the solution contains an arbitrary constant c. The meaning of this constant becomes clear if we specify the **initial value** N_0 of our population at time $t = 0$. By direct substitution, we see that

$$N_0 = ce^{k \cdot 0} = c, \tag{1.5}$$

so that the constant represents the initial value of the population. The solution of our equation may therefore be written as

$$N = N_0 e^{kt}. \tag{1.6}$$

It can be proved that no other function satisfies the differential equation and the initial condition $N(0) = N_0$. Therefore the solution is said to be **unique**. The solution tells us the population number N at any time.

What is the significance of the parameter k? We shall first discuss its **dimensions**. An equation such as (1.3) must in the first instance be dimensionally correct to be at all meaningful. If we let the bracket [ ] denote "the dimensions of" the quantity contained within them, we see from the preceding discussion that

$$\left[\frac{dN}{dt}\right] = \left[\frac{\Delta N}{\Delta t}\right] = \left[\frac{N}{t}\right]. \tag{1.7}$$

From equation (1.3), the dimensions of dN/dt must be the same as $[kN]$, and therefore $[k] = [1/t]$, or, k has the dimension of reciprocal time. We can also recognize this fact from equation (1.6), because the argument of the exponential function or exponent of e must be a pure number, whence $[kt] = [1]$. From (1.3), we see that the **fractional growth rate** at any time ($N^{-1}\, dN/dt$) is a constant, and this constant is k. It is also called the **specific growth rate**.

. Suppose now that k is positive. We ask, how long does it take the population to double itself? In the case of bacterial populations, this interval is called the **doubling time**. If there are no cell deaths, the doubling time is the same as the **mean generation time**, which is the mean lifetime of a single cell. Call this doubling time T. Then at $t = T$, $N = 2N_0$. Substituting this relationship into equation (1.6) leads to $2N_0 = N_0 e^{kT}$, or

$$k = \frac{\log 2}{T} = \frac{0.6931}{T}. \tag{1.8}$$

Thus, k is inversely related to the doubling time. Equation (1.6) has also been called the **law of Malthus**, since it was Malthus who called attention to the dangers of geometric growth characterized by positive k.

An important physical difference occurs when the parameter k is negative. For a population, k then represents a rate of decline, or **death rate**. Similarly, if N represents the mass of a chemical species, and k is negative, then k represents a **dissappearance rate**. More generally, it is called a **decay rate**. The time it takes for the species to decay to $1/2$ of its initial value is called the **half-life** and is denoted by $T_{1/2}$.

The use of radioactive substances as tracers has become widespread in biology and medicine. Because the radioactivity of atoms is easily measured, the incorporation of radioactive atoms into living organisms is readily traced. Although the decay of a particular radioactive atom is a statistical event subject to the laws of probability, a macroscopic amount of radioactive material made up of a very large number of atoms behaves in a quite predictable way. Thus, if N is the amount of radioactive material, its decay in time is governed by (1.3) with k negative.

Most of the neutrons produced in the atmosphere by cosmic rays are eventually absorbed by nuclei of nitrogen by the reaction

$$^{14}N + n \rightarrow {}^{14}C + {}^{1}H.$$

The ^{14}C atoms, which are radioactive, are formed high in the atmosphere (10–15 km altitude) and rapidly become oxidized to $^{14}CO_2$. This then mixes with natural CO_2 and participates in the same physical and biochemical processes. In particular, a steady state of exchange is set up between the atmosphere and organisms so that a certain proportion of the carbon in living plants and animals is in this radioactive form. This radioactivity amounts to 5.5 dis/min/g of carbon. When a plant or animal dies, the exchange ceases and the radioactive carbon remains embedded in the organism to decay according to its natural rate. For ^{14}C, the half-life is 5730 years. These facts form the essential basis of **radiocarbon dating**. The basic mathematical theory of it is expressed by equation (1.6) with k negative.

1.2. DETERMINATION OF GROWTH OR DECAY RATES

Cell populations that are described by equation (1.6) are said to be in "**log phase**," a terminology that appears unfortunate, since they are more correctly described as being in **exponential growth phase**, or simply **exponential phase**. The origin of the term log phase arises from the following considerations.

In biology it is the usual case that a growth curve $y(t)$ is experimentally determined. For example, we wish to know whether a bacterial population is in fact in exponential growth, and if so, what the growth constant k is. If we take logarithms of both sides of equation (1.6) (with y replacing N), we see that the relationship between $y(t)$ and t can be described in linear fashion as

$$\log y = \log y_0 + kt. \tag{1.9}$$

In other words, the function $\log y$, when plotted as a function of t, is described by a straight line with slope k and intercept $\log y_0$. Therefore, if we make such a plot and find that, within experimental error, the resulting curve is in fact a straight line, then we know that the population is indeed in exponential growth with a growth rate determined by the slope of the straight line. Because linear relationships are easy to comprehend and to visualize, it is often desirable to represent functional relationships in linear fashion, if possible (see problems 1 and 2).

Another important reason for using a logarithmic scale, that is to say, using $\log y$ instead of y as an ordinate, is that the values of y may be changing by orders of magnitude. Here we are conforming to the common vocabulary usage in the natural sciences for comparing two numbers (rather than two functions). One number is larger than the other by an **order of magnitude** if it is approximately 10 times as large, by two orders of magnitude if it is approximately 100 times as large, and so forth. When this is the case, it is obviously wiser to use a logarithmic scale rather than a linear scale for graphical purposes.

The use of $\log_{10} y$ for the ordinate instead of y is facilitated by the use of **semilog paper**. This is a kind of graph paper in which the abscissa utilizes a linear scale, equal intervals on the abscissa representing equal increments of the value of the abscissa variable. On the other hand, the ordinate utilizes a logarithmic scale; equal intervals on the ordinate scale represent equal intervals of $\log_{10} y$. However, the values of y, the argument of the logarithm function, are assigned to the ordinate scale, rather than the values of the logarithm itself. Therefore, if y is measured, the necessity of taking $\log_{10} y$ in making a plot is obviated. We try to make this clear in Figure 1.1 by assigning ordinate values to both y and $\log_{10} y$.

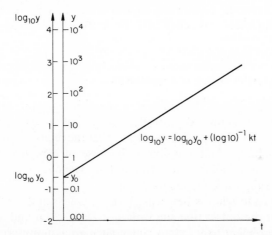

Figure 1.1. A semilog plot of the function $y = y_0 e^{kt'}$, or equivalently, the equation in the figure. Semilog paper utilizes $\log_{10} y$ as the ordinate (left-sided ordinate axis), but labels the ordinate distances with the values of y (right-sided ordinate axis).

Recall that the relationship between natural logarithms and logarithms to the base 10 is given by $\log y = \log_{10} y \cdot \log 10$. If then y represents a quantity in exponential growth as in equation (1.9), we may write

$$\log_{10} y = \log_{10} y_0 + (\log 10)^{-1} kt, \tag{1.10}$$

$$\log 10 = 2.3026, \frac{1}{\log 10} = 0.4343.$$

We have displayed this curve in the figure.

Assume that y has been determined experimentally as a function of the time t and plotted on semilog paper. Let the result be a straight line as shown in Figure 1.1. We wish to determine the values of y_0 and k. According to (1.10), the y intercept of the curve is $\log_{10} y_0$ (on the $\log_{10} y$ scale), and therefore the value of y_0 may be read off the curve directly from the y scale. The slope of the curve is given by

$$k(\log 10)^{-1} = \frac{\Delta\log_{10} y}{\Delta t}. \tag{1.11}$$

Now let t_1 and t_2 be the values of the abscissa at any two points on the curve, and let y_1 and y_2 be the corresponding values of y. Then, to determine k by means of (1.11), we require the logarithm to the base 10 of y_1 and y_2. Thus,

$$k = \log 10 \frac{\log_{10} y_2 - \log_{10} y_1}{t_2 - t_1}. \tag{1.12}$$

In the above example, convenient choices for y_1 and y_2 are 10 and 1, respectively, or any two successive powers of 10, whence $\Delta\log_{10} y = 1$, and $k = \log 10/\Delta t$.

1.3. THE METHOD OF LEAST SQUARES

A biologist who wishes to represent some experimental data in explicit quantified form is confronted with the problem of experimental error. For example, suppose he has reason to suspect that the relationship between two observed variables x and y is linear. Because of experimental error, the observations will not all fall on a straight line. The question therefore arises as to which straight line fits the data "best."

A criterion as to what is best was given by Gauss and is widely utilized. We assume at the outset that the values of x are given and ascribe all the error in measurement to y. (Many times, such an assumption is completely arbitrary, and in a second examination, we could equally well invert the roles of x and y.) Gauss' criterion of the "best" line is the one which minimizes the **error**, which is defined as the sum of the squares of the deviations of the observed values of y from the line. Let the equation of the line be $y = mx + b$, and let the observed value of y be denoted by y_i^* at the point x_i. The index i runs from 1 to n if there are n measurements altogether. Then the error E is given by the expression (see Figure 1.2)

$$E = \sum_{i=1}^{n} [y(x_i) - y_i^*]^2, \tag{1.13}$$

where $y(x_i)$ is the value of y at the point x_i. The error defined in this manner is seen to depend only on the absolute magnitude of the deviation of a given

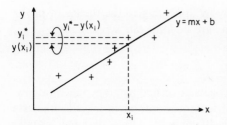

Figure 1.2. A set of experimental observations, indicated by the points marked +, to which the linear relationship $y = mx + b$ has been theoretically assigned. The squared error at the point x_i between the experimentally observed value y_i^* and the theoretical value $y(x_i)$ is $(y_i^* - y(x_i))^2$.

y_i^* from $y(x_i)$, and not on whether the point y_i^* falls above or below the line. Substituting $y(x_i) = mx_i + b$, the above equation becomes

$$E = \sum_{i=1}^{n} (mx_i + b - y_i^*)^2. \tag{1.14}$$

The quantities x_i and y_i^* are given numbers so that E is a function that depends on m and b, whose values we wish to choose in such a way that E is minimized. In order words, E is a function to be minimized with respect to the two variables m and b. The necessary conditions for this are that the partial derivatives of $E = E(m,b)$ with respect to m and b are set equal to zero, or

$$\frac{\partial E}{\partial m} = 2[m \sum (x_i)^2 + b \sum x_i - \sum y_i^* x_i] = 0,$$

$$\frac{\partial E}{\partial b} = 2[m \sum x_i + b \sum 1 - \sum y_i^*] = 0. \tag{1.15}$$

In these equations and the ones following, the summation sign $\sum$ is to be understood as extending from $i = 1$ to $i = n$.

Equations (1.15) are two linear inhomogeneous equations for the quantities m and b. Their solution is

$$m = \frac{n \sum y_i^* x_i - \sum y_i^* \sum x_i}{n \sum (x_i)^2 - (\sum x_i)^2},$$

$$b = \frac{\sum y_i^* \sum (x_i)^2 - \sum y_i^* x_i \sum x_i}{n \sum (x_i)^2 - (\sum x_i)^2} = n^{-1}(\sum y_i^* - m \sum x_i). \tag{1.16}$$

These values of m and b do minimize E because $E \to \infty$ as either $|m|$ or $|b| \to \infty$. The above illustrated procedure of fitting observations to an assumed functional form is called the **method of least squares**. The procedure does not depend on how many observations n have been made, as long as n is greater than the number of unknown parameters.

Such questions as, how "good" is the fit, what is the variance of this set of data as compared to another set, and so forth, are more properly statistical questions which can be answered only by appealing to the theory of probability. The answers to such questions are beyond our scope here.

1.4. NUTRIENT UPTAKE BY A CELL

As another example of a biological process governed by an exponential law, we shall consider the transport of a nutrient into some bacterial cells. **Bacteria** are microorganisms, very small single living cells, with a diameter usually not exceeding 3 μm. Reproduction is usually accomplished asexually by fission. Bacteria are termed **procaryotic** cells because they have a primitive type of nucleus that lacks a membrane. **Escherichiae** consititute a type of bacteria that has the shape of a short rod and is frequently motile. **Escherichia coli**, or **E. coli** for short, is a normal inhabitant of the intestinal tract of man and other animals. In its normal environment, it is nonpathogenic. *Escherichia coli* has been more thoroughly investigated than any other bacteria. Its length is about 2 μm and its mass is about 5×10^{-13} g. Many varieties or **strains** of *E. coli* exist.

Consider the mutant bacterial strain *E. coli* ML 32,400, which is **galactose-negative**, that is, incapable of growth on galactose. **Galactose** is one of the **simple sugars**, widely distributed in nature and over 100 in number, that are a source of carbon and energy for cells. Some other such abundant simple sugars that are used as energy sources for growing microorganisms are **glucose**, **sucrose**, **maltose**, **arabinose**, **lactose**, and **xylose**. The parent strain *E. coli* ML 30 does utilize galactose as a source of carbon. The cause of the inability to do this in the mutant strain is known to be the lack of an enzyme, **galactokinase**, which is necessary to metabolize galactose. Resting cells of this mutant strain will, however, take up large amounts of galactose when it is present in the cell medium.

The uptake of galactose by the strain can be readily observed by utilizing galactose that is labeled with ^{14}C in the growth medium (Horecker, Thomas, and Monod, 1960). Assume that the galactose concentration in the medium is constant, and let c denote the radioactive galactose concentration inside the cells. Initially, because the strain is galactose-negative, the concentration c is zero. We assume also that the rate of entry of galactose into the *E. coli* at any moment depends only on the difference between the concentration in the interior c and a final concentration $\bar{c}$, which is found to be much larger than the external concentration in the medium. Such an assumption no doubt represents a simplified view of the transport mechanism, although it leads to an accurate representation of observations. Figure 3.15 illustrates a similar transport process, the uptake of radioactively labeled thymine by a thymine-negative strain of *E. coli*. In that figure, the three curves represent the uptake of thymine in three different compounds within the cell. It follows from the assumptions that $c = c(t)$ satisfies the equation

$$\frac{dc}{dt} = k(\bar{c} - c), \tag{1.17}$$

where k is a positive constant and $c(0) = 0$. This is a slightly more complicated than our previous one. Before we solve it, we ask first what the **equilibrium** value of c is, or, when does $dc/dt = 0$? Clearly this occurs when influx $k\bar{c}$ equals outflux kc, or $c = \bar{c}$. The form of equation (1.17) suggests that we should investigate the quantity $(c - \bar{c})$. It is readily seen that equation (1.17) can be rewritten as

$$\frac{d(c - \bar{c})}{dt} = -k(c - \bar{c}).$$

This is an equation whose solution we already know to be

$$(c - \bar{c}) = (c - \bar{c})_0 e^{-kt} = -\bar{c}e^{-kt}. \tag{1.18}$$

The subscript 0 on the right-hand side means that we are to evaluate the quantity in parentheses at $t = 0$. The solution for c may therefore be written, by transposing $\bar{c}$ in equation (1.18), as

$$c = \bar{c}[1 - e^{-kt}], \tag{1.19}$$

which expresses the fact that c approaches its equilibrium value asymptotically. The curve based on equation (1.19) is illustrated in Figure 1.3a. For purposes of qualitative comparison, the observed intracellular galactose concentration is shown in Figure 1.3b. In the quoted experiment, some of the cells which had achieved the maximum or saturation value of intracellular concentration $\bar{c}$ were introduced suddenly at the time $t_0 = 21$ min (indicated by an arrow in Figure 1.3b) to a nonradioactive ^{12}C-galactose medium. The intracellular radioactive galactose then disappeared as shown by the decaying branch of the curve in Figure 1.3b. When representing such circumstances on the basis of the simple theory presented herein, equation (1.17) no longer contains the influx term $k\bar{c}$ and its solution becomes $c = \bar{c} \exp[-k(t - t_0)]$ for $t \geq t_0$. We see from Figure 1.3b that the decaying branch does have the qualitative features of such a decaying exponential.

How would we obtain the rate constant k from data represented by the curve shown in Figure 1.3a? If we plotted c versus t on semilog paper, we would find that the resulting curve is not a straight line. However, for large times, we would see that the curve is a horizontal straight line because then e^{-kt} is negligible, and $\log c \approx \log \bar{c}$. In other words, the asymptote for the semilog plot is $\log \bar{c}$. Knowing $\bar{c}$, we can then calculate $\bar{c} - c$ and make a semilog plot of the latter quantity versus time. Such a plot permits us to determine k in the usual way.

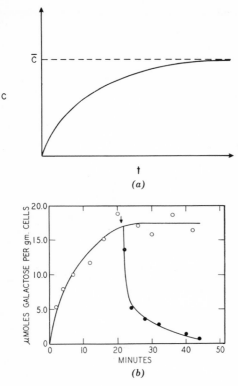

(a)

(b)

Figure 1.3. (*a*) Theoretical dependence of the concentration c as a function of the time t, according to equation (1.19). $\bar{c}$ is the stationary value of c that is approached asymptotically as $t \to \infty$. (*b*) The observed uptake of radioactive ^{14}C-galactose by the galactose-negative strain *E. coli* ML 32,400, following exposure to a radioactive ^{14}C-galactose medium at $t = 0$. The branch of the curve which decays to zero commencing at the time indicated by the arrow represents the ^{14}C-galactose content of some cells that were suddenly reexposed to a nonradioactive ^{12}C-galactose medium. From Horecker, Thomas, and Monod (1960), with permission of the authors and the American Society of Biological Chemists.

1.5. INHOMOGENEOUS DIFFERENTIAL EQUATIONS

In a formal way we have just solved the differential equation

$$\frac{dy}{dt} + ay = b, \tag{1.20}$$

where a and b are constants. Such an equation is called an **inhomogeneous** differential equation because of the presence in it of the inhomogeneous term b, which does not depend on y or its derivative. If b were equal to zero, the differential equation would be called **homogeneous**. What is the solution to this equation if a and b are functions of t? The answer is that it equals the general solution of the homogeneous equation (obtained by setting b equal to zero) plus any particular solution of the inhomogeneous equation.

To understand this last remark, we introduce some terminology from the theory of differential equations. A differential equation is said to be **ordinary** when it contains a single independent variable, as in equations (1.3) and (1.20). The **order** of a differential equation is the order of the highest derivative that occurs in it. For example, equation (1.20) is called a first order ordinary differential equation. The solution to an nth order ordinary differential equation contains n arbitrary constants, and is called the **general solution**. If particular values to the arbitrary constants are assigned, the resulting solution is called a **particular solution**.

In the solution (1.19) of equation (1.17), $c = \bar{c}$ is a particular solution, and de^{-kt} with d constant is the general solution to the homogeneous equation. We chose d equal to $\bar{c}$ in order to satisfy the **initial condition**. When a and b are time-dependent, the solution to equation (1.20), subject to the initial condition that $y(0) = y_0$, is

$$y(t) = \exp\left[-\int_0^t a(\xi)d\xi\right]\left\{y_0 + \int_0^t b(\tau)\exp\left[\int_0^\tau a(\xi)d\xi\right]d\tau\right\}. \qquad (1.21)$$

The solution is seen to consist of two terms. The first term derives from the general solution of the homogeneous equation, and the second term is a particular solution of the inhomogeneous equation.

1.6.   GROWTH OF A MICROBIAL COLONY

We know that a colony of bacteria or other microorganisms contained in a nutrient medium will not grow indefinitely. There are many possible reasons for this, to wit, lack of space, lack of oxygen, disappearance of nutrients, the appearance of toxic substances, or changes in ion concentration in the medium, especially pH. We attempt to represent microbial growth in a more realistic way by assuming that the population $N(t)$ at any time t is represented by the differential equation

$$\frac{dN}{dt} = kN - \beta N^2, \qquad k > 0, \beta > 0. \qquad (1.22)$$

The last term on the right-hand side of (1.22) is always negative, and we can see intuitively that N can never grow indefinitely. As it attempts to do so, dN/dt must ultimately become negative because N^2 is of a higher order than N.

We emphasize here that the term kN represents the net growth rate, or the excess of the birth rate over the death rate. If the limiting factor in growth is the appearance of toxic substances, say, we can provide a plausibility argument for its representation by the term $-\beta N^2$ as follows. A given cell detects the cumulative toxic effect of all N cells, if the toxic material diffuses freely throughout the intercellular medium. Thus, the toxic effect on a given cell is proportional to N. The toxic effect on all N cells is N times the effect on one cell and hence proportional to $N \cdot N$. In other words, the number of intercellular interactions of N cells is of the order of N^2. In any case, the representation of the growth process of the colony of microorganisms by equation (1.22) is a pure assumption whose validity is tested by the comparison between the deductions of the theory and experience.

Before solving (1.22) we obtain valuable information about the solution by inquiring about the **stationary state** when $dN/dt = 0$. Then

$$kN - \beta N^2 = 0.$$

This has two solutions, $N = 0$, and $N = N_e \equiv k/\beta$. These two solutions are called the **stationary states** or **equilibrium states** of the system. We can characterize the equilibrium states as being **stable** or **unstable**, by examining the behavior of these states in **response to a small perturbation**. That is to say, we shall consider the behavior of these states when N suffers a small displacement away from its equilibrium value. This method of analysis is called the **method of small perturbations**. The mechanical origin of this idea is illustrated by the consideration of a tiny ball balanced on a cylindrical shell in either of the following two equilibrium positions (see Figure 1.4). In case

Figure 1.4. Illustrating unstable (A) and stable (B) positions of a tiny ball balanced on a cylindrical shell. Stability means the ability of the ball to return to its equilibrium position in response to small perturbations.

A

B

A, the ball is unstable because a small displacement from equilibrium grows larger and larger. In case B, the ball is stable because the ball will return to its equilibrium position after a small displacement.

For equation (1.22), we saw that there were two equilibrium states, and their stability properties must be investigated separately. First, we examine the neighborhood of the equilibrium value $N = 0$. To do so, we expand the function $f(N)$, the right-hand side of (1.22), in a Taylor series about the value $N = 0$, and neglect all terms higher than the linear term. Thus,

$$f(N) = kN - \beta N^2 = f(0) + (N - 0)f'(0) + \cdots = kN + \cdots. \quad (1.23)$$

Because we are only inquiring about the behavior N in a small neighborhood of $N = 0$, we replace $f(N)$ in (1.22) by the expression (1.23). Then (1.22) becomes

$$\frac{dN}{dt} = kN \quad (1.24)$$

in the neighborhood of $N = 0$. The solution is $N = ce^{kt}$, which grows exponentially away from $N = 0$. Because k is positive, the equilibrium value $N = 0$ is said to be unstable.

In the neighborhood of $N = N_e$,

$$f(N) = f(N_e) + (N - N_e)f'(N_e) + \cdots = -k(N - N_e) + \cdots. \quad (1.25)$$

Then (1.22) becomes

$$\frac{dN}{dt} = -k(N - N_e) \quad (1.26)$$

in the neighborhood of $N = N_e$. The solution to (1.26) is already known to us to be

$$N = N_e + ce^{-kt}, \quad (1.27)$$

which approaches N_e as t increases. Because the coefficient of t in (1.27) is negative, the equilibrium value N_e is said to be stable. Thus, we see that stability is determined by the sign of the coefficient of t in the exponential term representing the displacement of N from its equilibrium value.

Let us return to the solution of equation (1.22). We shall solve it by the method of **separation of variables**. In this method, we try to write the equation so that a function of N appears on one side of the equation, and a function of t on the other side. If this can be done, the method is applicable. Thus, equation (1.22) can be rewritten as

$$\frac{dN}{dt} = kN \frac{(N_e - N)}{N_e}.$$

Treating dN and dt symbolically in an algebraic manner, we rewrite this equation as

$$\frac{N_e dN}{N(N_e - N)} = kdt.$$

We then integrate both sides. On the right, the time varies from $t = 0$ to an arbitrary time t. On the left, N varies from its initial value N_0 to its value N at time t:

$$\int_{N_0}^{N} \frac{N_e dN}{N(N_e - N)} = \int_{0}^{t} kdt. \quad (1.28)$$

Note that the choice of a lower limit in this equation imposes the initial condition and hence disposes of the arbitrary constant in the solution. The integration on the left-hand side of (1.28) is facilitated by rewriting the integrand as a sum of partial fractions, namely,

$$\frac{N_e}{N(N_e - N)} = \frac{1}{N} + \frac{1}{N_e - N}. \quad (1.29)$$

Substituting (1.29) into (1.28) and integrating both sides, we obtain

$$\left[\log N - \log(N_e - N)\right]\big|_{N_0}^{N} = kt\big|_0^t,$$

$$\log \frac{N}{N_e - N} - \log \frac{N_0}{N_e - N_0} = kt,$$

$$\frac{N}{N_e - N} = \frac{N_0}{N_e - N_0} e^{kt},$$

or

$$N = \frac{N_0 N_e}{N_0 + (N_e - N_0)e^{-kt}}. \tag{1.30}$$

This equation is called the **logistic law** of growth. Usually $N_0 < N_e$, in which case N never exceeds N_e and has a characteristic S-shaped appearance as shown in Figure 1.5. However, it is conceivable that $N_0 > N_e$. Then N is never less than N_e and has the appearance of a simple decay to equilibrium.

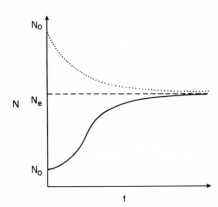

Figure 1.5. Behavior of the population N as a function of the time t obeying the logistic law of growth, equation (1.30). For the solid line, the initial value of N at $t = 0$, N_0, is less than the equilibrium value N_e. For the dotted line, N_0 is greater than N_e.

We note that when $N_e e^{-kt} \gg N_0$, the term $N_e e^{-kt}$ in the denominator is dominant or much larger than the other terms in the denominator, and (1.30) reduces to the Malthusian growth law,

$$N = N_0 e^{kt} + \cdots. \tag{1.31}$$

The dots on the right-hand side signify that we have neglected terms of lower order than the preceding term. After a sufficiently long time, however, the opposite extreme prevails when $N_e e^{-kt} \ll N_0$. Then the term N_0 in the denominator is dominant, and the solution can be expressed approximately as

$$N = \frac{N_e}{1 + ((N_e - N_0)/N_0)e^{-kt}} = N_e \left[1 - \frac{N_e - N_0}{N_0} e^{-kt} + \cdots \right]$$

$$= N_e - \frac{N_e}{N_0}(N_e - N_0)e^{-kt} + \cdots. \tag{1.32}$$

This equation indicates that the approach to the equilibrium value N_e is exponential.

A bacterial colony grown in a nutrient medium or "broth" usually displays a growth curve given by the logistic law, provided the initial inoculum N_0 is taken from an exponentially growing colony. If it is not, there is usually an initial phase of growth called the **lag phase** during which time the growth of the population is retarded. Figure 1.6, taken from the work of Monod (1949), displays the temporal growth of a colony of *E. coli* that was grown in two different media, one containing glucose and one containing

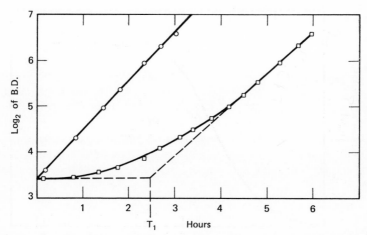

Figure 1.6. The growth of *E. coli* in synthetic media with glucose (circles) and xylose (squares) as organic nutrient source. The abscissa represents the time and the ordinate represents the logarithm to the base 2 of the bacterial density (B. D.), which is the dry weight of the cells per unit volume of the culture. The culture was maintained previously on an arabinose medium at temperature 37°C. The culture grown on xylose medium, unlike the one grown on glucose medium, exhibits a lag phase of growth. The time T_1 in the figure represents the lag time, according to a convenient definition as the point of intersection of two tangent lines, which is illustrated. Reproduced, with permission, from *The Growth of Bacterial Cultures* by J. Monod, Annual Review of Microbiology, volume 3. Copyright © 1949 by Annual Reviews, Inc. All rights reserved.

xylose as an essential organic nutrient source. The cells utilized to initiate the growth of the colonies were taken from a culture maintained on an arabinose medium. It is seen that the cells transferred to the glucose medium continue to grow without lag. The cells transferred to the xylose medium exhibit a lag phase of several hours duration, before a rate of growth is achieved which is the same as that in the glucose medium. This lag phase is believed to be due to biochemical adjustments in the cells, in particular, the formation of a specific enzyme system which must be made in order for the cells to be capable of growing in the presence of the nutrient xylose.

When the population of colony is at or very close to its equilibrium or stationary value N_e, it is said to be in its **stationary phase** of growth. Also, after a long time, the population usually decays to zero from its equilibrium value N_e. Obviously, new growth and death conditions have intervened, such as exhaustion of a necessary nutrient in the medium, say, oxygen, or the accumulation of a toxic metabolic product, which are not represented by equation (1.22).

Figure 1.7, taken from the work of Gause (1934), shows the growth in time

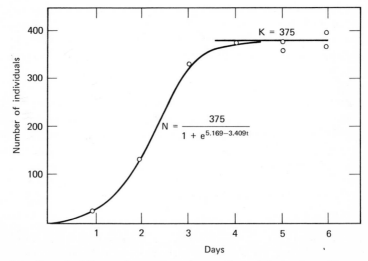

Figure 1.7. Growth of *Paramecium caudatum* in a medium of fixed volume. The number of cells at day 0 is small, but not zero. From Gause (1934), From G. F. Gause, *The Struggle for Existence*, © 1934, with permission of The Williams & Wilkins Co., Baltimore.

of a population of the infusorian **Paramecium caudatum**, and illustrates the logistic growth law. At $t = 0$ cells were placed in individual test tubes containing 0.5 cm³ of a nutrient medium. The average data of 63 subsequent population counts are displayed in the figure. These data points were then fitted by least squares to the logistic law, with parameter values as shown. **Infusoria** constitute a class of protozoans having hairlike protrusions called **cilia**, useful for locomotion and food ingestion. **Protozoa** are single-celled animals, the lowest division of the animal kingdom. They are microorganisms that are more complex and generally larger than bacteria. Their complexity derives from the possession of many organlike structures called **organelles**, such as cilia, gullet, contractile vacuole. Because protozoa possess well defined nuclear membranes, they are classified as **eucaryotic** cells. A typical protozoan volume is of the order of 10^4 μm^3, in contrast to a typical volume of a procaryotic cell which is of the order of 1 μm^3.

1.7. GROWTH IN A CHEMOSTAT

In any investigation of the underlying cellular processes governing growth, it is usually desirable to maintain a bacterial culture under study in a steady state of exponential growth. The previously described method of growth of a microorganism in a container with a fixed amount of nutrient medium is called the **batch culture** technique. Because of the lag phase of growth, and the stationary phase that is eventually achieved, there is really only an intermediate interval between these two phases during which the growth of the colony can be accurately described as being in steady exponential growth. As we have seen from the discussion in the preceding section, even this description is approximate.

An alternate solution to this experimental problem of maintaining cells in a steady state of growth is the introduction of the **continuous culture** technique for growing microorganisms. In this method, a growth chamber of fixed volume V contains a nutrient liquid medium. The chamber has a feeding system whereby the nutrient liquid is supplied at a volume rate per unit time Q, and an overflow provision to maintain the volume of the chamber constant. In addition, the culture is mixed continuously, and aerated continuously, if necessary (see Figure 1.8). Because of the overflow and the mixing, cells are washed out continuously at a rate which is proportional to the total number of cells in the chamber. Under these conditions, the population N in the chamber obeys the equation

$$\frac{dN}{dt} = kN - qN, \qquad (1.33)$$

where the first term on the right-hand side represents the growth rate of the population in the undisturbed culture under the given conditions when the cell density is not too great, and the second term represents the disappearance rate of cells from the culture due to dilution and overflow. Here q is

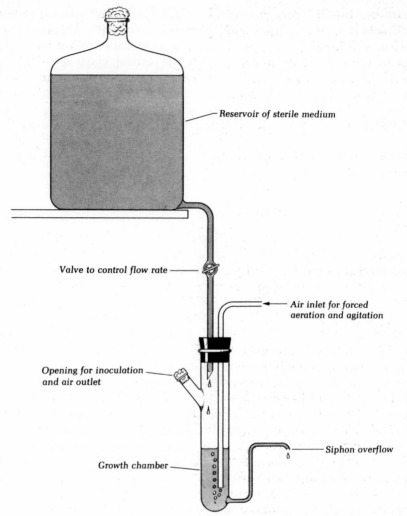

Figure 1.8. Diagram of a chemostat, a device for growing bacteria by continuous culture. From R. Y. Stanier, M. Doudoroff, & E. Adelberg, *The Microbial World*, 3rd edition, © 1970, p. 319. By permission of Prentice-Hall, Inc., Englewood Cliffs, New Jersey.

called the **dilution rate** and is defined as $q \equiv Q/V$. It represents the fractional rate of flow of medium into or out of the growth chamber per unit time.

It can be seen from equation (1.33) that if q is greater than k, the culture is being diluted out too quickly, and the population will decline to zero. If, however, q is less than k, the density of microorganisms in the chamber will increase. By monitoring the cell density continuously, by optical methods, say, and using the information so obtained to regulate the dilution rate q, the total population in the chamber can be maintained at a desired level N_0. When the cell density in the chamber is monitored by turbidity measurements, the continuous flow device is called a **turbidostat**.

A different principle for controlling the population is utilized in the continuous culture device called the **chemostat** (Novick and Szilard, 1950; Monod, 1950). Here the nutrient medium is composed of an excess of all required nutrients except one, the **limiting growth factor**, the concentration of which can be varied. An example of a limiting growth factor that is essential for the growth of almost all bacteria is an energy source containing carbon, such as a simple sugar. Clearly, if the concentration c of the limiting growth factor is zero, no growth can occur, while if it appears in excess, the growth rate is maximal. In general the growth rate k (in the absence of outflow) will depend on c, and the governing growth equation (1.33) is more properly written as

$$\frac{dN}{dt} = k(c)N - qN. \tag{1.34}$$

The dependence of k on the concentration of a typical limiting growth factor has the form shown in Figure 1.9 (Novick, 1955). In the figure, the

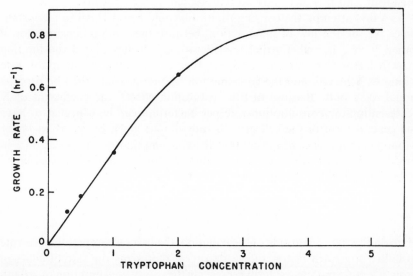

Figure 1.9. The growth rate of the tryptophan requiring strain *E. coli* B/1, *t* is shown as a function of the concentration of tryptophan. Reproduced, with permission, from *The Growth of Bacteria*, by A. Novick, Annual Review of Microbiology, volume 9. Copyright © 1955 by Annual Reviews, Inc. All rights reserved.

growth rate k is plotted as a function of the concentration of the nutrient **tryptophan**, for a strain of the bacterium *E. coli* B/1, t. This strain requires the essential amino acid tryptophan for its growth. **Amino acids**, which number about twenty, are the fundamental building blocks of all proteins. Monod (1950) proposed that the mathematical form of k is

$$k(c) = k_m \frac{c}{K + c},$$ (1.35)

which is suggested by experience and by the theoretical possibility that the uptake of the growth factor is enzyme-mediated. (This last remark will become understandable following the discussion of the Michaelis–Menten equation in the next chapter.) In equation (1.35) above, k_m and K are constants. Note that k_m represents the maximum possible value of k and is attained for large values of the concentration c. K has the dimensions of a concentration and is the value of c for which $k = k_m/2$.

Equation (1.35) must be supplemented by a rate equation for the determination of $c = c(t)$, the concentration of the limiting growth factor in the medium at any time t. Let the concentration of the limiting growth factor in the entering flow be designated by c_0. In the exit flow, the concentration is the same as it is in the growth chamber, $c(t)$. The limiting growth factor is disappearing from the medium as a result of consumption by the microorganism, and this fact must be represented mathematically.

As a first attempt, the consumption rate could be set equal to a constant times Nc, by the law of mass action, because there is no consumption if either N or c is null. Further consideration indicates that if the limiting growth factor were present in excess, then a maximum consumption rate would be achieved, and the consumption rate term would then be proportional to N only. Because of this "saturation effect," the coefficient of N representing the consumption rate per bacterium can be expected to have the general form of the cell growth rate $k(c)$, as given by equation (1.35). Monod (1950) in fact proposed that the consumption rate depends on c in precisely the same way as $k(c)$ does. Thus, the concentration of c is assumed to be determined by the equation

$$\frac{dc}{dt} = qc_0 - qc - \frac{1}{y} k(c)N.$$ (1.36)

Here on the right-hand side of the equation, the first term represents the rate of influx of growth factor, the second term represents the rate of efflux of growth factor, and the last term represents the rate of consumption of growth factor by the microorganisms. In the last term, y is a constant of proportionality. The reason y is placed in the denominator is that it thereby attains a biophysical meaning, which will become evident later.

We inquire now as to whether equations (1.34) and (1.36) have a stationary state, denoted by $c = \bar{c}$, and $N = \bar{N}$. The mathematical requirement for such a state is that dN/dt and dc/dt be zero. Thus, from equations (1.34) and (1.36) the existence of such a state requires that $\bar{c}$ and $\bar{N}$ constitute a solution of the equations

$$k(\bar{c}) - q = 0, \tag{1.37a}$$

$$qc_0 - q\bar{c} - \frac{k(\bar{c})\bar{N}}{y} = 0. \tag{1.37b}$$

The first equation above has a positive solution $\bar{c}$ provided $q < k_m$. If $q > k_m$, then the overflow rate is too rapid and the cells will gradually disappear from the chemostat. Hence the solution of (1.37a) is, with the aid of equation (1.35),

$$\bar{c} = \frac{Kq}{k_m - q}, q < k_m. \tag{1.38}$$

From (1.38) and equation (1.37b), we find the stationary population value $\bar{N}$ to be

$$\bar{N} = y(c_0 - \bar{c}) \tag{1.39}$$

where $\bar{c}$ is given by equation (1.38). A finite stationary population $\bar{N}$ is seen to exist, according to equation (1.39), provided $\bar{c} < c_0$. For a given value of c_0, if the dilution rate q is made to increase from zero, then $\bar{c}$ will increase in accordance with equation (1.38) until it attains the value c_0. At this value, $\bar{N} = 0$ by equation (1.39), and a finite stationary population ceases to exist. The maximum dilution rate q_m at which this occurs is obtained from equation (1.38) by setting $\bar{c} = c_0$ there and solving for $q = q_m$, whence

$$q_m = \frac{k_m c_0}{K + c_0}. \tag{1.40}$$

Note that q_m is always less than k_m. Also, the effect of decreasing the initial concentration c_0 is to decrease q_m.

In practice, it is often the case that $q \ll q_m$, so that we can neglect q with respect to k_m in equation (1.38). It also follows that we can neglect $\bar{c}$ with respect to c_0 in equation (1.39). Then, approximately, $\bar{c}$ and $\bar{N}$ are given by the expressions

$$\bar{c} = \frac{Kq}{k_m},$$
$$\bar{N} = yc_0, q \ll q_m. \tag{1.41}$$

The last equation states that $\bar{N}$ is independent of the fractional flow rate q, for q sufficiently small.

The significance of the constant y can now be clarified if we view the chemostat in the steady state as a bacterial mass production process. We ask what is the **yield**, defined as the mass output of bacteria per unit time $q\bar{N}m_0$ divided by the mass of growth factor consumed per unit time $Q(c_0 - \bar{c})$. Here, m_0 is the mass of a single bacterium. Thus, with the aid of equation (1.39),

$$\text{yield} \equiv \frac{q\bar{N}m_0}{Q(c_0 - \bar{c})} = \frac{ym_0}{V}. \tag{1.42}$$

The yield is seen to be proportional to the parameter y. Because of the above relation, the quantity ym_0/V is called the **yield constant**. The actual output of bacteria per unit time or **output rate** $q\bar{N}$ does of course depend on the dilution rate.

A quantitative test of the theory of the chemostat presented herein is provided by investigations of the growth of *Aerobacter cloacae* bacteria in

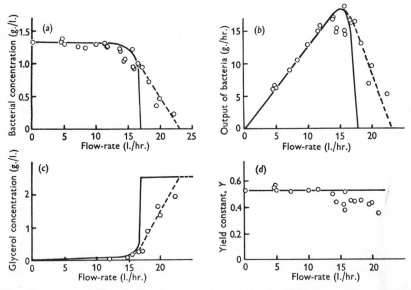

Figure 1.10. The circles indicate the observed dependence of (a) bacterial mass concentration $m_0\bar{N}/V$, (b) mass output rate $qm_0\bar{N}$, (c) concentration of limiting growth factor $\bar{c}$, and (d) yield constant $Y = ym_0/V$, as a function of the flow rate Q, for continuous culture in a chemostat of *Aerobacter cloacae* with glycerol as limiting growth factor. The solid lines are theoretical curves based on equations (1.38), (1.39) and (1.42). Here $V = 20$ liter, $c_0 = 2.5$ g/liter and $K = 12.3$ µg/ml, $k_m = 0.85$/hr, $ym_0/V = 0.53$, as determined by other experiments. Reproduced, with permission, from D. Herbert, R. Elsworth, & R. C. Telling (1956) *The Continuous Culture of Bacteria; a Theoretical and Experimental Study*, Journal of General Microbiology, volume 14, Cambridge University Press.

a chemostat with glycerol as the sole carbon source and limiting growth factor, at 21 different flow rates (Herbert, Elsworth, and Telling, 1956). **Aerobacter cloacae** is a coliform (colon inhabiting bacillus), similar to *E. coli*, and commonly found in sewage. Figure 1.10 shows the experimental dependence on the flow rate Q of the following: the bacterial mass concentration $m_0\bar{N}/V$, the mass output rate $q\bar{N}m_0$, the steady-state concentration of limiting growth factor $\bar{c}$, and the yield ym_0/V. The theoretical dependence of these quantities according to equations (1.38), (1.39), and (1.42) is also shown. The agreement between theory and experiment is seen to be good except when q is in the neighborhood of its maximum permissible theoretical value q_m. The discrepancy between theory and experiment is believed to be due to inhomogeneities in the medium which result from inadequate mixing in the chemostat chamber.

The chemostat has the important property that it is self-stabilizing. That is, the growth of a microorganism in it tends toward a stationary state $(c,N) = (\bar{c},\bar{N})$ for a given flow rate $q < q_m$, and this stationary state is stable in response to any small perturbation away from it. The mathematical demonstration of the above assertion follows from an examination of equations (1.34) to (1.36) in a small neighborhood of the stationary state (see Chapter 3, problem 13).

1.8. INTERACTING POPULATIONS: PREDATOR–PREY SYSTEM

We wish to consider a simple example (Lotka, 1925) of two interacting populations for which the level of one population depends in an intimate way on the level of another population. A parasite is born from an egg deposited in the host. The host is killed in this process. How do the host and parasite (or **prey** and **predator**) populations vary with time? Let x and y represent the number of hosts and parasites, respectively, at time t. The number of eggs deposited by the invading parasites depends on the probability of the parasites and hosts coming together, and is therefore assumed to be proportional to xy. We denote the constant of proportionality by α, so that the rate at which parasites kill the host is αxy. Denote the birth rate (the fractional rate of increase due to births per unit time) of the host population by k_b, and the natural death rate exclusive of parasite killing by k_d. The death rate k_d represents the deaths due to all causes except the parasite under consideration. Then the net rate at which the population increases, exclusive of parasite killing, is assumed to be $(k_b x - k_d x)$. We see that the constants k_b and k_d enter only in the combination $k_b - k_d$, so that we might just as well introduce a single symbol β for it, $\beta = k_b - k_d$. As long as $k_b > k_d$, β is positive.

When the birth rate, the natural death rate, and the parasite-killing rate are all taken into account, the host population is assumed to be governed by the equation

$$\frac{dx}{dt} = \beta x - \alpha xy. \tag{1.43}$$

Here α is a positive constant, and the minus sign accounts for the fact that αxy is a death rate. Similarly, the parasites are assumed to have a natural death rate δ due to all causes. Births are dependent on the number of eggs that are laid and that hatch. Assume for simplicity that an invading parasite lays only one egg in a host, and that the fraction of eggs f that hatch is a constant, where f is some number between zero and unity, $0 < f \leqq 1$. Then the parasite population is assumed to be governed by the equation

$$\frac{dy}{dt} = -\delta y + f\alpha xy, \qquad \delta > 0. \tag{1.44}$$

Equations (1.43) and (1.44) are the basic mathematical postulates of a host–parasite system and are known as the **Lotka–Volterra equations** (Lotka, 1925; Volterra, 1926). They are to be supplemented by an initial condition specifying the original population sizes:

$$x = x_0, \, y = y_0, \, \text{at } t = 0. \tag{1.45}$$

As in the previous section, we are again confronted with the problem of solving a **system of ordinary differential equations**. Furthermore, they are **nonlinear**, in that terms appear in them which are neither constant nor linear in the dependent variables and their derivatives. Without the presence of the nonlinear term xy, they would be linear and simple to solve. In retrospect, we see that equation (1.22), the logistic equation, was also nonlinear, although we were able to solve it by the method of separation of variables. Speaking generally, nonlinear equations are rather difficult to solve. However, approximation methods are available for obtaining approximate solutions which can be quite useful. These often depend for their success on a good understanding or biological–physical–chemical intuition about the process being represented mathematically, and on an ability to decide which simplifying approximations can usefully be made under the given circumstances.

Instead of trying to solve our equations as they stand, we prefer to utilize an approximation technique which has more general utility. We shall investigate the solution in the neighborhood of its stationary states by the method of small perturbations. We inquire as to the existence of stationary populations for the predator and prey, as we did before in the case of a single population and in the case of the chemostat equations, by seeking a solution to our equations for which the time derivatives are null, or when

$$x(\beta - \alpha y) = 0,$$
$$y(-\delta + f\alpha x) = 0. \tag{1.46}$$

Aside from the **trivial solution** $x = 0$, $y = 0$, which is not of much biological interest, it is seen that there is one nontrivial solution of these equations representing the equilibrium state (x_e, y_e), where

$$x = x_e \equiv \frac{\delta}{f\alpha},$$
$$y = y_e \equiv \frac{\beta}{\alpha}. \tag{1.47}$$

We now expand the right-hand sides of equations (1.43) and (1.44) in a Taylor series about the point (x_e, y_e) and retain only the linear terms. To accomplish this, we utilize Taylor's expansion for a function of two variables, that is,

$$f(x,y) = f(x_e, y_e) + (x - x_e)\left[\frac{\partial f(x,y)}{\partial x}\right]_{\substack{x=x_e \\ y=y_e}}$$
$$+ (y - y_e)\left[\frac{\partial f(x,y)}{\partial y}\right]_{\substack{x=x_e \\ y=y_e}} + \cdots. \tag{1.48}$$

For the functions given in (1.46), there results

$$x(\beta - \alpha y) = -\alpha x_e(y - y_e) + \cdots,$$
$$y(-\delta + \alpha x) = f\alpha y_e(x - x_e) + \cdots. \tag{1.49}$$

Now (1.43) and (1.44) become, when only the terms shown in (1.49) are retained in them,

$$\frac{dx}{dt} = -\alpha x_e(y - y_e),$$
$$\frac{dy}{dt} = f\alpha y_e(x - x_e). \tag{1.50}$$

These equations are the linearized forms of equations (1.43) and (1.44) in the neighborhood of the stationary state.

The mathematical analysis of equations (1.50) is somewhat simplified if we change the dependent variables from (x,y) to (ξ, η) which represent the displacements from equilibrium

$$\xi = (x - x_e),$$
$$\eta = (y - y_e). \tag{1.51}$$

Equations (1.50) become

$$\frac{d\xi}{dt} = -\alpha x_e \eta,$$
$$\frac{d\eta}{dt} = f\alpha y_e \xi. \tag{1.52}$$

We attempt to solve these equations by the method of separation of variables, which yields

$$dt = -\frac{d\xi}{\alpha x_e \eta}, \; dt = \frac{d\eta}{f\alpha y_e \xi}. \tag{1.53}$$

By combining the two equations in (1.53), we can write

$$\frac{-\xi d\xi}{x_e} = \frac{\eta d\eta}{f y_e}. \tag{1.54}$$

Integration of (1.54) leads to the relation

$$\frac{\xi^2}{x_e} + \frac{\eta^2}{f y_e} = c, \tag{1.55}$$

where c is a constant determined by the initial condition. Now, by using (1.51) to eliminate ξ and η, we impose the initial condition and determine the constant c to be

$$c = \frac{(x_0 - x_e)^2}{x_e} + \frac{(y_0 - y_e)^2}{f y_e}. \tag{1.56}$$

Equation (1.55) is the equation of an ellipse whose center is at the origin in the $\xi\eta$ plane. In the xy plane, the ellipse is centered at the equilibrium point (x_e, y_e). A family of such ellipses is shown in Figure 1.11. A particular ellipse in the figure is determined by the value of c that is assigned, that is, by the initial conditions. The arrowheads on the ellipse indicate the temporal course of events: the host and parasite undergo cyclic changes in population, alternately waxing and waning. Speaking qualitatively, the increase in the host population favors an increase in the parasite population. The latter increase at the expense of the host until they almost annihilate the host. But they too must begin to disappear because of a lack of food supply (for the parasite larva). This favors the recovery of the host population and the cycle begins all over again.

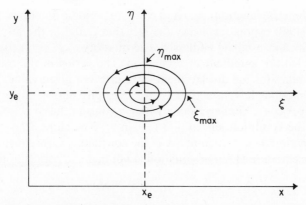

Figure 1.11. Family of ellipses representing displacements in time of a predator–prey system about the equilibrium point (x_e, y_e). The initial state places the system at a point on an ellipse, and the direction of increasing time is indicated by the arrowhead.

Knowing the interdependence of ξ and η, we can return to equations (1.53) and determine the explicit dependence of these variables on the time. It is convenient to integrate the first equation in (1.53) starting from the time t_0 when $\eta = 0$ and ξ is at its maximum value ξ_{max} (see Figure 1.11), given by equation (1.55) as

$$\xi = \xi_{max} \equiv (x_e c)^{1/2}. \tag{1.57}$$

Thus, we get from (1.53), (1.55), and (1.57)

$$\alpha x_e \int_{t_0}^{t} dt = -\left(\frac{x_e}{fy_e}\right)^{1/2} \int_{\xi_{max}}^{\xi} \frac{d\xi}{(\xi_{max}^2 - \xi^2)^{1/2}}.$$

Performing the integrations yields

$$\alpha(fx_e y_e)^{1/2}(t - t_0) = \cos^{-1}\frac{\xi}{\xi_{max}}\bigg|_{\xi_{max}}^{\xi} = \cos^{-1}\frac{\xi}{\xi_{max}}. \tag{1.58}$$

Finally, solving (1.58) for ξ leads to the relation

$$\xi = \xi_{max} \cos[\alpha(fx_e y_e)^{1/2}(t - t_0)]. \tag{1.59}$$

If t_0 is not given, but the value of ξ at $t = 0$ is given, then t_0 can be found by substituting that initial value of ξ in (1.58), setting $t = 0$, and solving for t_0. In utilizing (1.58), care must be exercised to determine in which quadrant of the $\xi\eta$ plane the initial point is located.

From (1.55) and (1.59) it follows that

$$\eta = \eta_{max} \sin[\alpha(fx_e y_e)^{1/2}(t - t_0)], \tag{1.60}$$

$$\eta_{max} \equiv (fy_e c)^{1/2}. \tag{1.61}$$

The trigonometric functions $\sin t$ and $\cos t$ are periodic functions. Equations (1.59) and (1.60) express explicitly the fact that ξ and η, the displacements from equilibrium oscillate endlessly. The quantity ξ_{max} in (1.59) or η_{max} in (1.60) is called the **amplitude** of oscillation. The **period** of the oscillation is the time it takes for the displacement to return to a given value. The trigonometric functions have a period 2π, for example, $\sin(t + 2\pi) = \sin t$, and $\cos(t + 2\pi) = \cos t$. Similarly $\sin ct$ with c a constant has a period T determined by the condition $\sin c(t + T) = \sin ct$. Therefore, $cT = 2\pi$, which yields the period $T = 2\pi/c$ in terms of the constant c. Consequently, we see from either equation (1.59) or equation (1.60) that the period of the cycle is given as

$$T = \frac{2\pi}{\alpha(fx_e y_e)^{1/2}}. \tag{1.62}$$

The period can also be expressed in terms of the parameters appearing in our fundamental equations, namely,

$$T = \frac{2\pi}{(\beta\delta)^{1/2}}. \tag{1.63}$$

Note that the period is independent of the initial conditions. This result is only approximately true, and is a consequence of our linearization procedure. In general, T does depend on the amplitude of the perturbation, that is, on the initial conditions.

Because of the well-known property of trigonometric functions that $\cos(t - \pi/2) = \sin t$, equation (1.60) can also be written as

$$\eta = \eta_{max} \cos[\alpha(fx_e y_e)^{1/2}(t - t_0) - \tfrac{\pi}{2}]. \tag{1.64}$$

This expression, when compared with equation (1.59), indicates that the maximum or peak of the predator population lags behind the peak in the prey population by an amount $\pi/2$, or $1/4$ the period of the oscillation. The constant quantity $\pi/2$ appearing in the argument of the trigonometric function in equation (1.64) is called a **phase angle**.

In the mathematical terminology of nonlinear equations, the equilibrium point (x_e, y_e) is called a point of **neutral stability** in that the trajectory of the point (x,y) as it evolves in time neither goes into the equilibrium point or far away from it. Such a neutrally stable point corresponds, in our previous discussion of the stability of equilibrium points for a system with only one dependent variable, to the case when the coefficient of the linear term is zero.

We have here neglected other population effects in the Lotka–Volterra equations (1.43) and (1.44), for example terms proportional to x^2 and y^2 such as appeared in our discussion of a single population. When such terms are included, it has been found that the concentric ellipses are replaced by spirals that wind all the way into the stationary point. The equilibrium point is then stable. It is possible that the trajectories move outward, which means that the equilibrium point is unstable. The trajectories may also remain closed.

The protozoan *Paramecium aurelia* feeds on the yeast *Saccharomyces exiguus.* **Yeast** cells are microorganisms that depend upon higher plants and animals for their source of energy. They normally live on dead or decaying plant or animal matter. They have played an historically useful role in making wine and beer, and in leavening bread. More recently, they have been used to synthesize B complex vitamins, and have been recognized as an easily cultivated, highly nutritive food supplement for animals, including man. The genus **Saccharomyces** is noted for its ability to vigorously ferment (anaerobically metabolize) the common sugars, producing ethyl alcohol and carbon dioxide.

When a mixed population of *P. aurelia* and *S. exiguus* was introduced into a nutrient medium, the fluctuations in population density as a function of time were observed, as shown in Figure 1.12 (Gause, 1935). The interaction between the two cells is a qualitative example of the predator–prey system we have discussed. Although the smooth curves drawn through the data points are not sinusoidal, the period of the fluctuations is approximately constant. Note also how the peaks of the predator population follow or lag behind the peaks of the prey population by an amount which is approximately $\frac{1}{4}$ the period, in accordance with theoretical expectations.

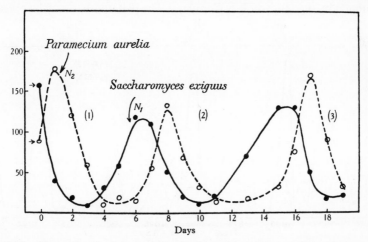

Figure 1.12. Fluctuations in the population densities of the predator cell *Paramecium aurelia* (per 15 cm^3) and of its prey, *Saccharomyces exiguus* (per 0.1 mm^3). From Gause (1935), with permission.

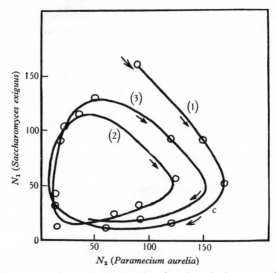

Figure 1.13. The variation in population density of predator is shown as a function of population density of prey for the same data shown in the previous figure. From Gause (1935), with permission.

The same data are shown plotted as a predator–prey diagram in Figure 1.13.

Further qualitative evidence for population fluctuations as a consequence of the "struggle for existence" is shown in Figure 1.14, which is a yearly record of the number of lynx and snowshoe hare pelts collected from hunters in Canada by the Hudson Bay Company. The pelt number may be presumed

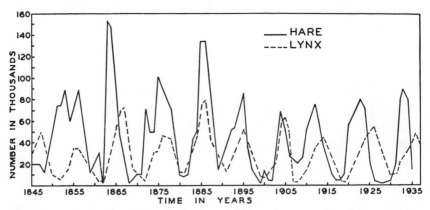

Figure 1.14. The yearly number of pelts of lynx and snowshoe hare collected by the Hudson Bay Company in Canada is shown as a function of the time. Reproduced, with permission, from E. P. Odum, *Fundamentals of Ecology*, 3rd edition. W. B. Saunders Company, Philadelphia, 1971. Redrawn from MacLulich (1937).

to be proportional to the population of the animals in a given year. The phase relationship between the two curves serves to identify the predator and prey.

Suppose a predator and prey are destroyed in a proportional manner by an external agent, such as overhunting, or a forest fire. Who recovers more rapidly, the predator or prey? When both predator population y and prey population x are reduced, the product term xy is proportionally reduced even more in the Lotka–Volterra equations. This latter term is the birth term for the predator, and it is the predator, therefore, that suffers the most. For example, if x and y are both reduced by half, the birth rate of the prey, βx, is half its former rate, but the birth rate of the predator, $f\alpha xy$, is one quarter its former rate. The consequence is that the prey always recovers more rapidly from such a catastrophic event. This somewhat surprising inference, which is easily made from the rate equations, is known as **Volterra's principle**.

A concrete manifestation of this principle is found in the work of entomologists who have observed that an attempt to control an insect pest by spraying with an insecticide is often followed by a new population surge of the pest, to a higher level than before. The effect is attributable to the simultaneous destruction by the insecticide of the natural predators of the pest species. The beneficial insects, the predator in this case, are not able to recover as rapidly from the effect of the spraying as the pest.

What is the effect of introducing a second prey into the system which feeds on the predator? We might naturally expect that this would protect the first prey and render it less vulnerable to the predator, but this is not necessarily the case. In fact, the presence of the second prey may prevent the waning of the predator sufficiently when the population of the first prey diminishes, so much so that the first prey is rendered extinct. When does this happen and under what conditions? These are the kinds of questions that can be answered by mathematical analysis.

1.9.   MUTATION AND REVERSION IN BACTERIAL GROWTH

As we have seen previously, the kinetics of the growth of a bacterial colony in a culture can be described by a linear rate equation under certain optimal growth conditions. This remains true even if **mutations** arise, that is, a variant type begins to grow in the cell culture. The result of mutations is to convert a certain proportion of the population under study from its initial form to a variant form. The standard form which is usually found in nature is called the **wild type**. The variant forms are called **mutants**. Some of the variant cells likewise get converted back to the original form by a process called **back-mutation** or **reversion**. The mutation rates of such a bacterial system can be deduced by a quantitative study of its growth characteristics.

Serratia marcescens is a common bacterial species which produces a distinct brick-red pigment intracellularly when grown in a culture at room temperature. The color is a useful property for identification in the laboratory. Cultures of the strain *Serratia marcescens* # 274 are unstable, producing daughter colonies of more than one type upon plating. **Plating** is the transfer of cells from a culture to a solid agar medium, on the surface of which colonies form. The variant colonies are recognizable by their color, the predominant variety being dark red and the predominant variant being bright pink. A small number of colonies colored pale pink and white are also produced, which will be considered here to be negligible. Regardless of the color of the variant in the inoculum used to initiate the culture, stable populations evolve containing about 97% dark red cells and 3% bright pink cells. When such cells were grown exponentially, the generation time was found to equal 64 min and to be independent of the color of the colony used as inoculum (Bunting, 1940). These results are illustrated in Figure 1.15.

Let $\psi(t)$ be the fraction of dark red cells at time t. Table 1.1 quotes the results of two experimental determinations of ψ, from the cited reference. These data were subsequently reanalyzed (Shapiro, 1946), and a semilog plot of

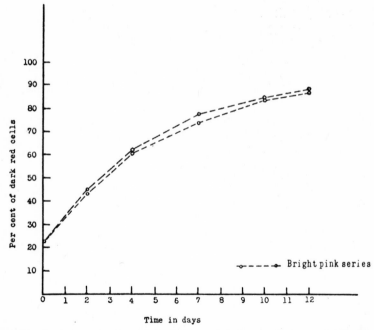

Figure 1.15. Variation with time of the fraction of dark red cells in a culture of *Serratia marcscens* for different initial inoculums. The remaining fraction of cells were bright pink cells. Reproduced, with permission, from Bunting (1940).

TABLE 1.1 THE CHANGE IN THE FRACTION
OF DARK RED CELLS AS A FUNCTION OF TIME.[a]

	Experiment	
	No. 5	No. 6
Time[b] (days)	ψ	ψ
0	0.23	0.23
2	0.45	0.44
4	0.62	0.61
7	0.78	0.74
10	0.84	0.85
12	0.87	0.88

[a] The results of two experiments of Bunting (1940).
[b] The time 0 is the time of the initial inoculum.

$\psi_\infty - \psi$ as a function of time was made, as shown in Figure 1.16. Here ψ_∞ is the asymptotic limit of $\psi(t)$ as $t \to \infty$, determined from other experiments as $\psi_\infty = 0.97$. It was found that such a plot could be adequately fitted by a straight line. The slope of the straight line was found to be $-0.0747/\text{day}$. The intercept of the straight line was constrained by the assumption that it equals $\psi_\infty - \psi(0)$. These results can be interpreted satisfactorily on the basis of the following theory.

Consider two bacterial types, X_1 and X_2, which are interrelated genetically, so that X_1 cells can arise from X_2 cells by mutation, and vice versa. Denote the number of cells of types X_1 and X_2 at time t by $n_1(t)$ and $n_2(t)$, respectively. We assume that the growth conditions in a culture are favorable for both types to grow at their maximal rates. The kinetic equations governing their growth are assumed to be of the form

$$\frac{dn_1}{dt} = k_1 n_1 + \beta n_2, \tag{1.65a}$$

$$\frac{dn_2}{dt} = \mu n_1 + k_2 n_2. \tag{1.65b}$$

Here k_1 and k_2 are the fractional growth rates for X_1 and X_2 type cells, respectively, μ is the fractional mutation rate per unit time at which X_1 cells are converted to X_2 cells, and β is the fractional mutation rate per unit time for reversion of X_2 type cells to X_1 type. The parameter μ is usually called

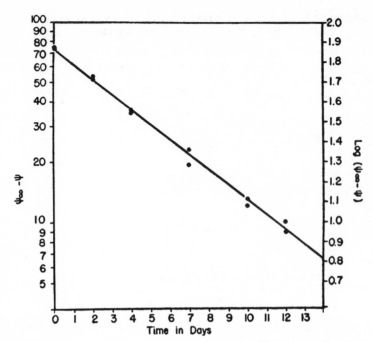

Figure 1.16. Semilog plot of $(\psi_\infty - \psi(t))$ for *Serratia marcescens* #274, where $\psi(t)$ is the fraction of dark red cells at time t, and ψ_∞ is a constant equal to 0.97. The data points represent the two separate experiments of Bunting quoted in the table. The scale of $\psi_\infty - \psi$ is in percent. Reproduced from A. Shapiro (1946), Symp. Quant. Biol. © 1947, p. 233, with permission of the Cold Spring Harbor Laboratory.

the **mutation rate**, while β is called the **back-mutation rate** or **reversion rate**. The quantities k_1, k_2, μ, and β are all assumed to be positive constants.

The solution to such an equation system is presented in a different context in Chapter 3. In the above-mentioned experiments, it was feasible to determine the relative amount of each mutant as a function of the time, because the relative amount can continue to be measured even though the cell culture is diluted at regular intervals with fresh culture medium. Hence, it is desirable to introduce new variables that represent the observable quantities. Thus, let $N(t)$ be the total number of cells of any type in the culture at time t, and let $\psi(t)$ be the fraction of cells that are of type X_1 at time t. Then, by definition,

$$N = n_1 + n_2, \psi = \frac{n_1}{n_1 + n_2},$$

or   (1.66)

$$n_1 = \psi N, n_2 = (1 - \psi)N.$$

By adding the two equations appearing in (1.65) and utilizing (1.66), we can readily derive the differential equation satisfied by $N(t)$. Using it together with equations (1.65) and (1.66), the equation satisfied by $\psi(t)$ can be found. Thus, equations (1.65) are equivalent to the equation system

$$\frac{dN}{dt} = (k_2 + \beta)N + (k_1 + \mu - k_2 - \beta)N\psi,$$

$$\frac{d\psi}{dt} = \beta + (k_1 - k_2 - 2\beta)\psi - (k_1 + \mu - k_2 - \beta)\psi^2.$$

(1.67)

It would appear that we have needlessly complicated the formulation of the problem in changing from the variables n_1 and n_2 to N and ψ, because equations (1.67), in contrast to (1.65), are nonlinear. However, we shall make the assumption that

$$k_1 + \mu = k_2 + \beta. \tag{1.68}$$

This has the effect of eliminating the nonlinear terms in the equation system (1.67), so that it now becomes a pair of linear ordinary differential equations. These equations are simpler than the pair of equations (1.65) because they are **uncoupled** (no interaction term appears in either equation), namely,

$$\frac{dN}{dt} = (k_1 + \mu)N, \tag{1.69}$$

$$\frac{d\psi}{dt} = \beta - (\mu + \beta)\psi. \tag{1.70}$$

In deriving equations (1.69) and (1.70), we have used equation (1.68) to eliminate k_2 from the coefficients of N and ψ. Thus only three parameters, k_1, μ, and β, appear in these equations. We also note that ψ has a steady state value ψ_∞ defined by the equation $d\psi/dt = 0$, or

$$\psi_\infty = \frac{\beta}{\mu + \beta}. \tag{1.71}$$

Consequently, equation (1.70) can be rewritten as

$$\frac{d\psi}{dt} = -(\mu + \beta)(\psi - \psi_\infty). \tag{1.72}$$

The solution to equations (1.70) and (1.72), which satisfies the initial condition

$$(N,\psi) = (N_0,\psi_0) \text{ at } t = 0, \tag{1.73}$$

is

$$N = N_0 e^{\lambda t},$$
$$\psi - \psi_\infty = (\psi_0 - \psi_\infty)e^{-\alpha t}, \tag{1.74}$$

where

$$\lambda = k_1 + \mu,$$
$$\alpha = \mu + \beta. \tag{1.75}$$

In equation (1.73) we have used vector notation to denote that $N = N_0$ and $\psi = \psi_0$ at $t = 0$. From the experimental determination of λ, α, and ψ_∞, and equations (1.68), (1.71), and (1.75), we can determine k_1, k_2, and the mutation rates μ and β as

$$k_1 = \lambda - (1 - \psi_\infty)\alpha, \qquad \beta = \psi_\infty \alpha,$$
$$k_2 = \lambda - \psi_\infty \alpha, \qquad \mu = (1 - \psi_\infty)\alpha. \tag{1.76}$$

The cited analysis of the data of Bunting yielded the following values of the parameters λ, α, and ψ_∞:

$$\lambda = \log 2/64 \text{ min} = 15.6/\text{day}$$
$$\alpha = \log 10 \cdot (0.0747)/\text{day} = 0.172/\text{day}, \tag{1.77}$$
$$\psi_\infty = 0.97.$$

Hence, from equations (1.76), it is inferred that

$$k_1 = 15.6/\text{day}, \qquad \mu = 0.0052/\text{day},$$
$$k_2 = 15.4/\text{day}, \qquad \beta = 0.167/\text{day}. \tag{1.78}$$

It is perhaps of greater biological significance to express the mutation rates in units of (generation time)$^{-1}$ because a mutation is the result of a genetic "event" that occurs once or not at all in a given cell generation. If it is assumed that no cell deaths occur, then the generation time is the doubling time, 64 min, and the mutation rate μ and the reversion rate β are expressible in units of (generation time)$^{-1}$ as follows:

$$\mu = 2.3 \times 10^{-4} \text{ per generation time,}$$
$$\beta = 7.3 \times 10^{-3} \text{ per generation time.} \tag{1.79}$$

Thus we see that mutation and reversion rates, as well as growth rates, can be inferred from quantitative kinetic studies of bacterial growth. The inferences drawn were based, inter alia, on the assumption (1.68), which was made for purposes of mathematical convenience. Actually, it is not strictly necessary to make the latter assumption, and the experiments can be better analyzed without it. For this purpose, it is simpler to work with the original equation system (1.65). The quantitative conclusions are slightly altered by such an analysis.

There is an alternative theory commonly utilized to measure mutation rates (Luria and Delbruck, 1943), which is complementary to the deterministic theory described in the text. This theory analyses mathematically the number of mutations observed and the associated fluctuations, in direct samplings from bacterial cultures.

SELECTED REFERENCE BOOKS

D'A. Thompson (1942), *On Growth and Form*, 2nd ed., Cambridge Univ. Press, Cambridge.

G. F. Gause (1934), *The Struggle for Existence*, Hafner, New York (reprinting 1969).

J. Monod (1942), *Recherches sur la croissance des cultures bactériennes*, Hermann and Cie, Paris.

A. J. Lotka (1925), *Elements of Physical Biology*, Williams and Wilkins, Baltimore (reprinted as *Elements of Mathematical Biology*, Dover, New York, 1956).

E. O. Wilson and W. H. Bossert (1971), *A Primer of Population Biology*, Sinauer Associates, Stamford, Conn.

R. Y. Stanier, M. Doudoroff, and E. A. Adelberg (1970), *The Microbial World*, 3rd ed., Prentice-Hall, Englewood Cliffs, N.J.

PROBLEMS

Problems marked with an asterisk (*) are difficult, and optional. Problems 1–3 are based on elementary calculus and not dependent on the text material.

1. Enzymatically controlled reactions. According to the Michaelis–Menten equation, the rate of product formation v of an enzymatically controlled reaction depends on the initial substrate concentration s in the following manner,

$$v = \frac{Vs}{s + K},$$

where K and V are constants. Given some experimental measurements of v and s for a certain reaction, how would you determine V and K?

2. The shortening of muscles. Hill's empirical relationship between the load P being raised by a muscle and the speed V with which the muscle shortens is given as

$$(P + a)V = (P_0 - P)b, \tag{1.80}$$

where a, b, and P_0 are constants characteristic of the given muscle (Hill, 1938). How should the data from an experiment be plotted so as to permit an easy determination of the constants a and b, assuming P_0 was known from an independent experiment?

3. The evolution of propulsive devices in primitive creatures. Consider a primitive organism living in the ocean, a nutrient fluid environment.

Assume that it can gather food if it is immobile at a fixed rate E_0 (due to simple diffusion say), but that it can increase its rate by evolving a propulsive device such as a cilium and thereby expose itself to a greater fluid volume (Carlson, 1962). Let this increased rate be $E_1 v$, where v is its propulsive or swimming speed and E_1 is a constant. At the same time the power expended by the creature, which was a fixed rate P_0 when it was at rest, is assumed to increase by an amount $P_1 v^2$, where P_1 is a constant (see Section 4.6). If the organism gains more energy per unit time than it loses in moving about, then the evolution of propulsive devices can be considered likely. Under what circumstances, and for what range of velocities, is it desirable for the organism to evolve a swimming apparatus? What is the optimal value of v for which the net rate of energy gain is maximized? Consider two cases:

(a) $E_0 > P_0$;
(b) $E_0 < P_0$.

Hint. Consider the net rate of energy gain as a function of the velocity, $E_{in} - E_{out} = f(v)$. What is its graphical representation?

4. For some radioactive material in exponential decay show that

$$k = -\frac{\log 2}{T_{1/2}}. \tag{1.81}$$

5. The dynamics of the process of **disinfection**, the killing of microorganisms by heat or cold, chemicals, ultraviolet irradiation, and so on, can usually be represented by equation (1.3) with N representing the number of survivors, and k a negative number. The disinfection process is illustrated in Figure 1.17, which shows the number of survivors of a pathogenic microorganism following exposure to phenol solution, a common chemical disinfectant. Replace k by $-K$, where K is positive. Suppose a suspension containing 8×10^5 organisms/ml is treated with 5% phenol solution, for which $K = 90\%$/min. How many survivors are expected to be present after 10 min? 20 min?

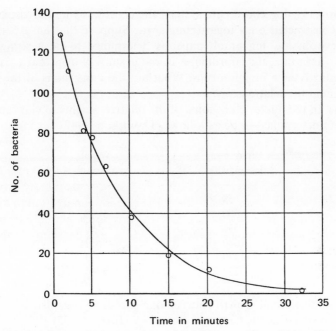

Figure 1.17. Disinfection of 3 hours' culture of *Bacillus paratyphosus* (*Salmonella paratyphi*), the pathogenic organism that causes paratyphoid fever, by a 0.6% phenol solution at 20° C. The circles represent the mean values of three samples. The ordinate is the number of viable cells remaining at time *t*, with the initial number of cells at time 1 min being 129. The solid line is a theoretical exponential curve. Reproduced, with permission, from Chick (1908).

6. A cubic centimeter of an old piece of wood taken from a museum has an activity of ^{14}C that is one third that of a cubic centimeter of wood taken from a modern piece of wood. How old is the museum piece?

7. A dated vial of sodium phosphate solution has a labeled activity of 500 μCi/ml. How many milliliters of this solution should be administered exactly 10 days after the original assay to provide an activity of 250 μCi? For ^{32}P, $T_{1/2} = 14.3$ days.

8. Show that the cell population described by equation (1.6) can equally well be described by the equation

$$N = N_0 2^{t/T}. \tag{1.82}$$

9. In a population of 1000 cells which is in exponential growth with a mean generation time $T = 106$ min, a mutation arises for which the mean generation time is 53 min. How long will it take the mutational cell type to overtake the original population? What is the total population at that time?

10. For a log–log coordinate system, the abscissa as well as the ordinate is constructed on a logarithmic scale. Suppose that an experimental curve on log–log graph paper is determined to be a straight line. (a) Denoting the coordinates in the usual way by x and y, express y explicitly as a function of x. What are the dimensions of the parameters in the relationship?

(b) In the fiddler crab, some of the relative weights of claw and body were determined to be as follows (Thompson, 1942).

Wt. of body less claw (g)	Wt. of claw (g)
58	5
300	78
536	196
1080	537
1449	773
2233	1380

Such differential growth between one part of an organism and another has been termed **heterogonic** growth by Huxley (1932).

Use the method of least squares to determine a straight-line fit to these data, and determine the constants. Plot the data and the straight line on log–log graph paper, to inspect your result.

11. Hypothetical measurements of a population of *Paramecium caudatum* growing in a nutrient medium are shown (Gause, 1934). Determine k for it, on the assumption that the population obeys the logistic law.

t(days)	0	1	2	3	4	5	6	7	8
N	2	20	135	320	374	378	380	381	381

12. Empirical formulas have described the relationship between mammalian fetal weight w and gestational age t in the form $w = at^n$, where a and n are constants. An improvement on such formulas was achieved by the suggestion that embryo age should be calculated not from conception but from the end of the "lag" phase of prenatal development t'. Thus the empirical relation (see Figure 1.18) becomes

$$w = a(t - t')^n. \tag{1.83}$$

The time t' presumably marks a boundary between two different mechanisms of nutrient supply to the fetus.

Assume that the limiting factor in the rate of fetal growth is nutrient

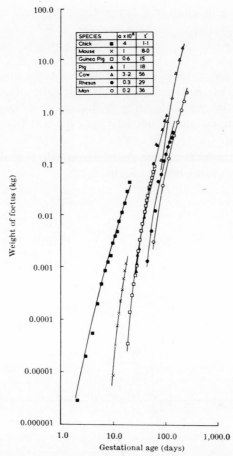

Figure 1.18. Fetal weight as a function of gestational age for a variety of species, based on data in Needham (1931). Reproduced from Payne and Wheeler (1967), with the permission of Nature.

flux, and that nutrients are supplied to the fetus across a surface corresponding to the vascular endothelium, the layer of cells lining the blood vessels. Assume that this surface is directly proportional to the total surface area of the fetus itself, or equivalently to $w^{2/3}$, for $t > t'$ (Payne and Wheeler, 1967). Write down the differential equation for $w = w(t)$, and show that the solution can be expressed in the form of the empirical relation quoted above.

13. One method of analyzing indicator dilution curves (see Chapter 3) is to assume that the observations of the concentration of an indicator

$c(t)$, which is a function of the time t, can be fitted by the functional form

$$c(t) = c_0(t - t_0)^v \, e^{-\lambda(t-t_0)}, \, t > t_0. \tag{1.84}$$

Let the observed values of the concentration at the times t_i, $i = 1, 2, \ldots,$ n, be denoted by $c^*(t_i)$, and assume that the value of t_0 is readily determined from the observations by inspection. Use the method of least squares to find equations that determine the best values of the parameters c_0, v, and λ.

14.   Many solid tumors have been shown empirically (Laird, 1965) to follow the **Gompertz growth law** (see Figure 1.19)

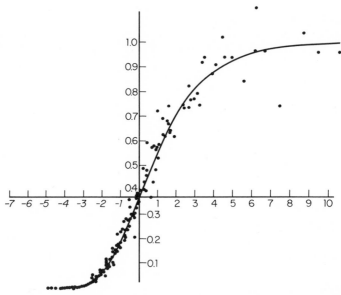

Figure 1.19   A "normalized" Gompertz plot of tumor size (ordinate) as a function of time (abscissa). The growth data for 19 examples of 12 different animal tumors have been superimposed by adjusting the units of the axes for each example. Reproduced from Laird (1965), with permission of the British Journal of Cancer.

$$V = V_0 \exp\left[\frac{k_0}{\alpha}(1 - e^{-\alpha t})\right], \tag{1.85}$$

where V is the volumetric size of the tumor of time t, and V_0, k_0, and α are constants. Derive this law from the following growth equations, which represent explicitly the fact that the growth "constant" $k(t) =$

$V^{-1}dV/dt$ decreases exponentially with time (Laird, Tyler, and Barton, 1965):

$$\frac{dV}{dt} = kV, \tag{1.86a}$$

$$\frac{dk}{dt} = -\alpha k, \tag{1.86b}$$

with $V(0) = V_0$, $k(0) = k_0$.

Show that an equivalent result is obtained by the assumption

$$\frac{dV}{dt} = \alpha V \log \frac{\bar{V}}{V}, \tag{1.87}$$

with $\bar{V} = V_0 e^{k_0/\alpha}$.

Hint. For the last demonstration set $\log(\bar{V}/V) = w$, and solve the equation for w.

15. Suppose that the flux Q in a chemostat is null so that the chemostat is being utilized for growth by the batch culture technique. Assume, in addition, that the initial concentration of limiting growth factor c_0 satisfies the inequality $c_0 \ll K$, so that the growth rate of the microorganism in the chemostat is approximately linear in c: $k = k_m c/K$. Show that equation (1.34) becomes the equation governing the logistic growth law. What is the value of the concentration of the limiting growth factor after a long time?

16. Discuss the host–parasite problem by finding dy/dx and sketching the trajectories in the xy plane,
    (a) for the case when: (1) the birthrate equals the death rate in the host population; (2) none of the parasites' eggs hatch.
    (b) Investigate the solution in the neighborhood of $x = y = 0$. Is this equilibrium state stable or unstable?
    (c)* Find the relationship between x and y in the general case, without a perturbation analysis.

17.* **Brucella abortus** is a pathogenic bacterium causing contagious abortion in cattle. In man, it causes the disease called brucellosis or undulant fever. When the bacterium *Brucella abortus* is grown on an agar medium, it forms a colony of smooth outline, and the constituents are called S-cells. During growth, mutations occur which form colonies having a rough or wrinkled appearance, and these cells are called R-cells. A flask of nutrient broth which is initially inoculated with S-cells contains a majority of R-cells by the time a stationary population size is attained. Analysis of the broth shows that the decline of S-cells coincides with the appearance of alanine, manufactured by S-cells, which kills S-cells but does not affect the growth of R-cells (Goodlow, Mika, and Braun, 1950). The growth curve exhibits a dip beginning about four days after inoculation (see Figure 1.20). Try to make up a

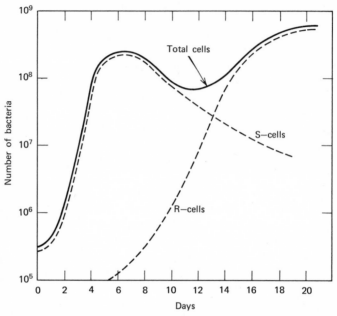

Figure 1.20. The solid line represents the growth in time of the pathogen *Brucella abortus* on an agar medium. From M. Sussman, *Growth and Development*, 2nd edition, © 1960, p. 84. By permission of Prentice-Hall, Inc., Englewood Cliffs, New Jersey. Based on the work of Goodlow, Mika, and Braun (1950).

theory for the kinetic behavior of S- and R-cells in interaction. What are the equilibrium states of the system?

18. The "unstable" variant strain M.T. 2CR of *Salmonella aertrycke* (*typhimurium*), when grown on broth and transferred to agar plates, was found to yield a diminishing fraction of the original rough type colonies, and an increasing fraction of small, smooth, dome-shaped colonies (Deskowitz and Shapiro, 1935). Ultimately, the fraction of rough type cells became zero, and the fraction of smooth type cells became unity. The smooth shaped colonies always yielded only smooth colonies on repeated plating. With $\psi(t)$ representing the fraction of smooth shaped colony forming cells at time t, a straight-line fit to the data points, plotted as $\log[1 - \psi(t)]$ versus time, yielded an exponential

decay constant of 0.059/hr. The fractional growth rate of the total culture, after the culture had reached its equilibrium composition, was found to have a constant value of 1.30/hr.

(a)   Find the mutation rate β for conversion of rough shaped colony forming cells to smooth shaped colony forming cells, and the growth rate constants k_1 and k_2 for smooth and rough cells, respectively.

(b)*   When the strain is grown on agar, different growth rates obtain for it. It is found that, after a long time, the fraction of smooth type cells obtained from plating attains the steady value 0.17, while the remaining fraction are rough type cells. It remains true that smooth type cells yield only smooth colonies on plating. The fractional growth rate of the total culture was found to have the constant value 1.22/hr. The observed exponential decay rate of $[\varphi(t) - \varphi_\infty]$ was found to be 0.061/hr, where $\varphi(t)$ is the ratio of smooth type cells to rough type cells at time t, and $\varphi(\infty) = \varphi_\infty$. Solve equations (1.65) directly with $\mu = 0$, and determine the values of k_1, k_2, and β. Do not assume that $k_1 = k_2 + \beta$.

2
ENZYME KINETICS

Virtually all chemical reactions in the cell involve the direct participation of **enzymes**, which are proteins that act as catalysts. **Proteins** are the major constituents of cellular matter. They are complex nitrogenous compounds, macromolecules composed of combinations of the twenty amino acids. The amino acids, which in combination are called **peptides**, are joined together in a linear fashion by **peptide bonds**. Hence a protein molecule is referred to as a **polypeptide chain** or, simply, as a **polypeptide**. The number of different proteins to be found in a bacterium of *E. coli* is estimated to be about 2×10^3 to 3×10^3, while for a human mammalian cell, the order of magnitude of this number is 10^6. A single protein molecule consists of anywhere from 50 to perhaps 10^4 amino acid molecules. Clearly, the diversity of proteins arises from the many possible ways of forming linear arrays or "words" (polypeptides) using twenty different "letters" (amino acids). The polypeptide chains of proteins are folded structures, which have two principal shapes, and are classified as being either **filamentous** proteins, also called **fibrous** proteins, or **globular** proteins. The latter are the most numerous.

The biological catalysts differ from all other catalysts known to chemistry in two essential ways. First, they are exceptionally efficient under the mild conditions of the normal physiological state: aqueous medium, standard pressure, and physiological temperature. Second, they exhibit **specificity**, acting selectively to bind small molecular species called **ligands**. A ligand that is acted upon by an enzyme to form a product is called a **substrate**. A single enzyme molecule can transform 10^3 to 10^6 molecules of substrate per minute. That is why their catalytic function in a cell can be performed rapidly and why extremely small quantities of enzyme suffice to carry out cellular processes. Enzymes also act as **regulators** of biochemical processes. Here, we wish to examine some aspects of the **kinetics** of enzymatic reactions, by which is meant their reaction rates, the temporal dependence of the reactants, and related properties.

46

2.1. THE MICHAELIS–MENTEN THEORY

The Michaelis–Menten equation has been successful in describing most enzymatically controlled reactions (Michaelis and Menten, 1913). We shall describe here its theoretical basis. Consider the simplest case of an enzymatic reaction, that between an enzyme and a single substrate. The fundamental assumption of the theory of Michaelis and Menten is that the enzyme and the substrate react reversibly to form a **complex** initially. The complex subsequently breaks down to form the free enzyme plus one or more **products**. The reactions are represented schematically as follows:

$$E + S \rightleftharpoons C, \tag{2.1a}$$

$$C \rightarrow E + P, \tag{2.1b}$$

Here E, S, C, and P stand for enzyme, substrate, complex, and product, respectively. The arrows in (2.1) indicate the possible directions of the reactions. The second assumption of the theory is that an equilibrium is rapidly established among E, S, and C in accordance with (2.1a) and in disregard of the reaction (2.1b). This approximation may be expected to be valid provided the reactions of (2.1a) proceed very rapidly as compared to the reaction (2.1b). We shall present here the mathematical representation of this theory, first given by Briggs and Haldane (1925).

In analytical chemistry, the concentration of a reactant in a solution is often specified, rather than the absolute amount. When the concentration represents the solute mass in moles per kilogram of solute, it is called the **molality**, and the unit is denoted by m. When the concentration is measured in moles per liter of solution, it is called the **molarity**, and the unit denoted by M. Here liter $= 10^3$ cm^3, a standard unit of volume. Thus cm$^3 =$ milliliter $=$ ml. We denote concentrations of E, S, C, and P by lower case letters. Naturally, the concentrations are time dependent functions. How do they vary? The answer is supplied by the rate equations corresponding to the reactions (2.1).

In accordance with the law of mass action, the rate equations are assumed to be the following:

$$\frac{ds}{dt} = -k_{+1}es + k_{-1}c, \tag{2.2a}$$

$$\frac{de}{dt} = -k_{+1}es + (k_{-1} + k_{+2})c, \tag{2.2b}$$

$$\frac{dc}{dt} = k_{+1}es - (k_{-1} + k_{+2})c, \tag{2.2c}$$

$$\frac{dp}{dt} = k_{+2}c. \tag{2.2d}$$

Equation (2.2a) states that the change Δs in the concentration of substrate in the infinitesimal time interval t to $t + \Delta t$ consists of two parts. One part is proportional to e, s, and Δt, and is a result of the **association** or formation of complex from enzyme and substrate. The other part is proportional to c and Δt, and is a consequence of the back reaction or **dissociation** of complex into enzyme and substrate. These proportionality relations become equalities with the introduction of the positive proportionality constants k_{+1} and k_{-1}, respectively. The minus sign in front of the term $k_{+1}es$ then indicates that the change in s resulting from complex formation is a decrease, while the plus sign in front of $k_{-1}c$ indicates that the change in s resulting from the back reaction is an increase. Similar interpretations are applicable to equations (2.2b)–(2.2d).

Note that all "growth" terms are positive and all "disappearance" terms are negative, so that the **rate constants** $k_{\pm i}$ are positive quantities by definition. As in the Lotka–Volterra equations, a nonlinear term, which in this case is proportional to es, appears. This term represents the collisions between enzyme and substrate molecules that lead to complex formation. Initially, it is assumed we start with enzyme and substrate only. Therefore,

$$(s,e,c,p) = (s_0,e_0,0,0) \qquad \text{at } t = 0, \tag{2.3}$$

where the subscript 0 denotes initial value. The temporal dependence of the concentrations is now completely known, provided we can solve the equation system (2.2)–(2.3).

The equations can be simplified by recognizing that the amount of enzyme is always conserved, existing either as free enzyme or as part of the complex. Mathematically, we observe, by adding equations (2.2b) and (2.2c), that $(d/dt)(e + c) = 0$, and therefore $e(t) + c(t)$ is a constant. The constant is determined by setting $t = 0$ when the values of e and c are known, from (2.3). Thus

$$e + c = e_0. \tag{2.4}$$

We utilize (2.4) to eliminate e from equations (2.2a) and (2.2c), which then become

$$\frac{ds}{dt} = -k_{+1}e_0s + (k_{+1}s + k_{-1})c, \tag{2.5a}$$

$$\frac{dc}{dt} = k_{+1}e_0s - (k_{+1}s + k_{-1} + k_{+2})c. \tag{2.5b}$$

This is a system of two nonlinear differential equations that is very similar to the host–parasite system already discussed. Clearly, if we solve these two equations, we can obtain p and e from equations (2.2d) and (2.4), respectively. However, we are not interested in the neighborhood of equilibrium states here, but wish to determine the velocity of the reaction, especially in the initial stages of the reaction. The initial velocity is usually measured.

In order to simplify the mathematical problem, we make use of the second assumption of the theory that a "quasi-steady state" is established very rapidly, so that the concentration of the complex C is changing very slowly with time. Thus, we assume that

$$\frac{dc}{dt} = 0. \tag{2.6}$$

With this assumption, equation (2.5b) becomes an algebraic equation that can be readily solved for c in terms of s, namely

$$c(t) = \frac{k_{+1}e_0s(t)}{k_{+1}s(t) + k_{-1} + k_{+2}}. \tag{2.7}$$

We now substitute (2.7) into (2.5a) and obtain the equation

$$\frac{ds}{dt} = -k_{+2}c = -\frac{k_{+1}k_{+2}e_0s}{k_{+1}s + k_{-1} + k_{+2}}. \tag{2.8}$$

This equation can be integrated directly to yield

$$s + \frac{k_{-1} + k_{+2}}{k_{+1}} \log \frac{s}{s_0} = s_0 - k_{+2}e_0t, \tag{2.9}$$

in which the initial condition has been imposed in order to determine the constant of integration.

Biochemists commonly wish to determine the **velocity of the reaction** v, which is usually defined as either the rate of appearance of the product, or the rate of disappearance of the substrate. These two rates are not strictly equal. However, with the approximation of the theory, it is seen from equations (2.2d) and (2.8) that these definitions are equivalent. Thus, the velocity of the reaction at time t is

$$v(t) = \frac{dp}{dt} = \left| \frac{ds}{dt} \right|. \tag{2.10}$$

Usually, reaction velocities are determined at $t = 0$. Denoting the initial value $v(0)$ by v_0, we obtain from equations (2.8) and (2.10) the relation

$$v_0 = \frac{Vs_0}{s_0 + K_m}, \tag{2.11}$$

where V and K_m are defined as

$$V \equiv k_{+2}e_0, \tag{2.12}$$

$$K_m \equiv \frac{k_{-1} + k_{+2}}{k_{+1}}. \tag{2.13}$$

Equation (2.11), the main result of the theory, is called the **Michaelis–Menten equation**. The constant K_m is called the **Michaelis constant**. The curve of v_0 as a function of s_0 is a rectangular hyperbola, and approaches its maximum value V asymptotically (see Figure 2.1). Thus V is the maximum value of the velocity of the reaction, or, simply, **v-max**. Because k_{+2} is the only reaction rate appearing in V, the **rate-limiting** step in the reaction is that represented by equation (2.1b). However, the rate-limiting constituent is the enzyme.

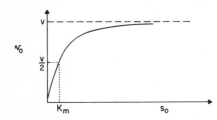

Figure 2.1. Illustrates the Michaelis–Menten equation (2.15).

For large values of s_0, we see that the velocity of the reaction is approximately V and is independent of the substrate concentration. The reaction velocity is then said to **saturate** with respect to the substrate or to exhibit the phenomenon of **saturation**. The introduction into a chemical reaction of a reactant in excess amount so as to make the reaction independent of the concentration of that reactant is a common biochemical technique for reducing the number of independent variables associated with the given reaction.

If k_{+2} is small compared to k_{-1}, as is the case for many reactions, it can be neglected in equations (2.2). The system then has a true equilibrium state, given by $ds/dt = de/dt = dc/dt = 0$, for which the interconversions among the molecular species leave the net number of molecules of a given species unchanging with time. Such a dynamical equilibrium state is also called a **steady state**. The equilibrium values of e, s, and c, denoted with an overbar, are found to be related by the equation

$$\frac{\overline{es}}{\overline{c}} = \frac{k_{+1}}{k_{-1}} \equiv K_d \approx K_m. \tag{2.14}$$

Hence in such a case, K_m is approximately the **equilibrium constant** or **dissociation constant** K_d for the reaction (2.1a). As the equations (2.2) stand, the only true equilibrium state for $e_0 \neq 0$ is given by $(e,s,c,p) = (e_0,0,0,s_0)$ and is not of interest. This state represents the state of total conversion of substrate to product. The back reaction $C \leftarrow E + P$ must exist on thermodynamic grounds, and complements the reaction (2.1b). Consequently, the complete conversion of substrate to product is never achieved. In practice, the rate constant for the back reaction is usually small. Therefore, the back reaction is usually neglected, as has been done here. When the back reaction is taken into account, the equilibrium constant relating the equilibrium concentrations of S and P can be expressed in terms of all the rate constants (see problem 1).

In experimental kinetic investigations, the velocity is measured for various values of the initial substrate s_0, and the curve shown in Figure 2.1 is determined. The kinetic properties of the enzyme in interaction with a particular substrate are characterized by the two parameters V and K_m. The value of K_m is determined according to equation (2.11) as the value of s_0 for which $v_0 = 1/2V$. A commonly employed equivalent representation of the data is the double reciprocal or **Lineweaver–Burk plot**, in which v_0^{-1} is plotted as a function of s_0^{-1} (Lineweaver and Burk, 1934). According to (2.11) this relationship is represented simply by a straight line,

$$\frac{1}{v_0} = \frac{1}{V} + \frac{K_m}{Vs_0}. \tag{2.15}$$

In Figure 2.2 it is seen that the ordinate intercept is V^{-1} and the abscissa intercept is $-K_m^{-1}$.

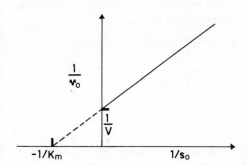

Figure 2.2. Lineweaver–Burk plot of the Michaelis–Menten equation for an enzyme reacting with a substrate.

2.2.*   EARLY TIME BEHAVIOR OF ENZYMATIC REACTIONS

If we examine more closely the solution we obtained for $c(t)$, equation (2.7), we observe that it does not satisfy the initial condition that $c(0) = 0$. In fact, the solution we have obtained is not valid in the neighborhood of $t = 0$. In spite of this fact, the inference regarding $v(0)$ is in good quantitative agreement with many enzymatically controlled reactions. In other words, $s(t)$ as given by (2.9) represents the experimental facts satisfactorily during the initiation of an enzymatically controlled reaction, but $c(t)$ as given by (2.7) does not. In order to understand this seemingly paradoxical result, we shall solve the rate equations (2.5) in a more formal albeit approximate manner, without making the restrictive assumption, equation (2.6).

We shall utilize a perturbation procedure to obtain a formal solution to equations (2.5) (Heineken, Tsuchiya, and Aris, 1967). To understand this procedure and the approximation implied by it, it is necessary to express our equations in nondimensional form by introducing the **nondimensional variables** $t' = k_{+1}e_0 t$, $s' = s/s_0$, and $c' = c/e_0$. Note that $k_{+1}e_0$, according to equations (2.5), has the dimensions of t^{-1} so that t' is dimensionless. It is called the "nondimensional time." If now equations (2.5) are divided by $k_{+1}e_0 s_0$ and the nondimensional variables are introduced, the equations take the form

$$\frac{ds'}{dt'} = -s' + \left(s' + \frac{k_{-1}}{k_{+1}s_0}\right)c', \tag{2.16a}$$

$$\frac{e_0}{s_0}\frac{dc'}{dt'} = s' - \left(s' + \frac{k_{-1} + k_{+2}}{k_{+1}s_0}\right)c'. \tag{2.16b}$$

In this form we see that e_0/s_0 is one of three nondimensional parameters characterizing the system, the other two being $k_{-1}/k_{+1}s_0$, and $(k_{-1} + k_{+2})/k_{+1}s_0$. In experimental investigations of enzyme kinetics it is usually the case the $s_0 \gg e_0$ and a value of $e_0/s_0 \sim 10^{-3}$ is not uncommon. We shall therefore treat the quantity e_0/s_0 as a very small quantity, and seek a solution to the above equation system in the form

$$s' = s'^{(0)} + \frac{e_0}{s_0}s'^{(1)} + \left(\frac{e_0}{s_0}\right)^2 s'^{(2)} + \cdots,$$

$$c' = c'^{(0)} + \frac{e_0}{s_0}c'^{(1)} + \left(\frac{e_0}{s_0}\right)^2 c'^{(2)} + \cdots. \tag{2.17}$$

Here $s'^{(j)} = s'^{(j)}(t')$ and $c'^{(j)} = c'^{(j)}(t')$, where j is an integer or zero, are functions of the time to be determined. Either of these series is called an **asymptotic expansion** because it tends to the true solution as the small parameter (e_0/s_0) tends to zero. The method of finding a solution to an equation system such as (2.16) by means of such a series is called a **singular perturbation procedure**. It is called a perturbation procedure because the solution is expanded as a power series in a small parameter. The procedure is singular when the small quantity or perturbation multiplies the highest derivative term appearing in the differential equations.

If we substitute equations (2.17) into (2.16) and neglect all terms that are of order e_0/s_0, we obtain two equations for $s'^{(0)}$ and $c'^{(0)}$. These equations are exactly the same as equation (2.5) with dc/dt set equal to zero. In other words, they are exactly the approximate equations we considered in the previous section. Thus, the solution we found there for $s(t)$ and $c(t)$ constitute the first term in the expansion (2.17). However, it is valid only for large values of t, and is called the **outer solution**.

To investigate the problem mathematically for early times, we cannot neglect the left-hand side of equation (2.16b) but must treat it on the same footing as the right-hand side. To accomplish this, we introduce a new nondimensional time $\tau = k_{+1}s_0 t = s_0 t'/e_0$ in place of the old one. Because the new nondimensional variable is much larger than the old one by a factor s_0/e_0, it is called a **stretched variable**. We introduce also the new functions $\tilde{s}(\tau)$ and $\tilde{c}(\tau)$ by the equations $\tilde{s}(\tau) = s'(\tau e_0/s_0)$, and $\tilde{c}(\tau) = c'(\tau e_0/s_0)$. In terms of these variables, equations (2.16) take the form

$$\frac{d\tilde{s}}{d\tau} = \frac{e_0}{s_0}\left[-\tilde{s} + \left(\tilde{s} + \frac{k_{-1}}{k_{+1}s_0}\right)\tilde{c}\right], \tag{2.18a}$$

$$\frac{d\tilde{c}}{d\tau} = \tilde{s} - \left(\tilde{s} + \frac{k_{+1} + k_{+2}}{k_{+1}s_0}\right)\tilde{c}. \tag{2.18b}$$

It is seen that the small parameter e_0/s_0 now multiplies a different term than the one it multiplied previously. Again, let us neglect terms of order e_0/s_0. Equation (2.18a) becomes $d\tilde{s}/d\tau = 0$, with the solution $\tilde{s} = $ constant. After imposing the initial condition, the solution becomes

$$\tilde{s} = 1, \text{ or } s = s_0. \tag{2.19}$$

Now we substitute this result into (2.18b), after which that equation is readily integrable. After imposing the initial condition, we obtain the concentration of the complex, which, expressed in terms of the original dimensional variables, is

$$c = \overline{c}[1 - e^{-(k_{+1}s_0 + k_{-1} + k_{+2})t}],$$

$$\overline{c} = \frac{k_{+1}e_0 s_0}{k_{+1}s_0 + k_{-1} + k_{+2}}. \tag{2.20}$$

According to (2.20), the complex concentration c increases from its value 0 at $t = 0$ to a final value $\overline{c}$ which we call the **quasi-steady-state value**. At the same time, according to (2.19), the substrate concentration remains unchanged. Of course, these conditions cannot apply to the state of affairs at "large" times, because we know, for example, that the substrate begins to get exhausted. The expressions given by (2.19) and (2.20) satisfy the initial conditions and represent the solution for "small" times. Together they comprise the **inner solution**. The inner solution, valid for short times, plus the outer solution, valid for long times, comprise the total solution to the problem, valid to the order e_0/s_0.

How long does it take for c to increase from zero to its quasi-steady-state value? The answer is the time it takes for the exponential term in (2.20) to become negligible, for which the characteristic decay time τ_0 appearing in the exponent is given by the expression

$$\tau_0 = (k_{+1}s_0 + k_{-1} + k_{+2})^{-1}. \tag{2.21}$$

For enzymes, this time is of the order of a second or less. The reason this time is so short is that k_{+1} for enzymatic reactions is usually found to be in the range $10^6/M$ sec to $10^8/M$ sec (see Table 2.1).

We note that the limiting value of the inner solution as $t \to \infty$, $(s,c) = (s_0,\bar{c})$, is the same as the limiting value of the outer solution as $t \to 0$, as is readily verified from equations (2.7), (2.9), (2.19), and (2.20). This fact indicates that the two solutions have a common time domain of validity, that is, the outer solution takes up where the inner solution leaves off. The term in the solution that is common to both solutions is called the **overlap term**. An alternative way of writing the total solution (up to a given order in e_0/s_0) is to write it as the sum of the inner solution and the outer solution, minus the overlap term. Thus, the total solution for s and c up to order e_0/s_0 can also be written in the form

$$s + K_m \log \frac{s}{s_0} = s_0 - Vt, \tag{2.22a}$$

$$c = e_0 \left[\frac{s}{s + K_m} - \frac{s_0}{s_0 + K_m} e^{-t/\tau_0} \right]. \tag{2.22b}$$

Graphs of s, c, and p are shown schematically in Figure 2.3. Note that

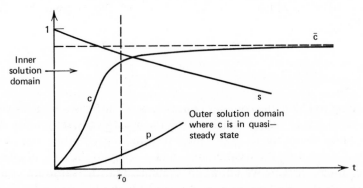

Figure 2.3. Schematic representation of the behavior with time of the concentrations s, c, and p of an enzyme–substrate system.

$\bar{c} \to e_0$ as $s_0 \to \infty$. Also, the total solution for s happens to be the same as the outer solution, to this order in e_0/s_0. The total solution given by (2.22) is not subject to the paradox set forth at the beginning of this section.

We are now in a position to understand better what occurs in an enzymatically controlled reaction. The reciprocal of the Michaelis constant is considered to be a quantitative measure of the **affinity** that the enzyme holds for a particular substrate. Other things being equal, we see that the affinity is greater, the larger the value of the forward rate constant k_{+1} (see Table 2.1). The further consequence of a large forward rate constant is that the complex rapidly attains a quasi-steady-state value. That is why the Michaelis–Menten equation, which utilizes an expression for v_0 that depends only on the outer solution, works so well. The time interval during which v_0 is usually measured is of the order of minutes. In fact, the student of biochemistry is admonished to determine the velocity of the reaction as close to the time origin as possible. However, using standard techniques of measurement, it is not possible to observe the initial build up of complex during the time τ_0. We may attribute the wide success of the Michaelis–Menten theory, inter alia, to these circumstances.

We can also understand why only two parameters serve to characterize the reaction velocity v_0 when three parameters k_{+1}, k_{-1}, and k_{+2} enter into the rate equations. The solution for s up to order e_0/s_0 is the same as the outer solution. The outer solution for s to this order depends on the parameters V and K_m only, as can be seen from equation (2.8) and the solution (2.22a).

The complete characterization of an enzymatically reaction obeying the rate equations (2.2) requires the determination of all three rate constants in the rate equations. The knowledge of only K_m and V is insufficient for this purpose. One way to supplement this knowledge is to determine the temporal dependence of s or p during the first few seconds of the reaction. These observations require special techniques. The results could be compared to the inner solution, and would provide an additional relationship that the rate constants must satisfy. However, it would then be necessary to develop the inner solution for s, say, to include the term of order e_0/s_0 in equation (2.17). This term is found by solving equation (2.18a) for s with c in it replaced by the expression (2.20). An alternate scheme for determining rate constants is the method of **chemical relaxation spectrometry** (Eigen, 1954; Eigen and DeMaeyer, 1963), discussed in Chapter 3, problem 14. Rate constants for formation and dissociation of complex, associated with some enzyme-substrate reactions, are shown in Table 2.1.

It is common to refer to the approximation whereby the Michaelis–Menten equation was obtained as the steady-state approximation. However, only dc/dt was set equal to zero, and not ds/dt. Furthermore, we see from the solution (2.7) or (2.22b) that c is not strictly constant for $t \gg \tau_0$, but varies

TABLE 2.1   RATE CONSTANTS FOR FORMATION AND DISSOCIATION
OF COMPLEX, OF SOME ENZYME–SUBSTRATE SYSTEMS.[a]

Enzyme	Substrate	k_{+1} $(M^{-1}sec^{-1})$	k_{-1} (sec^{-1})
Fumarase	Fumarate	$> 10^9$	$> 4.5 \times 10^4$
Glutamic-aspartic transaminase (aminic)	Oxalacetate	7×10^7	1.4×10^2
Glutamic-aspartic transaminase (aminic)	Ketoglutarate	2.1×10^7	70
Peroxidase	H_2O_2	9×10^6	< 1.4
Hexokinase	Glucose	3.7×10^6	1.5×10^3
Old yellow enzyme	FMN (flavin mononucleotide)	1.5×10^6	$\sim 10^{-4}$
Liver alcohol dehydrogenase	NAD (nicotinamide adenine dinucleotide)	5.3×10^5	74

[a] From the review of Eigen and Hammes (1963).

slowly, or in a **quasistatic** manner. Therefore, it is more appropriate to characterize the theory as a **quasi-steady-state** approximation.

2.3.   ENZYME–SUBSTRATE–INHIBITOR SYSTEM

The principal result of Michaelis–Menten theory, equation (2.11) or equivalently equation (2.15), did not really require the solution to any differential equations. The result followed formally from the underlying differential equations (2.5) by setting dc/dt equal to zero in (2.5b), denoting ds/dt in equation (2.5) by v_0, and eliminating c between these two equations. We did solve these equations in Section 2.2 in order to understand better the implications of the theory.

We shall now derive the result of applying Michaelis–Menten theory to the case in which two ligands compete for the same enzymatic site. Such a reaction is said to be **fully competitive**. The reactant that is singled out for measurement is called the **substrate**. The second reactant is called the **inhibitor**. Clearly, their roles may be reversed in subsequent studies. The reactions are represented schematically as

$$S + E \underset{k_{-1}}{\overset{k_{+1}}{\rightleftharpoons}} C_1 \overset{k_{+2}}{\rightarrow} P_1 + E,$$
$$I + E \underset{k_{-3}}{\overset{k_{+3}}{\rightleftharpoons}} C_2 \overset{k_{+4}}{\rightarrow} P_2 + E,$$

$$(2.23)$$

where I represents the inhibitor. Note that each allowed transition has been labeled with its associated rate constant. Again, we have neglected the reversibility of the reactions that form products in (2.23). For the "short" times during which such reactions are usually investigated, the amount of product formed is small, and the importance of the back reaction is thereby diminished. The kinetic equations corresponding to these reactions are the following:

$$\frac{ds}{dt} = -k_{+1}e_0 s + (k_{+1}s + k_{-1})c_1 + k_{+1}sc_2,$$

$$\frac{dc_1}{dt} = k_{+1}e_0 s - (k_{+1}s + k_{-1} + k_{+2})c_1 - k_{+1}sc_2,$$

$$\frac{di}{dt} = -k_{+3}e_0 i + (k_{+3}i + k_{-3})c_2 + k_{+3}ic_1, \qquad (2.24)$$

$$\frac{dc_2}{dt} = k_{+3}e_0 i - (k_{+3}i + k_{-3} + k_{+4})c_2 - k_{+3}ic_1,$$

$$e = e_0 - c_1 - c_2.$$

The concentrations are subject to the following initial condition,

$$(s,c_1,i,c_2,e) = (s_0,0,i_0,0,e_0), \text{ at } t = 0. \qquad (2.25)$$

In deriving equations (2.24), we have followed the same procedure utilized in deriving equations (2.5) from (2.2). We have omitted the equations representing the concentrations of the products P_1 and P_2. Now assume that e_0/s_0 is very small, and that both k_{+1}/k_{+3} and s_0/i_0 behave like constants as e_0/s_0 tends to zero. We express these assumptions in mathematical notation as follows:

$$\frac{e_0}{s_0} \ll 1, \frac{k_{+1}}{k_{+3}} \sim 1, \frac{s_0}{i_0} \sim 1. \qquad (2.26)$$

These assumptions are the mathematical requirements of the quasi-steady-state hypothesis.

Formally, in correspondence to the quasi-steady-state consideration of equations (2.5) of the single substrate-enzyme system, we set dc_1/dt and dc_2/dt equal to zero in (2.24). The two algebraic equations which result permit us to solve for c_1 and c_2 in terms of s and i. Hence, c_1 and c_2 can be eliminated from the differential equations for s and i.

Denoting the velocity of the reaction by $v(t) = |ds/dt|$, it follows from (2.24) that, in the quasi-steady-state approximation,

$$\frac{1}{v(t)} = \frac{1}{V^s}\left[1 + \frac{K_m^s}{s(t)}\left(1 + \frac{i(t)}{K_m^i}\right)\right], \qquad (2.27)$$

where

$$K_m^s = \frac{k_{-1} + k_{+2}}{k_{+1}}, \ K_m^i = \frac{k_{-3} + k_{+4}}{k_{+3}}, \ V^s = k_{+2}e_0. \qquad (2.28)$$

The result (2.27) is called the **time-dependent Michaelis–Menten equation** for an enzyme–substrate–inhibitor system. The differential equations for $s(t)$ and $i(t)$ in the quasi-steady-state approximation can be readily solved, in the same manner as described in the previous section for the case when the inhibitor was absent (Rubinow and Lebowitz, 1970). The result is

$$s(t) - s_0 - \frac{V^s}{V^i} i_0 \left\{ 1 - \left[\frac{s(t)}{s_0} \right]^\delta \right\} + K_m^s \log \left[\frac{s(t)}{s_0} \right] = -V^s t,$$

$$\frac{i(t)}{i_0} = \left[\frac{s(t)}{s_0} \right]^\delta, \qquad (2.29)$$

where δ is defined as

$$\delta = \frac{V^i K_m^s}{V^s K_m^i} = \frac{k_{+4}(k_{-1} + k_{+2})}{k_{+2}(k_{-3} + k_{+4})}, \ V^i = k_{+4}e_0.$$

To obtain the usual time-independent form of the Michaelis–Menten equation, set $t = 0$ in equation (2.27), and denote by v_0 the initial velocity of the reaction $v(0)$. Then

$$\frac{1}{v_0} = \frac{1}{V^s} \left[1 + \frac{K_m^s}{s_0} \left(1 + \frac{i_0}{K_m^i} \right) \right]. \qquad (2.30)$$

The result is that a Lineweaver–Burk plot of v_0^{-1} versus s_0^{-1} is again a straight line, except that the slope depends on the particular value of i_0 (see Figure 2.4). Note that the intercept on the v_0^{-1} axis is a fixed point, regardless of the value of i_0. From a set of such straight lines, both K_m^s and K_m^i

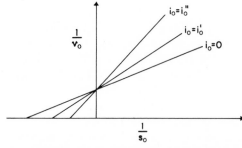

Figure 2.4. Lineweaver–Burk plot for a fully competitive enzyme–substrate–inhibitor system for various initial inhibitor concentrations i_0

can be determined. To determine V^i, however, an additional experiment in which the roles of inhibitor and substrate are reversed would have to be performed.

One word of caution is necessary. The result (2.30) depends on the tacit assumption, *not* covered by the assumptions in (2.26) above, that i_0 does not change (decrease) very much during the course of the experiment in which v_0 is measured. From (2.29) it is seen that the role of the parameter δ is critical in determining whether the fractional decrease of inhibitor is small or large. If δ is of the order of 1 or less, the change in inhibitor concentration will be slight during the course of the experiment. However, if δ is large compared to unity, the fractional decrease in the inhibitor concentration may be quite large even if $s(t)$ changes by 1–2% during the experiment. In such a case the Michaelis–Menten equation (2.30) is no longer valid and the time-dependent form of the theory of the reaction is needed, equation (2.27). It then becomes more difficult to infer the values of the characteristic parameters, defined in (2.28), from experiment.

Investigations of enzymatic reactions are utilized, among other things, for the purpose of making inferences concerning the nature of the reactions. Thus, the existence of intersecting straight lines as shown in the previous figure is sufficient to characterize the reaction of an enzyme with two reactants as "fully competitive." However, we infer from the mathematical analysis of such reactions that the converse is not true: We cannot be sure that two ligands are not fully competitive, because of the lack of appearance of straight lines intersecting on the v_0^{-1} axis in a Lineweaver–Burk plot. To check whether the lack of straight lines is not a "fully competitive" result, the parameter δ should be determined to see if it has an anomalous large value. If it does, then the two substrates may yet be fully competitive. If they are, a reversal of the roles of substrate and inhibitor will yield a straight line because then the value of the parameter δ is the reciprocal of its previous value (i.e., from (2.29), $s(t)/s_0 = [i(t)/i_0]^{1/\delta}$).

An example of such an anomalous enzyme–substrate–inhibitor system is that of the enzyme **asparaginase**, which reacts with the amino acid asparagine (and water) to yield the product aspartate (and ammonia). Enzymes are frequently named by adding the suffix-*ase* to the name of the substrate acted upon. It has been found that asparaginase derived from the bacterium *E. coli* is capable of reacting with the two substrates asparagine and glutamine. When reacting with the amino acid glutamine (and water), asparagine yields the product glutamate (and ammonia).

In this system $V^{\mathrm{asp}}/V^{\mathrm{glu}} \approx 16$, while $K_m^{\mathrm{asp}}/K_m^{\mathrm{glu}} \approx 0.1$, so that $\delta \approx 160$ when the "fast" substrate asparagine is the "inhibitor" and the "slow" substrate glutamine is the "substrate". In this case, Lineaweaver–Burk plots yield a family of curves that are not straight lines (Miller and Balis, 1969). When the roles of glutamine and asparagine are reversed, however, so that asparagine is the substrate and glutamine is the inhibitor, $\delta = 1/160$, and Lineweaver–Burk plots do yield a family of straight lines, as shown in Figure 2.4.

2.4. COOPERATIVE PROPERTIES OF ENZYMES

The macromolecular structure of enzymes permits them to engage in a variety of interactions involving many kinds of bonds. In general, however, an enzyme catalyzes only a single reaction. This great specificity is believed to be a **steric** property, that is to say, it depends on the shape of the enzyme molecule in a local region. A local region of interaction or "contact" between the enzyme molecule and a reacting substrate molecule is called an **active center** or **active site**.

Many enzymes consist of **subunits**, which are globular polypeptides. A protein consisting of many subunits is called an **oligomer**. The term is not usually applied to proteins consisting of more than about twenty subunits. The oligomer is called a **dimer**, **trimer**, **tetramer**, and so on, when the number of subunits is respectively 2, 3, 4, and so on. Subunits which are identical are called **protomers**. A protomer obtained by dissociation of the oligomer may also be referred to as a **monomer**, or **monomeric unit**. In some enzymes, each subunit contains an active center. We wish to discuss the kinetic behavior of enzymes with more than one active centers.

We shall first consider the theory for an idealized protein which is an oligomer consisting of n identical protomers, each containing one active center (see Figure 2.5). The active sites are assumed to be **independent** of each

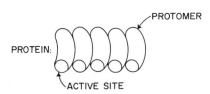

PROTOMER

PROTEIN:

ACTIVE SITE

Figure 2.5. Schematic representation of an idealized protein consisting of several identical protomers, each containing one active site.

other in their interaction with the molecules of a ligand, which we shall again call the substrate. Denote the substrate by S, and the complex of the protein combined with j ligand molecules by C_j, where j runs from 0 (meaning the protein is bare) to n (meaning the protein is fully reacted with n substrate molecules). The individual reactions are represented as follows,

$$S + C_j \rightleftharpoons C_{j+1}, \qquad j = 0, 1, 2, \ldots, n - 1. \qquad (2.31)$$

Denote the rate constant for binding the substrate ligand to a particular site of the protein as k_{+1} for association and k_{-1} for dissociation. We shall further assume that there is an abundance of substrate, and its concentration is varying so slowly that we may treat it as constant and set its value equal to s_0.

As an alternative to the representation (2.31) of the reactions, we shall introduce Figure 2.6. Such a diagram has a one-to-one correspondence with the rate equations for the concentrations of C_j. Thus, a distinct protein state is represented by an appropriately labeled circle. Such a circle is called a node or vertex, and a reaction pathway connecting two states is represented by a directed line segment or branch. Because of this correspondence, such diagrams have been studied in their own right and are called **graphs**. The mathematical study of such graphs is called **graph theory**.

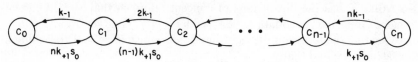

Figure 2.6. Representation by a graph of the system of reactions (2.31).

Each vertex has a rate equation associated with it. To find it, choose a vertex and set the time rate of change of the concentration of material associated with that vertex equal to a sum of terms, one for each branch entering and one for each branch leaving the vertex. A term is the product of the branch label with the concentration of the node from which the branch emanates. The sign of the term is positive or negative, accordingly as the arrowhead points towards or away from the chosen vertex.

Note that the rate constant in going from C_0 to C_1 is $nk_{+1}s_0$ because there are n unoccupied sites in the state C_0. It is $k_{+1}s_0$ in going from C_{n-1} to C_n because only one site is unoccupied. Similarly, in going from C_1 to C_0, there is only one site occupied, and the rate constant for dissociation is therefore k_{-1}. In going from C_n to C_{n-1}, there are n sites occupied by ligand molecules, and the rate constant is nk_{-1}. We see that the assumption of independence of binding sites leads to particularly simple interrelations among the rate constants for the various transitions.

By following the above prescription for inferring the rate equations from the graph, we obtain

$$\frac{dc_0}{dt} = -nk_{+1}s_0c_0 + k_{-1}c_1,$$

$$\frac{dc_j}{dt} = (n + 1 - j)k_{+1}s_0c_{j-1} - jk_{-1}c_j \qquad (2.32)$$

$$- (n - j)k_{+1}s_0c_j + (j + 1)k_{-1}c_{j+1}, \qquad j = 1, 2, \ldots, n - 1,$$

$$\frac{dc_n}{dt} = k_{+1}s_0c_{n-1} - nk_{-1}c_n.$$

We shall now make the steady-state hypothesis, setting the time derivatives in (2.32) equal to zero, and solve for the steady-state or equilibrium values of c_j. Define the equilibrium constant K as

$$K \equiv \frac{k_{-1}}{k_{+1}}. \tag{2.33}$$

Consider first the steady-state condition $dc_n/dt = 0$. It follows that the equilibrium values of c_n and c_{n-1} are related by the equation

$$c_n = \frac{1}{n} \frac{s_0}{K} c_{n-1}. \tag{2.34}$$

Note that K has the dimensions of concentration, so that s_0/K is dimensionless. Going next to equations (2.32) with $j = n - 1$ and setting $dc_{n-1}/dt = 0$, we see that the terms on the right-hand side vanish in pairs. Proceeding in this fashion, we find that

$$c_{n-1} = \frac{2}{n-1} \frac{s_0}{K} c_{n-2},$$

$$\cdots\cdots\cdots\cdots\cdots\cdots\cdots\cdots\cdots$$

$$c_2 = \frac{n-1}{2} \frac{s_0}{K} c_1, \tag{2.35}$$

$$c_1 = n \frac{s_0}{K} c_0.$$

By combining the equations appearing in (2.34) and (2.35), we see that all the equilibrium values of c_j for $j \geq 1$ may be expressed in terms of c_0 in a regular fashion. Thus

$$c_2 = \frac{n(n-1)}{1 \cdot 2} \left(\frac{s_0}{K}\right)^2 c_0,$$

$$c_3 = \frac{n(n-1)(n-2)}{1 \cdot 2 \cdot 3} \left(\frac{s_0}{K}\right)^3 c_0,$$

and so forth. In general

$$c_j = b_j^n x^j c_0, \qquad j = 0, 1, 2, \ldots, n, \tag{2.36}$$

where the nondimensional substrate concentration x is defined as

$$x \equiv \frac{s_0}{K}. \tag{2.37}$$

The quantity b_j^n is the **binomial coefficient** defined as

$$b_j^n = \frac{n!}{j!(n-j)!}, \qquad j, n \text{ integers}, j \quad n, \tag{2.38}$$

and $n!$ is the **factorial function**, $n! \equiv 1 \cdot 2 \cdot 3 \cdots (n-1) \cdot n$. The b_j^n is called a binomial coefficient because it first arose in connection with the binomial expansion of the quantity $(x + y)^n$, namely,

$$(x + y)^n = \sum_{j=0}^{n} b_j^n x^{n-j} y^j. \tag{2.39}$$

Note that, by definition, $0! \equiv 1$, so that $b_0^n = b_n^n = 1$.

The fraction of sites actually bound by the ligand is denoted by $Y(s_0)$ and is called the **saturation function**. It is defined by the equation

$$Y(s_0) \equiv \frac{\sum_{j=1}^{n} j c_j}{n \sum_{j=0}^{n} c_j}. \tag{2.40}$$

From equations (2.36) and (2.39) with $y = 1$, it follows that

$$\sum_{j=0}^{n} c_j = c_0 \sum_{j=0}^{n} b_j^n x^j = c_0 (1 + x)^n. \tag{2.41}$$

Also, because the operations of differentiation and summation can be commuted,

$$\sum_{j=1}^{n} j c_j = c_0 \sum_{j=1}^{n} j b_j^n x^j = c_0 x \frac{d}{dx} \sum_{j=0}^{n} b_j^n x^j = c_0 x \frac{d}{dx} (1 + x)^n,$$

or

$$\sum_{j=1}^{n} j c_j = c_0 x n (1 + x)^{n-1}. \tag{2.42}$$

From the results (2.41) and (2.42), the saturation function Y defined by (2.40) becomes

$$Y(s_0) = \frac{x}{1 + x} = \frac{s_0}{K + s_0}. \tag{2.43}$$

As is easily verified from equation (2.40) with $n = 1$, this is exactly the same result that would be obtained if the protein were a monomer!

By the same token, suppose that the protein is an enzyme, and that in addition to the reactions (2.31), the reactions

$$C_i \rightarrow C_{i-1} + P, \qquad i = 1, 2, \ldots, n, \tag{2.44}$$

where P is a product, are also occurring. Let k_{+2} be the rate constant for conversion of a ligand molecule to a product molecule at a particular site. The effect of the inclusion of (2.44) on the rate equations (2.32) for c_i is to replace k_{-1} wherever it appears by $(k_{-1} + k_{+2})$. Then, with the assumption of a quasi-steady state for the conversion of substrate to product, the Michaelis constant $K_m = (k_{-1} + k_{+2})/k_{+1}$ replaces the equilibrium constant K in equation (2.33) and those following it. The velocity of the reaction v_0 is given by the expression

$$v_0 \equiv k_{+2} \sum_{j=1}^{n} jc_j = V \frac{\sum_{j=1}^{n} jc_j}{n \sum_{j=0}^{n} c_j}, \tag{2.45}$$

where, by definition, $V = nk_{+2}e_0$, and e_0 is the total initial enzyme concentration,

$$e_0 = \sum_{j=0}^{n} c_j. \tag{2.46}$$

Hence, from (2.40) and (2.43), with K_m replacing K,

$$v_0 = \frac{Vs_0}{K_m + s_0}. \tag{2.47}$$

From this result, too, we see that, as far as the reaction velocity with respect to a given substrate is concerned, an oligomeric enzyme that consists of an arbitrary number of independent identical protomeric units is indistinguishable from an enzyme that is a monomer. An oligomeric enzyme that possesses a reaction velocity that is different from the monomeric result (2.47) is said to be a **cooperative system**. Hence, the idealized oligomeric enzyme or protein described above can be called a **noncooperative system**, or its behavior can be described as **noncooperative**. Alternatively, it can be said to display **zero cooperativity**.

It has been noted that in many cases of enzymatically controlled reactions, the curve of v_0 versus s_0 has an inflection point and is S-shaped before attaining its maximum (see Figure 2.7). In fact, the curve sometimes decreases after attaining a maximum value. The explanation of this phenomenon that is considered to be the most plausible is that the enzyme possesses several active centers with respect to the given substrate. Furthermore, the rate constants for association and dissociation or the equilibrium constant at one active site apparently depend upon whether or not ligands are bound at other active sites. In other words, the enzyme (or more generally, protein) is behaving as a cooperative system.

The recognition of cooperativity in the kinetic behavior of an enzyme is a matter of some concern. Usually, when cooperativity is suspected, a **Hill plot** is made. This procedure depends essentially on the assumption that v_0

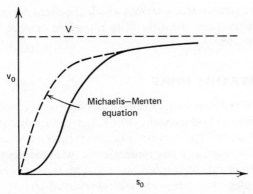

Figure 2.7. The solid line illustrates the sigmoidal shape of the reaction rate v_0 as a function of initial substrate concentration s_0 for a cooperative system. The dashed line contrasts the curve for $v_0(s_0)$ when the system is not a cooperative one.

can be represented by the Hill equation, which is equation (2.11) with s_0 replaced by s_0^n (Hill, 1910). The exponent n is determined from a plot of $\log(v_0/(V - v_0))$ versus $\log s_0$, called a Hill plot. The resulting curve is a straight line with slope n, if the Hill equation is applicable. The commonly accepted definition of positive, zero, or negative cooperativity is that n is greater than, equal to, or less than unity, respectively. Unfortunately, the Hill equation, if derived from differential rate equations in the same manner as the Michaelis–Menten equation was derived, depends on the somewhat restrictive assumption that the reaction between the substrate and the enzyme is of the nth order with respect to the substrate. That is to say, to obtain the Hill equation, the concentration s appearing in equations (2.2) must be replaced by s^n, where n is the number of active sites on the enzyme molecule. This means that intermediate states between the bare enzyme molecule and the complex formed from it and n molecules are neglected.

When a Lineweaver–Burk plot is made of the Hill equation, it is found that for values of n greater than, equal to, or less than unity, the resulting curves display positive, zero, or negative curvature, respectively (Levitzki and Koshland, 1969). This result suggests the following quantitative operational test for cooperativity of an oligomeric enzyme in interaction with a substrate, which is independent of the Hill equation. Make a Lineweaver–Burk plot of the reciprocal reaction velocity $1/v_0$ versus the reciprocal initial substrate concentration $1/s_0$. If the resulting curve is a straight line (see Figure 2.9a), the enzyme displays noncooperative behavior. If the resulting curve possesses positive curvature (concave upwards), the enzyme is said to display **positive cooperativity**. If the resulting curve possesses negative curvature (concave downwards), the enzyme is said to display **negative**

cooperativity. We remark that enzymes which are characterizable as cooperative by these criteria may be difficult to recognize as such from the curve $v_0(s_0)$ (see Figure 2.9b).

2.5. THE COOPERATIVE DIMER

We shall consider the simplest example of an oligomeric cooperative system, a cooperative dimer (Volkenshtein, 1969). Let each subunit of the dimer contain an active center. The enzyme molecule, in interaction with a substrate, has three possible states: E, a free molecule; C_1, with one center free and the other occupied by a substrate molecule; C_2, with both centers occupied by substrate molecules. The reactions are represented as

$$S + E \underset{k_{-1}}{\overset{k_{+1}}{\rightleftharpoons}} C_1 \overset{k_{+2}}{\to} E + P,$$
$$S + C_1 \underset{k_{-3}}{\overset{k_{+3}}{\rightleftharpoons}} C_2 \overset{k_{+4}}{\to} C_1 + P.$$

$$(2.48)$$

Because the sites are no longer independent, there is no simple relationship between k_{+1} and k_{+3}, or between k_{-1} and k_{-3}, as there was for a noncooperative system. We shall invoke the quasi-steady-state hypothesis as in Michaelis–Menten theory. Hence, we treat the substrate concentration as constant and equal to its initial value s_0. The graph representing these reactions is then as follows (Figure 2.8).

By following the prescription for writing down the rate equations from such a graph (see equations (2.32)), we find that the rate equations for the concentrations e, c_1, c_2, and p are

$$\frac{de}{dt} = -k_{+1}s_0 e + (k_{-1} + k_{+2})c_1,$$

$$\frac{dc_1}{dt} = k_{+1}s_0 e - (k_{-1} + k_{+2} + k_{+3}s_0)c_1 + (k_{-3} + k_{+4})c_2, \quad (2.49)$$

$$\frac{dc_2}{dt} = k_{+3}s_0 c_1 - (k_{-3} + k_{+4})c_2,$$

$$\frac{dp}{dt} = k_{+2}c_1 + k_{+4}c_2. \quad (2.50)$$

In addition, because of conservation of enzyme molecules,

$$e_0 = e + c_1 + c_2. \quad (2.51)$$

To apply Michaelis–Menten theory to this system, we impose the steady-state condition on the enzyme and its complexes, that is, we set $de/dt = dc_1/dt = dc_2/dt = 0$. In addition, we denote dp/dt by v_0, the quasi-steady-

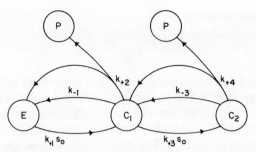

Figure 2.8. The graph for a dimeric enzyme E interacting with a substrate of concentration s_0 to form a product P. The complexes C_1 and C_2 contain, respectively, one and two substrate molecules.

state initial reaction velocity. Equations (2.49) become a set of three homogeneous linear algebraic equations for e, c_1, and c_2. Because of homogeneity, these three equations can only determine e, c_1, and c_2 up to a constant factor. We therefore replace one of the set (2.49) by (2.51), which contains the inhomogeneous term e_0. We can then solve for the steady-state values of e, c_1, and c_2, uniquely obtaining the result

$$e = e_0 \left[1 + \frac{s_0}{K_m} + \frac{s_0^2}{K_m K_m'} \right]^{-1}, \tag{2.52a}$$

$$c_1 = \frac{e_0 s_0}{K_m} \left[1 + \frac{s_0}{K_m} + \frac{s_0^2}{K_m K_m'} \right]^{-1}, \tag{2.52b}$$

$$c_2 = \frac{e_0 s_0^2}{K_m K_m'} \left[1 + \frac{s_0}{K_m} + \frac{s_0^2}{K_m K_m'} \right]^{-1}, \tag{2.52c}$$

where

$$K_m = \frac{k_{-1} + k_{+2}}{k_{+1}}, \tag{2.52d}$$

$$K_m' = \frac{k_{-3} + k_{+4}}{k_{+3}}. \tag{2.52e}$$

By substituting equations (2.52b) and (2.52c) into (2.50), the reaction velocity v_0 is determined to be

$$v_0 = \frac{e_0 s_0 (k_{+2} K_m' + k_{+4} s_0)}{[K_m K_m' + K_m' s_0 + s_0^2]}. \tag{2.53}$$

Let $V = k_{+2} e_0$, $k_{+4} = \alpha k_{+2}$, $K_m' = \beta K_m$, and $K_m/s_0 = \xi$. Then equation

(2.53) can be rewritten as

$$\frac{1}{v_0} = \frac{\beta \xi^2 + \beta \xi + 1}{V(\beta \xi + \alpha)}. \tag{2.54}$$

Hence, cooperativity is accounted for by the assignment of suitable values to the two parameters α and β. Speaking qualitatively, the parameter β is to some extent a relative measure of the binding of substrate to enzyme in the two states C_1 and C_2, while α is a relative measure of product formation from the two states C_1 and C_2. The dependence of v_0 on the rate constants associated with the state C_2 appears above in these two parameters only. By differentiation of equation (2.54) with respect to ξ, we compute

$$\frac{d}{d\xi}\left(\frac{1}{v_0}\right) = \frac{\beta}{V} \frac{(\beta \xi^2 + 2\alpha \xi + \alpha - 1)}{(\beta \xi + \alpha)^2},$$

$$\frac{d^2}{d\xi^2}\left(\frac{1}{v_0}\right) = \frac{2\beta}{V} \frac{[\alpha^2 + \beta(1 - \alpha)]}{(\beta \xi + \alpha)^3}.$$

In the last expression on the right the denominator is always positive, and we see that the curvature depends on the sign of the quantity in parenthesis in the numerator on the right-hand side. Thus, if $\alpha < 1$, the curvature is always positive, but if $\alpha > 1$, the curvature can be positive, zero, or negative. When

$$\beta = \frac{\alpha^2}{\alpha - 1}, \alpha > 1. \tag{2.55}$$

the curvature is zero. Substitution of (2.55) into (2.54) makes the denominator on the right-hand side of equation (2.54) a factor of the numerator, and reduces (2.54) to the simpler expression

$$\frac{1}{v_0} = \frac{\alpha \xi + 1}{\alpha V} \tag{2.56}$$

which is the Michaelis–Menten or zero cooperative result. In particular, comparison of the branch labels in the graphs of Figures 2.6 and 2.8 indicates that if we set $k_{+1} = 2k_{+3}$, $k_{+4} = 2k_{+2}$, and $k_{-3} = 2k_{-1}$, then $\alpha = 2$ and $\beta = 4$. These latter two values are a particular set which satisfies the condition (2.55) for reducing (2.54) to the zero cooperative result.

Furthermore, we see that if

$$\beta < \frac{\alpha^2}{\alpha - 1} \text{ or } \beta > \frac{\alpha^2}{\alpha - 1}, \alpha > 1, \tag{2.57}$$

the second derivative of $d^2(1/v_0)/d\xi^2$ and therefore the cooperativity of the enzyme is positive or negative, respectively, for all values of ξ. For example, if $k_{+3} = 2k_{+1}$, $k_{-3} = 2k_{-1}$, and $k_{+4} = 2k_{+2}$, then $\alpha = 2$ and $\beta = 1$. Hence

$\beta < \alpha^2/(\alpha - 1)$, the curve of $1/v_0$ versus $1/s_0$ (proportional to ξ) possesses positive curvature, and the interpretation is made that the enzyme displays positive cooperativity. In general the larger k_{+3} is compared to k_{+1}, the smaller is the value of K'_m/K_m, other things being equal, and the more readily is the criterion of positive cooperativity satisfied. This conclusion agrees with our intuitive concept of positive cooperativity as representing enhanced ligand binding in the state C_2 as compared to the state C_1. However, we see that positive cooperativity depends as well on the relative rates with which product is formed (the parameter α).

By contrast, if $k_{+3} = k_{+1}/10$, $k_{-3} = 2k_{-1}$, and $k_{+4} = 2k_{+2}$, which represents decreased ligand binding in the state C_2 as compared to the state C_1, then $\alpha = 2$ and $\beta = 20$. Hence $\beta > \alpha^2/(\alpha - 1)$, the curve of $1/v_0$ versus $1/s_0$ now possesses negative curvature, and the interpretation is made that the enzyme displays negative cooperativity.

These curves, along with other examples based on equation (2.54), are illustrated in Figure 2.9(a). In Figure 2.9(b) are shown the corresponding

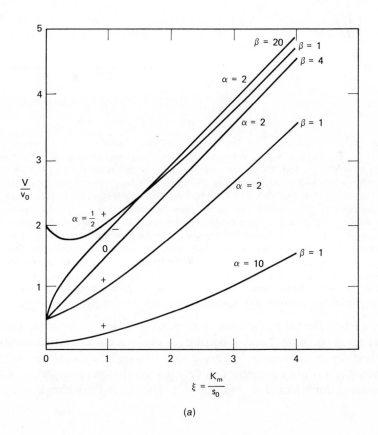

(a)

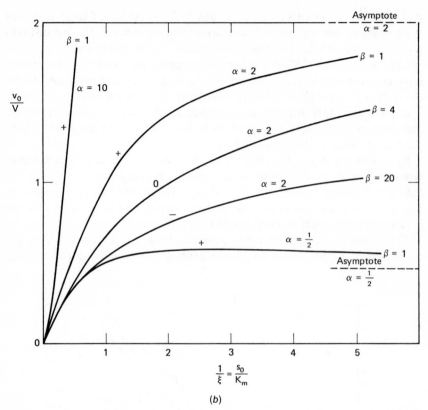

(b)

Figure 2.9. (a) A nondimensional Lineweaver–Burk plot of equation (2.54) for several values of $\alpha = k_{+4}/k_{+2}$ and $\beta = K'_m/K_m$. Alternatively, the abscissa can be thought of as representing $1/s_0$ in units of $1/K_m$, while the ordinate represents the reciprocal velocity $1/v_0$ in units of $1/V$. The three curves with positive curvature display positive cooperativity and are labeled with a + sign, the straight line with $\beta = 4$ displays zero cooperativity and is labeled with a zero, and the curve for $\beta = 20$ with negative curvature, labeled with a $-$ sign, displays negative cooperativity. (b) The curves corresponding to those in figure (a) and labeled in the same manner, are here plotted in a nondimensional version of the function $v_0(s_0)$. In dimensional terms, the abscissa represents s_0 in units of K_m, and the ordinate represents v_0 in units of V. According to equation (2.54), the asymptotic value of v_0/V for large s_0 is α. The asymptote for the curve labeled $\alpha = 10$ is off the scale. Note that the curves for $\alpha = 10$ and $\alpha = 2, \beta = 1$, are sigmoidal close to the origin, and the curve for $\alpha = 1/2$ has a maximum at a finite value of the abscissa. All the curves have a common slope at the origin equal to unity.

curves when plotted in the form of $v_0 = v_0(s_0)$. It should be noted that the cases for α equaling 2 and β equaling 1 or 20 are cooperative according to their representations in Figure 2.9(a), although their representation in Figure 2.9(b) is not qualitatively different from the curve displaying no cooperativity for which $\alpha = 2$ and $\beta = 4$. The curve for which $\alpha = 10$ and

$\beta = 1$ displays a sigmoidal shape near the origin, and the curve for which $\alpha = 2$ and $\beta = 1$ is slightly sigmoidal near the origin. Furthermore, the case $\alpha = 1/2$ and $\beta = 1$ is not sigmoidal in appearance and in fact displays a maximum near $\xi^{-1} = 2.7$, although, according to the classification scheme based on Figure 2.9(a), it is an example of positive cooperativity.

Let us expand equation (2.53) in a Taylor series about $s_0 = 0$, retaining only the first two terms in the expansion. The result is

$$v_0 = \frac{k_{+2}e_0}{K_m} s_0 + \frac{k_{+2}e_0 s_0^2}{K_m K'_m}\left(\frac{k_{+4}}{k_{+2}} - \frac{K'_m}{K_m}\right) = \frac{V}{\xi} + \frac{V}{\xi^2}\left(\frac{\alpha}{\beta} - 1\right). \quad (2.58)$$

From the above expression, we see that $d(v_0/V)/d(1/\xi)$ evaluated at $1/\xi = 0$ is unity, and that is why the illustrative curves in Figure 2.9(b) all have a common slope at the origin. It is beyond the capability of the model described herein to alter the slope of $(K_m/V)(dv_0/ds_0)$ at the origin. A model that does so will be described in the next section. We find also by evaluating $d^2 v_0/d(1/\xi)^2$ from equation (2.58), that positive curvature of the curve $v_0(s_0)$ in the neighborhood of $s_0 = 0$, which is the qualitative manner of recognizing cooperativity, illustrated in Figure 2.7, requires that $\alpha > \beta$. It can be seen in Figure 2.9(b) that the two curves that appear sigmoidal satisfy the latter condition.

2.6. ALLOSTERIC ENZYMES

Consider an enzyme that contains a ligand binding site which is distinct and distant from the catalytically active site of the enzyme. A proposed mechanism of control of the catalytic activity of the enzyme is that ligand binding at the distant site affects the activity of the enzyme at the active site (Monod, Changeux, and Jacob, 1963). The distant site is called an **allosteric site** and the effect is called an **allosteric effect**, because the ligand that binds at the distant site can be "structurally unlike" or **allosteric** with the subtrate that binds at the active site. An enzyme exhibiting such an effect is called an **allosteric enzyme**. By contrast, when substrate and inhibitor molecules compete for a single active site on the enzyme as in the enzyme–substrate–inhibitor system discussed in Section 2.3, the substrate and inhibitor molecules are structurally similar or **isosteric**.

The allosteric effect is presumed to arise because of a reversible **conformational change** in the enzyme: A change in the folding of the polypeptide chain, called an **allosteric transition**. The ligand that binds at the allosteric site is called an **effector** or **modifier**. If the modifier increases the activity or binding of the substrate, it is said to be an **activator**. If the modifier decreases the activity of the substrate, it is called an **inhibitor**.

If the modifier molecule is the same as the ligand that binds at the active site, the allosteric effect is said to be **homotropic**. If the modifier and substrate molecules are different, the effect is said to be **heterotropic**. Most allosteric proteins are oligomers. Furthermore, allosteric enzymes exhibiting cooperative (homotropic) effects almost invariably exhibit heterotropic effects as well.

Because allosteric enzymes are recognized by their anomalous kinetic behavior, in particular the appearance of a sigmoidal curve for $v_0(s_0)$, the descriptive term allosteric is often used to mean that the enzyme displays such a sigmoidal curve. Clearly, in the case when the ligands are identical, the allosteric effect is a mechanism by which an oligomer can achieve cooperative behavior. In other words, conformational change in the enzyme explains how distant active sites can interact during binding with substrate.

One of the principal quantitative theories of allosteric proteins at the present time is that due to Monod, Wyman, and Changeux (1965), and we shall now present some of its main features. We shall refer to allosteric proteins rather than enzymes because the theory is applicable to a wider class of proteins than enzymes. The theory assumes the following properties for the allosteric protein:

1. The protein is an oligomer made up of n identical protomers.
2. Each protomer has only one active site that is able to combine with a given ligand.
3. The active sites associated with a given ligand are independent of each other in their interactions with the ligand.
4. The protomer exists in two states, each of which has a different affinity for the ligand.
5. The protein can exist only in either of two states, in which the subunits are all in one or the other of their two states.

The allosteric effect is a consequence of the fact that the protein can exist in two different states. This is presumed to be due to a conformational change in the protein.

We denote the two states of the protein by R and T and the ligand molecule by S. The subscript j attached to R or T will denote the presence of j ligands attached to the protein in either state. Thus, j can vary from 0 to n, and the subscript 0 represents the protein with no occupied sites. We now impose the steady state hypothesis, and assume that the protein complexes are in equilibrium with a large number of ligand molecules of concentration s_0, which is treated as constant. The possibility of transformation of ligand molecules into product molecules is initially ignored, so that a true steady state is assumed to exist. Furthermore, it is assumed that the states R_0 and T_0 can convert from one to the other and are in equilibrium, with an equilibrium constant L, the **allosteric constant**. Thus, the possible reactions are written as

$$\left.\begin{array}{l} S + R_j \rightleftharpoons R_{j+1} \\ S + T_j \rightleftharpoons T_{j+1} \end{array}\right\} \quad j = 0, 1, 2, \ldots, n-1, \qquad (2.59)$$

$$R_0 \rightleftharpoons T_0.$$

Let us denote the rate constant for binding the ligand molecule to a particular site of the protein in the R state by k_{+1} for association, and k_{-1} for dissociation. Similarly, we denote by k_{+3} and k_{-3} the corresponding rate constants for the binding of the ligand molecule to the protomer of the protein in the T state. The rate constants for transformation for the R_0 to the T_0 state and back are denoted by k_{+0} and k_{-0}, respectively.

The graph for the process is shown in Figure 2.10. We see that it is composed of two subgraphs representing the R and T systems, each consisting of $n + 1$ vertices, which are joined together via the interaction between the states R_0 and T_0. We remark that the theory is presented with the conformational change from the R state to the T state occuring only when there are no ligand molecules attached, that is, $R_j \rightleftharpoons T_j$ for $j \neq 0$ is not allowed. However, this simplification in the theory is unnecessary, as we shall show later.

We can now write the rate equations corresponding to the graph, according to the prescription given in Section 2.4. Denoting concentrations by lower case letters, we obtain

$$\frac{dr_0}{dt} = -k_{+0}r_0 + k_{-0}t_0 - nk_{+1}s_0r_0 + k_{-1}r_1,$$

$$\frac{dr_j}{dt} = (n + 1 - j)k_{+1}s_0r_{j-1} - jk_{-1}r_j$$
$$\qquad - (n - j)k_{+1}s_0r_j + (j + 1)k_{-1}r_{j+1}, \qquad j = 1, 2, \ldots, n-1,$$

$$\frac{dr_n}{dt} = k_{+1}s_0r_{n-1} - nk_{-1}r_n,$$

$$\frac{dt_0}{dt} = k_{+0}r_0 - k_{-0}t_0 - nk_{+3}s_0t_0 + k_{-3}t_1, \qquad (2.60)$$

$$\frac{dt_j}{dt} = (n + 1 - j)k_{+3}s_0t_{j-1} - jk_{-3}t_j$$
$$\qquad - (n - j)k_{+3}s_0t_j + (j + 1)k_{-3}t_{j+1}, \qquad j = 1, 2, \ldots, n-1,$$

$$\frac{dt_n}{dt} = k_{+3}s_0t_{n-1} - nk_{-3}t_n.$$

At equilibrium, we set the time derivatives in (2.60) equal to zero and solve for the equilibrium values of r_j and t_j. We define the equilibrium constants

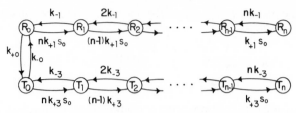

Figure 2.10. Graph for the allosteric reaction system of Monod, Wyman, and Changeux (1965), equations (2.59).

K_R, K_T, and L as

$$K_R = \frac{k_{-1}}{k_{+1}},$$

$$K_T = \frac{k_{-3}}{k_{+3}}, \tag{2.61}$$

$$L = \frac{k_{+0}}{k_{-0}}.$$

We recognize that either the R states or the T states by themselves constitute a noncooperative system, the theory of which we discussed in Section 2.4. In combining these two systems, only the transition between R_0 and T_0 has been added. By referring back to the previous results, equations (2.33)–(2.38), we see that

$$r_j = b_j^n x^j r_0,$$
$$t_j = b_j^n (cx)^j t_0,$$
$$t_0 = L r_0, \tag{2.62}$$
$$x \equiv \frac{s_0}{K_R},$$
$$c \equiv \frac{K_R}{K_T}.$$

Suppose now that the allosteric transition can occur, regardless of the number of ligand molecules attached to the protein. Then equations (2.59) are supplemented by the reactions

$$R_j \rightleftharpoons T_j, \qquad j = 1, 2, \ldots, n, \tag{2.63}$$

and branches connecting R_j and T_j directly appear in the graph of Figure 2.10. Consider for definiteness the transition

$$R_1 \underset{k_{-5}}{\overset{k_{+5}}{\rightleftharpoons}} T_1, \tag{2.63a}$$

in which we have assigned the rate constant k_{+5} for the forward reaction and k_{-5} for the backward reaction. Permitting this transition to occur would appear to have the consequence that the relationship between the equilibrium values of r_1 and t_1 would be altered, and the additional equilibrium constant $K_5 \equiv k_{+5}/k_{-5}$ would appear in the theory, but that is not the case. The justification for this statement is a consequence of the thermodynamic **principle of detailed balancing**. This principle, applied to a chemical reaction in a state of equilibrium, states that the frequency of transitions from one molecular equilibrium state to another equilibrium state is equal to that in the reverse direction. It follows that, for a cyclic set of reactions in equilibrium, the product of the rate constants taken in a clockwise direction is equal to the product of the rate constants taken in a counterclockwise direction. (See, for example, Hearon et al., 1959.)

Refer now to the graph of Figure 2.10 with the transitions given by (2.63a) included, and consider the loops formed by the branches connecting the states R_0, T_0, R_1, and T_1. We infer from the above principle that $k_{+0}k_{+3}k_{-5}k_{-1} = k_{-0}k_{+1}k_{+5}k_{-3}$, or equivalently with the aid of equations (2.61) and (2.62), that

$$\frac{k_{+5}}{k_{-5}} = \frac{k_{+0}k_{-1}k_{+3}}{k_{-0}k_{+1}k_{-3}}, \quad K_5 = Lc. \tag{2.64}$$

Similar considerations apply to the transitions $R_j \rightleftharpoons T_j$ for $j > 1$. Therefore the relationship between the equilibrium values of r_1 and t_1 given by equations (2.62) as

$$\frac{t_1}{r_1} = Lc \tag{2.65}$$

is unaltered by the inclusion of the reaction (2.63a), and is the same as the equilibrium condition $t_1/r_1 = K_5$. The same considerations apply to the reactions (2.63) for $j > 1$, so that the equilibrium relation between r_j and t_j, which according to equations (2.62) is

$$\frac{r_j}{t_j} = Lc^j, \quad j = 0, 1, \ldots, n, \tag{2.66}$$

is unaltered by the inclusion of the reactions (2.63). Consequently, the above equilibrium relations have a more general significance than the simplified reaction scheme (2.59) indicates, and the graph of Figure 2.10 could equally well have included the transitions (2.63).

The saturation function, which is the fraction of all sites to which ligands are bound, is defined by the expression

$$Y = \frac{\sum\limits_{j=1}^{n} j(r_j + t_j)}{n \sum\limits_{j=0}^{n} (r_j + t_j)}. \tag{2.67}$$

By using (2.62) in (2.67) and proceeding as in the derivation of equation (2.45), we find that

$$Y = \frac{Lcx(1 + cx)^{n-1} + x(1 + x)^{n-1}}{L(1 + cx)^n + (1 + x)^n}. \tag{2.68}$$

The curve for $Y(x)$ may be sigmoidal. If so, it is taken to be evidence for the allosteric behavior of the system.

When $c = 1$ (the affinity of both states towards the ligand is the same), or when $L \to 0$, or $L \to \infty$, the protein exists exclusively in one state consisting of n identical, independent monomeric units, namely, a noncooperative system. In the first two of these limiting cases, it is seen from (2.68) that the saturation function reduces to the simple expression

$$Y = \frac{x}{1 + x} = \frac{s_0}{K_R + s_0}. \tag{2.69}$$

In the third case, for which $L \to \infty$, Y again reduces to the above expression with K_R replaced by K_T. Equation (2.69) is exactly the same result that is obtained if the protein is a monomer, as we expect.

When c is negligible, (2.68) reduces to the expression

$$Y = \frac{x(1 + x)^{n-1}}{L + (1 + x)^n}. \tag{2.70}$$

Numerical calculations based on equation (2.68) show that $Y(x)$ is sigmoidal when L assumes a large value. If L is large, it means that the bare protein prefers to exist in the T state rather than the R state. The sigmoidal character of Y is accentuated when $c \ll 1$. To see this explicitly, let us calculate dY/dx and d^2Y/dx^2, evaluated at $x = 0$. Expanding equation (2.68) in a Taylor series about $x = 0$, and retaining the first two terms only, it is readily found that

$$\left(\frac{dY}{dx}\right)_{x=0} = \frac{Lc + 1}{L + 1}, \tag{2.71a}$$

$$\left(\frac{d^2Y}{dx^2}\right)_{x=0} = \frac{2}{(L + 1)^2}[nL(1 - c)^2 - (Lc^2 + 1)(L + 1)]. \tag{2.71b}$$

From (2.71a), we see that the slope of $Y(x)$ at $x = 0$ is small only if $L \gg 1$. Further, since the slope is then approximately $L^{-1} + c$, it is also necessary that $c \ll 1$. From (2.71b) we see that the curvature of $Y(x)$ is positive at $x = 0$ provided that

$$n > \frac{(Lc^2 + 1)(L + 1)}{(1 - c)^2 L}. \tag{2.72}$$

In particular, if L is large and c is small compared to unity, this condition becomes, approximately, $n > Lc^2 + 1$. For example, if $c = 0$, then positive cooperativity requires only that $n > 1 + 1/L$, and if $c = 0.04$ and $L = 1000$, as illustrated in Figure 2.11, the condition reads $n > 2.8$.

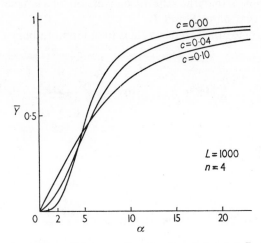

Figure 2.11. The dependence of the saturation function Y (labeled $\bar{Y}$ along the ordinate axis) on $x = s_0/K_R$, given by equation (2.68) and denoted by α in the figure, for $n = 4, L = 1000$, and several values of the parameter c. Reproduced, with permission, from Monod, Wyman, and Changeux (1965).

The chemical significance of $c \ll 1$ or $K_R \ll K_T$ is that there is virtually no binding of ligand molecules in the T state, and the R state has a greater affinity for the ligand. Combined with the condition $L \gg 1$, we can understand qualitatively why the saturation function is sigmoidal, as follows. At low concentrations of substrate, the protein exists preferentially in the T_0 state, so that Y increases very slowly with substrate concentration. However, with increasing substrate concentration, more and more of the protein gets tied up in the bound forms of the R state, and the conversion of the bound form of the R state to the T state is no longer so favored, according to equations (2.66). Hence the saturation function rises rapidly with substrate concentration. Figure 2.11 shows the computed dependence of $Y(x)$ for L large and various values of c.

It is legitimate to inquire, if the protein were an enzyme, whether its reaction velocity would possess the same sigmoidal character as $Y(x)$, and whether it could be sigmoidal under different conditions than those governing the sigmoidal form of Y. To answer these questions, assume that the following additional reactions are possible:

$$\left.\begin{array}{c} R_j \xrightarrow{k+2} R_{j-1} + P \\ T_j \xrightarrow{k+4} T_{j-1} + P \end{array}\right\} \quad j = 1, 2, \ldots, n. \tag{2.73}$$

Here it is assumed that the rate of conversion of enzyme to product is different in the two states R and T.

The expression for the reaction rate is then found to be

$$v_0 = k_{+2} \sum_{j=1}^{n} jr_j + k_{+4} \sum_{j=1}^{n} jt_j = \frac{V \sum_{j=1}^{n} jr_j + V' \sum_{j=1}^{n} jt_j}{n \sum_{j=0}^{n} (r_j + t_j)}, \tag{2.74}$$

where, by definition,

$$\begin{aligned} V &\equiv nk_{+2}e_0, \\ V' &\equiv nk_{+4}e_0, \end{aligned} \tag{2.75}$$

and e_0 is the total initial enzyme concentration,

$$e_0 \equiv \sum_{j=0}^{n} (r_j + t_j). \tag{2.76}$$

By analogy with the derivation of equation (2.68) from (2.67), it follows that in the quasi-steady state,

$$v_0 = \frac{V'Lcx(1 + cx)^{n-1} + Vx(1 + x)^{n-1}}{L(1 + cx)^n + (1 + x)^n}. \tag{2.77}$$

Here, however, $x = s_0/K_m^R$, where K_m^R is the Michaelis constant for the R state, equal to $(k_{-1} + k_{+2})/k_{+1}$, and $c = K_m^R/K_m^T$, where K_m^T is the Michaelis constant for the T state, equal to $(k_{-3} + k_{+4})/k_{+3}$. Furthermore, equation (2.77) is not valid if the transitions (2.63) occur. From this result we see that if $V' = V$, then $v_0 = VY(x)$, so that v_0 then has all the qualitative features of the curve $Y(x)$. However, when $V' \neq V$, the necessary conditions for the appearance of a sigmoidal curve $v_0(x)$ are slightly different from those for $Y(x)$. Thus,

$$\frac{1}{V}\left(\frac{dv_0}{dx}\right)_{x=0} = \frac{\alpha Lc + 1}{L + 1}, \tag{2.78a}$$

$$\frac{1}{V}\left(\frac{d^2v_0}{dx^2}\right)_{x=0} = \frac{2}{(L + 1)^2}[(n - 1)(L + 1)(\alpha Lc + 1) - n(Lc + 1)^2], \tag{2.78b}$$

where, by definition, $\alpha \equiv V'/V = k_{+4}/k_{+2}$. According to equation (2.78a), the derivative at the origin is small if $L \gg 1$, and $\alpha c \ll 1$. In addition, the condition for positive curvature of the function $v_0(x)$ in the neighborhood of $x = 0$ is, from equation (2.78b),

$$\frac{n-1}{n} > \frac{(Lc + 1)^2}{(L + 1)(\alpha Lc + 1)}. \tag{2.79}$$

This condition is equivalent to equation (2.72) when $\alpha = 1$. It can be seen that the condition will be readily satisfied if $n \geqq 2$, $L \gg 1$, and $c \ll 1$. However, if $\alpha \sim 0$, it is also necessary that Lc^2 be small compared to unity, in order for the curvature to be positive.

2.7.   OTHER ALLOSTERIC THEORIES

The allosteric theory of Monod, Wyman, and Changeux has been criticized on the grounds that it cannot account for negative cooperativity. Indeed, it can be verified from equation (2.68) that the curvature of $1/Y$ in a Lineweaver–Burk plot is positive near the origin, namely, $d^2(1/Y)/d(1/x)^2$ is always positive at $1/x = 0$. An alternative theory of allosteric proteins, the "induced fit" model, has been put forth by Koshland, Nemethy, and Filmer (1966). This theory is also based on the idea that proteins are composed of subunits. Like the Monod–Wyman–Changeux theory, it assumes that each subunit exists in either of two conformational states, A and B say, and that the ligand has a greater affinity for a subunit in one of its two conformations. However, the theory proposes that hybrid conformational states can exist, consisting of some subunits in the state A, while other subunits are in the state B. Thus, the basic generalization introduced by the induced fit model to the Monod–Wyman–Changeux model is that the subunits of the protein do not necessarily change their conformation in an "all-or-none" fashion.

The theory is not presented on the basis of rate equations, as was done in the previous section. Instead, the possible forms of the equilibrium constants are decided on the basis of energetic considerations of the interactions between subunits and their nearest neighbors. Such considerations depend on the number and kind of nearest neighbors a given subunit has. These in turn depend on the assumed spatial configuration of the subunits, and lead to different realizations of the theory, for example, "linear, square, tetrahedral, and concerted" models. With the two conformational states of the subunits denoted by circles and squares, the number of possible protein states in interaction with a ligand S is illustrated in Figure 2.12 for the case when the protein is a tetramer.

T - Form   R - Form

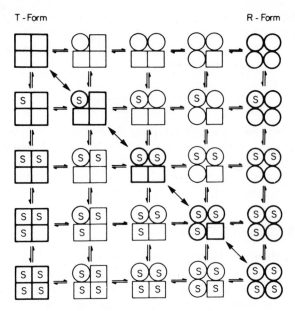

Figure 2.12. The "induced fit" model (Koshland, Nemethy, and Filmer, 1966) permits the protein to exist in hybrid configurations of subunits, which are in either of two states A and B, here represented as circles and squares. Ligand binding to a particular subunit is denoted by the letter S. The figure represents the protein as a tetramer, and shows the various states of the protein that are possible, together with their transitions. In the idealized form in which the theory was presented, only transitions along the main diagonal (upper left to lower right) were considered. Reproduced, with permission, from Gutfreund (1972), after Eigen (1967).

In order to reduce the number of parameters that enter into the theory, it is simplified by the additional assumption that a subunit is induced to change its conformation by the binding of a ligand, but does not change otherwise. Hence, the only allowed states of the protein are those along the diagonal marked with transitions in Figure 2.12. The Monod–Wyman–Changeux model is a special case of the general scheme presented in the figure in which the only allowed protein states are those represented in the extreme left and extreme right columns.

There are many other allosteric models that have been proposed. It is fair to state that the sigmoidal shape of $v_0(s_0)$ or $Y(s_0)$ is an insufficient basis for distinguishing between one theory and another, because they all have a number of adjustable parameters at their disposal to achieve such a shape. Further information regarding the conformation and other properties of enzymes and proteins is needed before a proper quantitative theory of the cooperative behavior of proteins can be found.

2.8. HEMOGLOBIN

Although hemoglobin is not an enzyme, it is a good example of a cooperative system, and therefore serves as a model with which to test theoretical concepts of cooperative behavior.

More is known about **myoglobin** and **hemoglobin**, iron-bearing macromolecules, than any other proteins. Myoglobin is found in muscle, and its function is to transport and store oxygen for subsequent use in an oxidative process. Hemoglobin exists normally in the blood, entirely within the red cell. The principal function of the red cell in fact is to contain hemoglobin. The chemical action of hemoglobin is to combine with oxygen in the lungs to form oxyhemoglobin. The latter gives up oxygen in the tissue to form hemoglobin again. Hemoglobin also combines with CO_2 in the tissues and releases it to the lungs. These properties are better understood by examining the saturation function representing the fractional oxygenation of hemoglobin as a function of the partial pressure of oxygen in the air surrounding it. The saturation function is a directly observable quantity. Whereas the saturation curve for myoglobin has the form of a rectangular hyperbola, the curve for hemoglobin is S-shaped, as shown in Figure 2.13.

Thus, the partial pressure of oxygen in the lungs where oxygen is taken up by hemoglobin is about 100 mm Hg. At this pressure, the hemolgobin is about 98% saturated, and its saturation is but slightly affected by changes in oxygen content of the air, because the saturation curve is relatively flat there. In the veins or tissue, the oxygen partial pressure is about 40 mm Hg, which is near the steep portion of the saturation curve, and represents a saturation value of about 75%. Furthermore, any abnormal demands for oxygen on the part of the tissue will deplete the oxygen partial pressure there to, say, 20 mm Hg, and the oxyhemoglobin will readily give up its oxygen, reducing its saturation to about 35%. At this value of the oxygen partial pressure, the myoglobin saturation value is about 90%. Hence, if the tissue is muscle, the oxygen will transfer itself from the hemoglobin to the myoglobin. In other words, the relative positions of the saturation curves illustrate the fact that, for a given partial pressure of oxygen, the affinity of myoglobin for oxygen is very much greater than that of hemoglobin.

The hemoglobin molecule consists of a protein part called **globin** and four disk-shaped molecular rings called **heme groups**. Globin is composed of four polypeptide subunits, divided equally into two types of structures, called α and β. Each subunit has a heme group attached, and the latter is the active center of the subunit. At the center of each heme group is an iron atom. The iron atom is capable of combining with oxygen, which accounts for the functioning of hemoglobin. Myoglobin consists of only one polypeptide chain and one heme group, and the molecular weight of

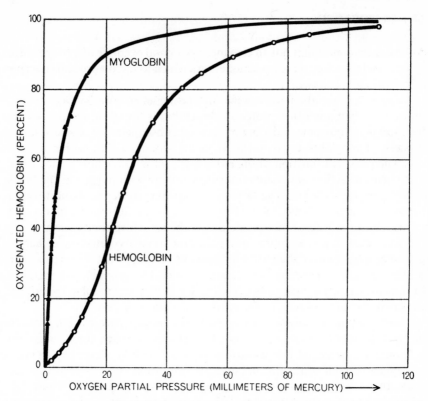

Figure 2.13. The curve labeled "hemoglobin" is the observed saturation function representing the fraction of oxygenated hemoglobin as a function of the partial pressure of oxygen in the surrounding air. Similarly, the curve labeled "myoglobin" represents the saturation function for myoglobin. From J. P. Changeux, *The control of biochemical reactions.* Copyright ©️ 1965 by Scientific American, Inc. All rights reserved.

myoglobin is about one fourth that of hemoglobin. The theoretical prediction of the observed shape of the hemoglobin saturation curve is an unsolved problem.

Perhaps the simplest theory of hemoglobin proposed is that it behaves like a cooperative tetramer. Thus, we assume the hemoglobin exists in five possible states H_j, where j runs from 0 to 4 and represents the number of oxygen molecules attached to the hemoglobin molecule. In other words, H_1 represents HbO_2, H_2 represents HbO_4, and so forth. The hemoglobin is assumed to be in equilibrium with oxygen, as is shown by the graph in Figure 2.14. In the graph, the quantity s_0 again represents the ligand molecule, which in the present instance is oxygen. Each transition between states possesses a different rate constant, as indicated by the branch values of the

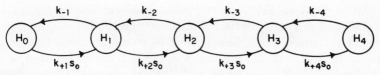

Figure 2.14. Graph of states of hemoglobin interacting with ligand of concentration s_0. The state H_j represents hemoglobin with j ligand molecules bound to it, $j = 0, 1, 2, 3, 4$.

graph. In equilibrium, only the branch values will enter into the determination of the saturation function.

Let the equilibrium constants be defined as

$$K_j \equiv \frac{k_{-j}}{k_{+j}}, \qquad j = 1, 2, 3, 4. \tag{2.80}$$

It is not necessary for us to write down the rate equations for H_j, as we can readily write down the equilibrium equations for the concentrations h_j of the hemoglobin complexes as follows:

$$K_j = \frac{s_0 h_{j-1}}{h_j}, \qquad j = 1, 2, 3, 4. \tag{2.81}$$

Therefore, the equilibrium concentrations are expressible in terms of h_0 as

$$h_1 = \frac{s_0}{K_1} h_0, \qquad h_3 = \frac{s_0^3}{K_1 K_2 K_3} h_0,$$
$$h_2 = \frac{s_0^2}{K_1 K_2} h_0, \quad h_4 = \frac{s_0^4}{K_1 K_2 K_3 K_4} h_0. \tag{2.82}$$

The saturation function $Y(s_0)$, representing the fractional saturation of the available oxygen sites, is defined as in equation (2.40) by the expression

$$Y(s_0) = \frac{\displaystyle\sum_{j=1}^{4} j h_j}{4 \displaystyle\sum_{j=0}^{4} h_j}. \tag{2.83}$$

Substituting equation (2.82) into (2.83) leads to the final result

$$Y(s_0) = \frac{A_1 s_0 + 2 A_2 s_0^2 + 3 A_3 s_0^3 + 4 A_4 s_0^4}{4\{1 + A_1 s_0 + A_2 s_0^2 + A_3 s_0^3 + A_4 s_0^4\}}, \tag{2.84}$$

where

$$\begin{aligned}
A_1 &= K_1^{-1}, \\
A_2 &= (K_1 K_2)^{-1}, \\
A_3 &= (K_1 K_2 K_3)^{-1}, \\
A_4 &= (K_1 K_2 K_3 K_4)^{-1}.
\end{aligned} \tag{2.85}$$

Equation (2.84) was first suggested on an empirical basis by Adair (1925), who advanced this "intermediate compound" theory after discovering that hemoglobin contained four hemes. Since O_2 concentration is directly proportional to oxygen partial pressure p, s_0 in (2.84) can be replaced by p so as to make a direct comparison with the experimental curve shown in Figure 2.13. The proportionality constant can be absorbed into the A_j. Thus, if $s_0 = \alpha p$ where p is the partial pressure of oxygen in the gas surrounding the hemoglobin solution, and α is a constant of proportionality called the **solubility coefficient**, then $Y(p)$ is given by equation (2.84) with s_0 replaced by p and

$$\begin{aligned}
A_1 &= \alpha K_1^{-1}, \\
A_2 &= \alpha^2 (K_1 K_2)^{-1}, \\
A_3 &= \alpha^3 (K_1 K_2 K_3)^{-1}, \\
A_4 &= \alpha^4 (K_1 K_2 K_3 K_4)^{-1}.
\end{aligned} \tag{2.86}$$

With four disposable constants to fit one curve, it is not surprising that the experimental curve can be adequately represented by a judicious choice of the K_j. We emphasize that the above theory is a cooperative theory in that the binding of the ligand molecule at one site facilitates or inhibits the binding of the ligand molecule at a second site, and the K_j are not interrelated in a simple way.

The most recent measurements of $Y(p)$ for sheep hemoglobin (Roughton, DeLand, Kernohan, and Severinghaus, 1972) yield the values, in units of mm Hg = Torr = 1333.22 dyne/cm^2,

$$\begin{aligned}
A_1 &= 2.18 \times 10^{-2}, A_3 = 3.75 \times 10^{-6}, \\
A_2 &= 9.12 \times 10^{-4}, A_4 = 2.47 \times 10^{-6}.
\end{aligned} \tag{2.87}$$

From these values and equation (2.86), it is readily inferred that the values of the equilibrium constants are

$$\begin{aligned}
(K_1^{-1}, K_2^{-1}, K_3^{-1}, K_4^{-1}) &= 10^{-2}(2.18, 4.18, 0.411, 65.9)\alpha^{-1} \text{ Torr} \\
&\quad 2.18 \times 10^{-2}(1, 1.92, 0.189, 30.2)\alpha^{-1} \text{ Torr.}
\end{aligned} \tag{2.88}$$

If the four sites were independent and had the same intrinsic affinity for association and dissociation with the ligand, then by comparing equations (2.82) and (2.36) with $n = 4$, we infer that the theoretical values of the equilibrium constants would be given as

$$\begin{aligned}
(K_1^{-1}, K_2^{-1}, K_3^{-1}, K_4^{-1}) &= (4K^{-1}, \tfrac{3}{2}K^{-1}, \tfrac{2}{3}K^{-1}, \tfrac{1}{4}K^{-1}) \\
&= 4K^{-1}(1, \tfrac{3}{8}, \tfrac{1}{6}, \tfrac{1}{16}),
\end{aligned} \tag{2.89}$$

where K is the intrinsic equilibrium constant for a given site. The expression for $Y(s_0)$ would reduce to (2.69), which has no inflection point and is therefore incorrect. The preceding argument was already put forth by Hartridge and Roughton (1925) at about the same time the Adair equation was proposed.

A comparison of the last expressions in (2.88) and (2.89) makes it evident that the really large discrepancy between observation and the independent site theory occurs for the ratio K_1^{-1}/K_4^{-1}. Apparently, there is a very great affinity for binding of an oxygen molecule to hemoglobin once three oxygen molecules are already bound to it.

Monod, Wyman, and Changeux did apply their theory of allosterism to some careful observations of $Y(p)$ for equine hemoglobin (Roughton and Lyster, 1965). They found that these data could be fitted for $n = 4$ with $L = 9054$ and $c = 0.014$. However, a good fit to the more recently quoted data has not yet been demonstrated (Roughton, DeLand, Kernohan, and Severinghaus, 1972) for the Monod-Wyman-Changeux model, the induced fit model, or for older models that have been proposed such as those of Pauling (1935), and Margaria (1963) (see problem 7). The data can be fitted by the Adair model.

At the present time it seems fair to state that a theoretical explanation of the very large observed value of K_1^{-1}/K_4^{-1} is still needed. An alternative suggestion has been made regarding the binding of hemoglobin to oxygen that would permit hemoglobin to dissociate into dimers and/or monomers.

2.9. GRAPH THEORY AND STEADY–STATE ENZYME KINETICS

We see that the reaction of enzymes or proteins with ligands can be very complex affairs with the number of rate constants characterizing the reaction process becoming very large, as the number of intermediate states increases. We have also seen that, with the hypothesis of a steady state, the determination of the velocity of the reaction is reduced to the problem of solving a system of linear algebraic equations. Furthermore, explicit algebraic expressions are desired for the solution, rather than numerical ones. The problem of achieving such expressions is substantially simplified by the use of graph theory. We shall present the method of doing so here (Volkenstein and Goldstein, 1966a; Volkenshtein, 1969).

Assume that the enzyme or protein exists in various states of combination with ligands, substrates, inhibitors, effectors, and so on, which shall be designated as complexes of concentration c_i where $i = 1, 2, \ldots, n$. Thus there are n states or complexes in which the protein molecule can be. In a quasi-steady-state enzymatic reaction, the conversion of substrate to product is taking place so slowly that the substrate concentration can be treated as constant in the rate equations for the complexes. Then the equilibrium values of the concentrations of the complexes satisfy the equations

$$c_i \sum_{j=1}^{n}{}' A_{ji} = \sum_{j=1}^{n}{}' A_{ij}c_j, \qquad i = 1, 2, \ldots, n, \tag{2.90}$$

where the prime signifies that the summation excludes the value $j = i$. Here, the left-hand side represents transitions of the protein from the state i to the state j, and the right-hand side represents transitions from the state j to the state i, with $i \neq j$.

The coefficients A_{ij} are constants that depend on the rate constants and the ligand concentrations. They are elements of the **coefficient matrix** A. Equations (2.90) can also be written in the form

$$Ac = 0, \tag{2.91a}$$

or

$$\sum_{j=1}^{n} A_{ij}c_j = 0, \qquad i = 1, 2, \ldots, n. \tag{2.91b}$$

Here, the coefficient matrix A and the **column vector** C represent the square array and linear array, respectively,

$$A = \begin{bmatrix} A_{11} & A_{12} & A_{13} & \cdots & A_{1n} \\ A_{21} & A_{22} & A_{23} & \cdots & A_{2n} \\ A_{31} & A_{32} & A_{33} & \cdots & A_{3n} \\ \cdots\cdots\cdots\cdots\cdots\cdots\cdots \\ A_{n1} & A_{n2} & A_{n3} & \cdots & A_{nn} \end{bmatrix}, c = \begin{bmatrix} c_1 \\ c_2 \\ . \\ . \\ . \\ c_n \end{bmatrix}. \tag{2.92}$$

The **diagonal elements** A_{ii} of the matrix A are defined as

$$A_{ii} \equiv -\sum_{j=1}^{n}{}' A_{ji}, \qquad i = 1, 2, \ldots, n. \tag{2.93}$$

The matrix A has n rows and n columns, and is therefore designated as an $n \times n$ (row $\times$ column) matrix, or **square matrix**. Note that the elements A_{ii} lie on the principal diagonal of A, running from upper left to lower right. The other elements of A are designated as **off-diagonal elements**. The matrix A has the special property that the diagonal elements are expressible in terms of the off-diagonal elements, in view of (2.93). The vector c is **n-dimensional**, and has **components** c_i, $i = 1, 2, \ldots, n$. The product Ac is again an n-dimensional vector whose ith component is given by the left-hand side of equation (2.91b) (see Appendix B).

Because the equations (2.91) are homogeneous, they are not all linearly independent, and one of them is a linear combination of the others. Consequently, in order to solve for the c_i, one of the equations can be dispensed with. It does not matter which one this is. It is replaced by the additional equation

$$e_0 = \sum_{i=1}^{n} c_i, \qquad (2.94)$$

which expresses the fact that the total amount of enzyme at any time is conserved. We do not make the replacement explicit because the method of solution we shall employ does not require it.

According to the theory of graphs, we can construct a diagram, called a graph, to correspond to equations (2.91) and (2.93). Thus, we assign a **node** or **vertex** to each c_i and label it accordingly. The quantity c_i is called the **nodal value**. Corresponding to each nonzero off-diagonal element A_{ij} of A, we draw a **directed line segment** leading from node j to node i. The directed line segment is called a **branch**. Note that A_{ji} is represented by a directed line segment from node i to node j, and is not the same as A_{ij}. The quantity A_{ij} is called the **branch value**. According to equation (2.93), the diagonal element A_{ii} equals the sum of all the coefficients representing the directed line segments leaving the node i. The totality of nodes plus branches is called a **graph**. The number of nodes in the graph is called the **order** of the graph.

It is readily seen that there is a one-to-one correspondence between the graph and equations (2.91). Therefore, knowing the graph, we can write down the equilibrium equations. In the study of enzymes, we want to express the c_i explicitly in terms of rate constants and ligand concentrations. Therefore it is desirable to label the directed line segment corresponding to A_{ij} by the explicit expression for A_{ij} in terms of the rate constants and the ligand concentrations.

If we want to consider time-dependent properties of c_i, we merely replace the right-hand side of equation (2.91b) by dc_i/dt, the time derivative of $c_i(t)$. In this manner, the graphs are in one-to-one correspondence with the rate equations. This correspondence was already discussed in Section 2.4. Finally, if we wish to consider the ligand concentrations as time-dependent, then we must replace the constant initial values of the ligand concentrations by their time-dependent values. The graph can then be utilized to determine by inspection the rate equations for the complexes. They must be supplemented by rate equations for the time-dependent ligand concentrations, which do not, however, follow from the graph in a transparent manner.

We shall now show how the graph is a great aid to the explicit solution of the steady state equations, equations (2.91) and (2.94). Of course, it is well known from the study of algebra how to solve such linear equations, for example, by Cramer's rule. However, the latter method requires the calculation of determinants, which can be a rather laborious procedure when the number of equations is large. In order to make use of graphs, it is necessary to introduce some additional mathematical definitions.

A **path** is a continuous sequence of branches along which no node is encountered more than once. The **path value** is the product of the branch values belonging to the path. A path which ends where it begins is called a **loop**. A particular node may be selected for consideration and called the **basic node** or **base**. A **basic tree** associated with a given base is a set of branches which touches all the nodes of the graph but does not form a loop, and which is always directed towards the base. Thus, a basic tree consists of $n - 1$ branches, where n is the order of the graph, unless the graph consists of two (or more) subgraphs that are totally disconnected. The **tree value** is the product of the branch values of a basic tree. The sum of all basic trees associated with a given node i is called the **basic determinant** D_i of the graph.

The solution to the set of linear equations (2.91) plus (2.94) is expressed in terms of basic determinants as

$$c_i = \frac{e_0 D_i}{\sum\limits_{j=1}^{n} D_j}, \qquad i = 1, 2, \ldots, n. \qquad (2.95)$$

We will accept without proof the assertion that the above expression is equivalent to the solution obtained by Cramer's rule. It can be recognized that the denominator in (2.95) is just the determinant of the coefficient matrix formed from $n - 1$ of equations (2.91) together with equation (2.94). Knowing the values of c_i, we can then calculate the velocity of the reaction v, which is given as

$$v = \frac{dp}{dt} = \sum_{i=1}^{n} k_i c_i, \qquad (2.96)$$

where p is the concentration of product formed from the complexes. Of course, if a given complex c_i does not form any product, then the associated rate constant k_i will be zero. For example, if c_1 is chosen to represent the free protein molecule with no ligands attached, k_1 is identically zero and the term $k_1 c_1$ vanishes in the summation in (2.96).

There are some operations that can be performed with graphs, which can simplify the calculation of the basic determinants. We shall mention in particular two of these.

1.   Parallel branches can be added: Any two branches between two nodes that have the same direction, may be replaced by a single branch whose branch value is the sum of the values of the previous two branches. The permissibility of this operation follows obviously from the correspondence between equations and graphs. Because of this operation, the number of trees that are possible is decreased.

2. The number of nodes can be reduced by selecting an **auxiliary node** j, the choice of which is arbitrary. Let all the paths directed from j to i have path values $P_{ij}^{(1)}$, $P_{ij}^{(2)}$, and so forth. By compressing one of these paths with path value $P_{ij}^{(k)}$, say, into a point, the node j is merged with the node i, all branches and nodes between i and j along the given path are eliminated, and a graph results that is of lower order than the original graph. The basic node of this new graph of lower order is taken to be the merged node. Its basic determinant $D_{ij}^{(k)}$ is computed. This process is repeated for all possible paths directed from j to i. Then the basic determinant D_i of node i of the original graph can be calculated by means of the formula

$$D_i = \sum_k P_{ij}^{(k)} D_{ij}^{(k)}, \tag{2.97}$$

where the summation extends over all possible values of k. This formula corresponds to the expansion of a determinant by minors. Any basic determinant can be calculated in this manner. This method is useful when the order of the graph is large.

The above graph-theoretic procedures are best understood by working out some examples. We shall now calculate reaction velocities with these procedures for some enzymatically controlled reactions involving two or more ligands.

2.10. ENZYME–SUBSTRATE–MODIFIER SYSTEM

As a first example of the use of graph theory, we shall consider a hypothetical enzyme–substrate–modifier system. In such a system, it is assumed that, in addition to the free enzyme C_0 combining with its substrate S to form a complex C_1, an effector or modifier ligand M is present that can also combine with the enzyme to form a complex C_2. This system was analyzed first by Botts and Morales (1953), and subsequently by King and Altman (1956; see also King, 1956). The latter investigators also introduced a simple graphical algorithm for solving problems of steady-state enzymatic reactions involving two or more ligands, not unrelated to that of the preceding section. The effect of the binding of modifier is allosteric, so that the rate constants for C_2 combining with S are different from the rate constants for C_0 combining with S. Let the enzyme–modifier–substrate complex be denoted by C_3. In conventional biochemical notation, the reactions are as follows,

$$
\begin{aligned}
C_0 + M &\underset{k_{-0}}{\overset{k_{+0}}{\rightleftharpoons}} C_2, \\
C_0 + S &\underset{k_{-1}}{\overset{k_{+1}}{\rightleftharpoons}} C_1 \overset{k_{+2}}{\rightarrow} C_0 + P, \\
C_2 + S &\underset{k_{-3}}{\overset{k_{+3}}{\rightleftharpoons}} C_3 \overset{k_{+4}}{\rightarrow} C_2 + P, \\
C_1 + M &\underset{k_{-5}}{\overset{k_{+5}}{\rightleftharpoons}} C_3 \overset{k_{+4}}{\rightarrow} C_2 + P.
\end{aligned}
\tag{2.98}
$$

We now make the steady-state hypothesis for the concentrations of the free enzyme and its complexes C_i, which we shall denote as usual by the corresponding lowercase letters. The graph representing the reactions for these quantities is shown in Figure 2.15, with the letters m_0 and s_0 representing the initial concentrations of modifier and substrate, respectively. In the graph, we have purposely omitted, in the interest of clarity, the nodes corresponding to product formation, because these nodes do not enter into the solution of the equations for the steady-state values of the concentrations of the complexes. They do serve, however, to indicate the expression for the reaction velocity, equation (2.96).

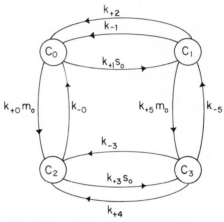

Figure 2.15. Graph representing the reactions of the enzyme–substrate–modifier system, equations (2.98). The formation of product is not indicated.

Because of the rule that parallel branches can be added, the graph can be further simplified, as shown in Figure 2.16. Our next task is to find the basic determinants of the graph. We shall first calculate D_0, the basic determinant associated with the base C_0. We draw all the basic trees associated with the node C_0. There are exactly four basic trees, given in Figure 2.17 with their associated tree values. Hence, the basic determinant D_0 of the graph is

$$D_0 = (k_{-1} + k_{+2})[k_{-0}(k_{-3} + k_{+4} + k_{-5}) + k_{+3}k_{-5}s_0] \\ + k_{+5}m_0k_{-0}(k_{-3} + k_{+4}). \tag{2.99}$$

We shall now calculate D_1, the basic determinant associated with the node C_1, utilizing an auxiliary node that we choose to be C_2. There are only two paths which lead from C_2 to C_1, $2 \to 0 \to 1$, and $2 \to 3 \to 1$. These have path

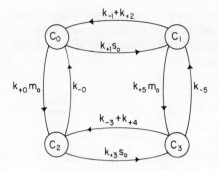

Figure 2.16. Simplification of the graph of Figure 2.15 by use of the rule that parallel branches can be added.

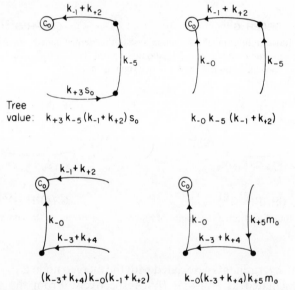

Figure 2.17. The basic trees and the tree values associated with the node C_0 of the graph of Figure 2.16.

values designated as $P_{12}^{(1)}$ and $P_{12}^{(2)}$, that is,

$$P_{12}^{(1)} = k_{-0}k_{+1}s_0,$$
$$P_{12}^{(2)} = k_{+3}k_{-5}s_0. \tag{2.100}$$

When we merge the nodes C_1 and C_2 along path (1), we obliterate the intermediate node C_0 and all the branches connecting C_0 to C_1 and C_2. Thus, we form the new graph $G_1^{(1)}$ of order 2, consisting of the node C_3 and the merged node C_{1-2}. Similarly, by merging C_1 and C_2 along path (2), we form the new graph $G^{(2)}$ of order 2, consisting of the node C_0 and the merged node C_{1-2} (see Figure 2.18). Now by using the additivity of parallel branches, the graphs further simplify as shown in Figure 2.19.

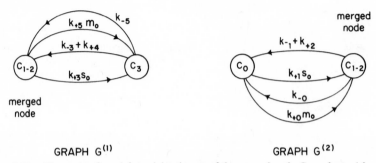

GRAPH $G^{(1)}$ GRAPH $G^{(2)}$

Figure 2.18. Illustrating from left to right, the use of the merged node C_{1-2}, formed from the merger of nodes C_1 and C_2 (see Figure 2.16) along the paths (1): $2 \to 0 \to 1$, and (2): $2 \to 3 \to 2$, respectively, for the calculation of the basic determinant D_1.

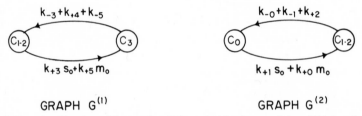

GRAPH $G^{(1)}$ GRAPH $G^{(2)}$

Figure 2.19. Simplification of the graphs of Figure 2.18 by the use of the additivity of parallel branches.

The basic determinants associated with the merged node C_{1-2} for $G^{(1)}$ and $G^{(2)}$ are designated as $D_{12}^{(1)}$ and $D_{12}^{(2)}$, respectively. From the graphs, it is obvious that

$$D_{12}^{(1)} = k_{-3} + k_{+4} + k_{-5},$$
$$D_{12}^{(2)} = k_{+1}s_0 + k_{+0}m_0. \tag{2.101}$$

Inasmuch as we only wished to determine $D_{12}^{(1)}$ and $D_{12}^{(2)}$, we realize now that we could have saved ourselves some trouble in deriving the graph $G^{(1)}$ (or $G^{(2)}$) by neglecting all paths which lead from C_1 or C_2 to the node C_0 (or C_3). From equation (2.97), we calculate the value of D_1 to be

$$D_1 = P_{12}^{(1)}D_{12}^{(1)} + P_{12}^{(2)}D_{12}^{(2)}$$
$$= k_{-0}k_{+1}s_0(k_{-3} + k_{+4} + k_{-5}) + k_{+3}s_0k_{-5}(k_{+1}s_0 + k_{+0}m_0). \tag{2.102}$$

The advantage of utilizing auxiliary nodes to calculate basic determinants becomes more apparent as the order of the graph becomes larger.

By either of the above two procedures, we readily find that

$$D_2 = m_0\{k_{-3} + k_{+4})[k_{+0}(k_{-1} + k_{+2}) + k_{+0}k_{+5}m_0 + k_{+1}k_{+5}s_0]$$
$$+ k_{+0}k_{-5}(k_{-1} + k_{+2})\}, \tag{2.103}$$
$$D_3 = m_0 s_0[k_{+1}k_{+5}(k_{-0} + k_{+3}s_0) + k_{+0}k_{+3}(k_{-1} + k_{+2} + k_{+5}m_0)].$$

From (2.96),

$$v = \frac{dp}{dt} = k_{+2}c_1 + k_{+4}c_3. \tag{2.104}$$

From (2.95),

$$c_1 = \frac{e_0 D_1}{D_0 + D_1 + D_2 + D_3},$$
$$c_3 = \frac{e_0 D_3}{D_0 + D_1 + D_2 + D_3}, \tag{2.105}$$

or, after substituting equations (2.99), (2.102), and (2.103), and dividing numerator and denominator of (2.105) by $k_{+0}k_{+3}k_{+5}$,

$$c_1 = e_0 s_0 \left[\frac{k_{+1}}{k_{+0}} K_0' s_0 + K_0 \left(\frac{k_{+1}}{k_{+3}} K_0' + \frac{k_{+1}}{k_{+5}} K_m' \right) + m_0 K_0' \right] D^{-1},$$
$$c_3 = e_0 m_0 s_0 \left(\frac{k_{+1}}{k_{+0}} s_0 + \frac{k_{+1}}{k_{+3}} K_0 + \frac{k_{+1}}{k_{+5}} K_m + m_0 \right) D^{-1}, \tag{2.106}$$

where

$$D = s_0^2 \frac{k_{+1}}{k_{+0}} (K_0' + m_0)$$

$$+ s_0 \left\{ \frac{k_{+1}}{k_{+0}} K_0' K_m + K_0 \left(\frac{k_{+1}}{k_{+3}} K_0' + \frac{k_{+1}}{k_{+5}} K_m' \right) \right.$$

$$\left. + m_0 \left(K_0' + \frac{k_{+1}}{k_{+0}} K_m' + \frac{k_{+1}}{k_{+3}} K_0 + \frac{k_{+1}}{k_{+5}} K_m \right) + m_0^2 \right\}$$

$$+ K_m K_0 \left(\frac{k_{+1}}{k_{+3}} K_0' + \frac{k_{+1}}{k_{+5}} K_m' \right) \tag{2.107}$$

$$+ m_0 \left[K_m \left(\frac{k_{+1}}{k_{+3}} K_0' + \frac{k_{+1}}{k_{+5}} K_0 \right) + K_0 K_m' \right] + m_0^2 K_m',$$

$$K_m = \frac{k_{-1} + k_{+2}}{k_{+1}}, \; K_0 = \frac{k_{-0}}{k_{+0}},$$

$$K_m' = \frac{k_{-3} + k_{+4}}{k_{+3}}, \; K_0' = \frac{k_{-5}}{k_{+5}}.$$

Substituting (2.106) and (2.107) into (2.104), we obtain the final expression for the reaction velocity as

$$
\begin{aligned}
v = s_0 \Bigg[& k_{+2}e_0 \left\{ \frac{k_{+1}}{k_{+0}} K_0' s_0 + K_0 \left(\frac{k_{+1}}{k_{+3}} K_0' + \frac{k_{+1}}{k_{+5}} K_m' \right) + m_0 K_m' \right\} \\
& + k_{+4}e_0 m_0 \left\{ \frac{k_{+1}}{k_{+0}} s_0 + \frac{k_{+1}}{k_{+3}} K_0 + \frac{k_{+1}}{k_{+5}} K_m + m_0 \right\} \Bigg] D^{-1}
\end{aligned}
\tag{2.108}
$$

Because of the principle of detailed balancing, the rate constants are not all independent. Consider the system in equilibrium, which would result if the reactions leading to product formation are not permitted. The graph representing this equilibrium state is obtained from the graph of Figure 2.15 by omitting the branches labeled k_{+2} and k_{+4}. From the condition that the product of the branch values of the clockwise loop equals that of the counterclockwise loop, we obtain the constraint equation

$$
k_{+0}k_{+3}k_{-5}k_{-1} = k_{-0}k_{-3}k_{+5}k_{+1}.
\tag{2.109}
$$

Equation (2.108) reduces to the ordinary Michaelis–Menten result when the modifier concentration $m_0 \rightarrow 0$. The curve $v = v(s_0)$ represented by the above equation can exhibit cooperative behavior as shown in Section 2.5.

2.11. ENZYME–SUBSTRATE–ACTIVATOR SYSTEM

In a hypothetical enzyme–substrate–activator system, assume that the ligand A reacts competitively at the same site of the enzyme as the substrate S. In addition, when the ligand A is bound at one site, it enhances allosterically the binding of the ligand S at a neighboring site. Such an effector molecule is called a **competitive activator**. As an example of such a system (Volkenstein and Goldstein, 1966), we consider an enzyme which is a dimer. The graph representing the reaction scheme is shown in Figure 2.20. The graph also indicates those complexes that break down to form the product. Because the nodes representing the product and their connecting branches are irrelevant to the determination of the reaction velocity, the connecting branches are shown as dashed lines. For simplicity, it is assumed that the sites act equivalently in product formation so that $k_{+2} = k_{+4} = k_{+8}$. As in the enzyme–substrate–modifier system, the complex C_1 could in principle undergo a transition to the state C_4 by the addition of an activator molecule, but this possibility has been ignored.

Although the graph is of order six, the basic determinants are each determined by only one tree because of the connectivity of the graph. Thus, it may

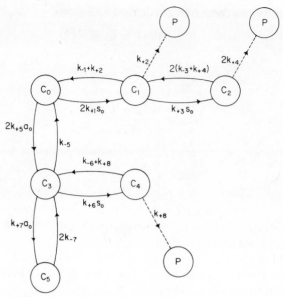

Figure 2.20. Graph illustrating an enzyme–substrate–activator system for which the allosteric activator of concentration a_0 reacts competitively at the same active site of the enzyme as does the substrate of concentration s_0.

be seen directly from the graph that

$$D_0 = 4(k_{-1} + k_{+2})(k_{-3} + k_{+2})k_{-5}(k_{-6} + k_{+2})k_{-7},$$
$$D_1 = 8(k_{-3} + k_{+2})k_{-5}(k_{-6} + k_{+2})k_{-7}k_{+1}s_0,$$
$$D_2 = 4k_{-5}(k_{-6} + k_{+2})k_{-7}k_{+1}k_{+3}s_0^2,$$
$$D_3 = 8(k_{-1} + k_{+2})(k_{-3} + k_{+2})(k_{-6} + k_{+2})k_{-7}k_{+5}a_0,$$
$$D_4 = 8(k_{-1} + k_{+2})(k_{-3} + k_{+2})k_{-7}k_{+5}k_{+6}s_0a_0,$$
$$D_5 = 4(k_{-1} + k_{+2})(k_{-3} + k_{+2})(k_{-6} + k_{+2})k_{+5}k_{+7}a_0^2.$$

$$(2.110)$$

The velocity of the reaction is determined from the figure with $k_{+2} = k_{+4} = k_{+8}$, and equation (2.96) becomes in this case

$$v = 2k_{+2}e_0s_0\left[\frac{1}{K_m}\left(1 + \frac{s_0}{K_m'}\right) + \frac{a_0}{K_A K_{mA}}\right]$$
$$\times \left[1 + 2\frac{s_0}{K_m} + \frac{s_0^2}{K_m K_m'} + 2\left(1 + \frac{s_0}{K_{mA}}\right)\frac{a_0}{K_A} + \frac{a_0^2}{K_A K_A'}\right]^{-1},$$

$$(2.111)$$

where the equilibrium constants are defined as follows:

$$K_m = \frac{(k_{-1} + k_{+2})}{k_{+1}},$$

$$K'_m = \frac{(k_{-3} + k_{+2})}{k_{+3}},$$

$$K_{mA} = \frac{(k_{-6} + k_{+2})}{k_{+6}},$$

$$K_A = \frac{k_{-5}}{k_{+5}},$$

$$K'_A = \frac{k_{-7}}{k_{+7}}.$$

(2.112)

When the activator A is absent, the expression for v is that for a cooperative dimer such as we considered in Section 2.5.

Consider the ability of the activator to enhance allosterically the binding of the substrate ligand at neighboring sites. This effect should be observable at low concentrations of a_0. However, at large concentrations of a_0, the competitive activator will exert an inhibitory effect on product formation, because it occupies the sites at which the substrate binds, and ties up the enzyme in the inactive complex C_5. In fact, we see that $v \to 0$ as $a_0 \to \infty$, according to (2.111). Because of these opposing effects of the competitive activator, the possibility exists that at some intermediate value of the activator concentration, a maximum in the velocity of the reaction will be observed. The quantitative expression of these considerations is as follows.

The necessary condition for a maximum in the curve $v = v(a_0)$ is that $dv/da_0 = 0$, or, from (2.111),

$$\frac{a_0^2}{K_A K'_A} + \frac{2K_{mA}}{K_m}\left(1 + \frac{s_0}{K'_m}\right)\left(1 + \frac{a_0}{K'_A}\right) + \frac{s_0^2}{K_m K'_m} - 1 = 0. \quad (2.113)$$

The presence of both positive and negative terms permits this equation to have a real positive root. If the activator enhances substrate binding at neighboring sites, then it follows that $k_{+6} \gg k_{+1}$, or, from (2.112), $K_{mA} \ll K_m$. Then we may neglect the term of order K_{mA}/K_m in (2.113). If it is further the case that s_0 is not too large so that $s_0^2 < K_m K'_m$, then the real positive root of (2.113) is given approximately as

$$a_0 = (K_A K'_A)^{1/2}\left(1 - \frac{s_0^2}{K_m K'_m}\right)^{1/2} \quad (2.114)$$

Speaking generally, kinetic analysis of enzyme reactions has made a major contribution to the elucidation of the mechanisms of the catalytic behavior of enzymes. It is to this end that graph theory plays a role in obtaining readily the results of a steady-state analysis of such reactions. There still remains the general question of analyzing the results of experimental studies involving two or more substrates so as to infer which of many possible theoretical mechanisms is governing a given reaction. Some theoretical efforts in this direction have already been made, especially with respect to the understanding of enzyme-catalyzed reactions in the presence of two or more substrates. The interested reader is directed especially to the work of Cleland (1963, 1967).

2.12. ASPARTATE TRANSCARBAMYLASE

The principal constituents of the cell nucleus are the **nucleic acids**, which are chains of **nucleotides**, or **bases** attached to sugar phosphates. The backbone of the chain consists of alternating sugar and phosphate groups. There are two kinds of nucleic acids, namely **ribonucleic acid (RNA)**, which directs protein synthesis, and **deoxyribonucleic acid (DNA)**, which is the heritable and genetic material of a cell. RNA and DNA are each linear polymers constructed from four kinds of nucleotides. DNA molecules are of enormous length, and the number of nucleotides in a single molecule may exceed 3×10^5. By contrast, the number of nucleotides in a single DNA molecule of a very small **bacteriophage**, a virus that attacks bacteria, is about 5000.

The bases found in nucleic acids are classified as being either **purines** (adenine and guanine), or **pyrimidines** (cytosine, thymine, and uracil). Pyrimidine biosynthesis is initiated by the combination of aspartate with carbamyl phosphate to form carbamyl aspartate. The latter reaction is catalyzed by the enzyme **aspartate transcarbamylase (ATCase)**, an extensively studied enzyme that behaves as a regulator and about which some structural oligomeric features are known.

In practice, carbamyl phosphate is added in excess amounts, and the enzymatic reaction is studied as if there were a single substrate. The curve for the velocity of the reaction as a function of aspartate concentration displays a typical sigmoidal shape expected of cooperative systems (see the curve labeled "Control" in Figure 2.21(a)). One of the end products in the biosynthetic reactions initiated by ATCase is **cytidine triphosphate (CTP)**, which is used in RNA synthesis. Figure 2.21(a) also shows that the form of the curve is affected by the presence of the product CTP, essentially shifting it to the right. Consequently, more substrate is needed to achieve a given reaction velocity in the presence of product than that needed when the product is absent. In other words, as long as cytidine is being incorporated into nucleic acid, ATCase will continue its catalytic function. When incorporation stops,

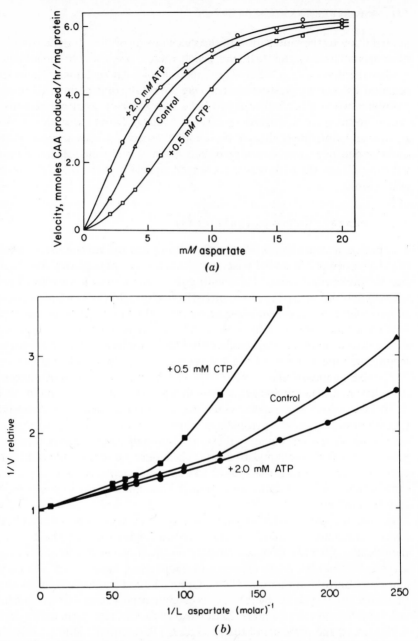

Figure 2.21. (*a*) The velocity of the formation of carbamyl aspartate (CAA) by ATCase is shown as a function of substrate concentration, and in the presence of the inhibitor CTP or the activator ATP. Reproduced, with permission, from Gerhart (1970), based on data of V. Pigiet, Y. Yang, and H. K. Schachman (unpublished). (*b*) The same data shown in (*a*) are replotted in the Lineweaver–Burk manner. The ordinate is the reciprocal reaction velocity, and the abscissa (labeled $1/L$ aspartate) is the reciprocal of the aspartate concentration. The ordinate unit is the maximum velocity or asymptote of the curves displayed in (*a*). Reproduced, with permission, from Jacobson & Stark (1973), after Collins (1971). © Copyright 1972 by Kim D. Collins, Ph.D. Thesis, Stanford University (1971).

the cytidine accumulates and terminates the catalytic activity of ATCase. This effect is known as **product inhibition**. Thus CTP is an allosteric effector, or feedback inhibitor of the enzyme. By means of this effect, regulation of nucleotide biosynthesis is achieved.

When a mercurial is added to ATCase taken from *E. coli*, the affinity for aspartate increases, and the sigmoidal shape of the velocity curve reverts to a classical hyperbolic form. Structural studies (Gerhart and Schachman, 1965) indicate that the mercurial causes ATCase to dissociate into two kinds of subunits. One of them, called the **catalytic subunit** catalyzes the aspartate reaction but is not inhibited by CTP. The other subunit, called the **regulatory subunit**, has a high affinity for CTP, but displays no catalytic activity. Thus, CTP is an inhibitor only in the presence of the native ATCase molecule.

Additional studies (Weber, 1968; Rosenbusch and Weber, 1971) indicate that the there are six catalytic sites and six regulatory sites on each native ATCase molecule. Furthermore, when **adenosine triphosphate (ATP)** is added in the presence of CTP, the effect of CTP shown in Figure 2.21(*a*) is nullified, and the curve for the velocity of the reaction changes to a parabolic nonsigmoidal shape. In fact it is seen that the velocity of the reaction at a given substrate concentration is increased beyond that observed for native ATCase alone with the substrate. In this sense, ATP behaves like an activator. The interpretation of this effect that is made is that ATP is a competitor of CTP at the regulatory site. If ATP binds at the regulatory site, the catalytic site behaves in a noncooperative manner. If CTP binds at the regulatory site, a conformational change in the monomer takes place and the catalytic site is inhibited.

The data shown in Figure 2.21(*a*) have been replotted in the Lineweaver–Burk manner in Figure 2.21(*b*). On the basis of the curves shown and the criteria for cooperativity discussed in Section 2.4, it is inferred that normally ATCase is a positive cooperative system (see the curve labeled "Control" in Figure 2.21(*b*), but that it loses its cooperativity in the presence of ATP. These same conclusions also follow from the curves of Figure 2.21(*a*). We note from Figure 2.21(*b*) that, viewed as a Michaelis–Menten enzyme, v_0-max of ATCase does not change in the presence of inhibitor or activator, but the Michaelis constant K_m does.

In sum, ATCase displays activator characteristics discussed theoretically in Section 2.11, and modifier characteristics discussed theoretically in Section 2.10. Early data for ATCase (which indicated four instead of six active sites) was quantitatively analyzed in terms of the Monod–Wyman–Changeux model (Changeux and Rubin, 1968). Future work will no doubt further clarify the understanding of this and other cooperative enzymes.

SELECTED REFERENCE BOOKS

M. Dixon and E. C. Webb (1964), *Enzymes*, 2nd ed., Academic, New York.

A. L. Lehninger (1970), *Biochemistry*, Worth, New York.

H. R. Mahler and E. H. Cordes (1971), *Biological Chemistry*, 2nd ed., Harper and Row, New York.

M. V. Volkenshtein (1969), *Enzyme Physics*, Plenum Press, New York.

H. Gutfreund (1972), *Enzymes: Physical Principles*, Wiley, London.

PROBLEMS

1. Assume that the back-reaction $C \overset{k_{-2}}{\underset{}{\rightleftharpoons}} P + E$ is included with the reactions of the enzyme-substrate system (2.1). Find the modification of the Michaelis–Menten equation for this case (Haldane, 1930). Show that after a very long time, when equilibrium is established, the equilibrium concentrations $\bar{s}$ and $\bar{p}$ of the substrate and product, respectively, are related by **Haldane's relation**,

$$\frac{\bar{p}}{\bar{s}} = \frac{k_{+1}k_{+2}}{k_{-1}k_{-2}}. \tag{2.115}$$

2. Reaction velocities are commonly measured by measuring the substrate concentration s_1 at a time t_1 shortly after the onset of the reaction at $t = 0$. The velocity is then determined by the secant approximation to the derivative of s, which we shall denote by w_0, namely,

$$w_0 = \left| \frac{\Delta s}{\Delta t} \right| = \frac{s_0 - s_1}{t_1}, \tag{2.116}$$

where s_0 is the initial substrate concentration. According to equations (2.15) and (2.22a), what is the theoretical error $(1/w_0 - 1/v_0)$, as a function of s_0 and s_1? What are the two leading terms in the error if the power series representation of $\log (1 - \varepsilon)$, for $\varepsilon \ll 1$, is utilized?

    Hint. Eliminate t_1 from equation (2.116) by means of equation (2.22a).

3. Show that the substrate concentration $s(t)$ in a substrate–enzyme system is given for very short times, to order e_0/s_0, by the expression

$$s(t) = s_0 \left\{ 1 - \frac{Vt}{s_0 + K_m} - \frac{e_0 - V\tau_0}{s_0 + K_m} \left[1 - e^{-t/\tau_0} \right] \right\}. \tag{2.117}$$

    Hint. It is necessary to solve equation (2.5a) (equivalent to equation (2.18a)) with s and c on the right-hand side given by equations (2.19) and (2.20), respectively.

4. In **autocatalysis**, the product P of a reaction catalyses its formation from a substrate S. Assume that initially, the concentrations of substrate and product have the values s_0 and p_0, respectively. Write down the rate equations for the concentrations and find their functional depen-

dence on the time. Do not assume that an intermediate complex is formed in the reaction.

5. According to a current proposal (Rosenberg, Cavalieri, and Ungers, 1969), the initiation of DNA synthesis is under **negative control**, a principal mechanism of genetic control in the Jacob–Monod theory of protein synthesis (Jacob and Monod, 1961). That is to say, it is assumed that a **repressor** protein R, synthesized at the beginning of the replication cycle, combines with DNA to prevent synthesis. An **antirepressor** protein A, capable of reacting with R to form a complex, is synthesized continuously and slowly accumulates during cell growth until, at a critical concentration, the affinity of repressor for antirepressor overcomes that of repressor for DNA and permits the initiation of synthesis. Can this happen abruptly, that is, in an "off–on" manner? (Posed by L. Cavalieri.)

This question can be answered in the following manner.

(*a*) Write down the rate equations for the concentrations of D, R, A, and associated complexes, where D stands for DNA (see Figure 2.22). Ignore the fact that production of A is slowly taking place, and the fact that there is only one DNA molecule.

(*b*) Assume a steady state and show that the equilibrium concentrations of antirepressor and "free" DNA are solutions of certain cubic equations.

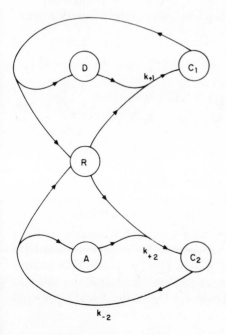

Figure 2.22. Interaction of repressor (R) with DNA (D) and antirepressor (A), to form the complexes C_1 and C_2, respectively.

$(c)^*$ Assuming $K_1/d_0 \approx 10^{-10}$ and $K_2/d_0 \gtrsim 10^{-13}$, show that, approximately, $a/a_0 \gtrsim 1 - d_0/r_0$, where $K_1 \equiv k_{-1}/k_{+1}$ and $K_2 \equiv k_{-2}/k_{+2}$ are the equilibrium constants associated with C_1 (repressor–DNA complex) and C_2 (repressor–antirepressor complex), respectively, a, d, and r denote the equilibrium concentrations of A, D, and R, respectively, and the subscript 0 denotes initial value.

6.* Show that when $c = 0$ and $L \gg 1$, the saturation function $Y(x)$ given by equation (2.70) has a point of inflection at approximately the value $x = [(n - 1)L/(n + 1)]^{1/n} - 1$.

   Hint. Set $x = \xi - 1$, and find the inflection point for Y considered as a function of ξ. In the equation determining the value of ξ at which the inflection point exists, assume ξ varies as some power of L and retain only the two dominant terms.

7. Assume that the first three oxygen molecules combine with hemoglobin in an independent manner, but because of an allosteric effect which follows the binding of the third oxygen molecule, the hemoglobin molecule has a very great affinity for the fourth oxygen molecule. Thus, let the rate constants for binding the first three oxygen molecules be as shown in the graph of Figure 2.6, and let the equilibrium constant for binding any of the first three oxygen molecules at a given site be K, as in equation (2.33). For the fourth oxygen molecule, let the rate constants for binding and dissociating be $\lambda_+ k_{+1} s_0$ and $4\lambda_- k_{-1}$, respectively, and let $\lambda_+/\lambda_- \equiv L + 1$, where L is then a measure of the positive cooperative effect. Show that

$$(K_1^{-1}, K_2^{-1}, K_3^{-1}, K_4^{-1}) = 4K^{-1}(1, \tfrac{3}{8}, \tfrac{1}{6}, \tfrac{1}{16} + \tfrac{L}{16}), \qquad (2.118)$$

and
$$Y(x) = \frac{x(1 + x)^3 + Lx^4}{(1 + x)^4 + Lx^4}, \qquad (2.119)$$

where
$$x = \frac{s_0}{K}.$$

8. Consider the fully competitive enzyme–substrate–inhibitor system of Section 2.3. Draw the associated graph, and derive the expression for the velocity of the reaction by the graph-theoretic method.

9. In **noncompetitive** inhibition, the inhibitor I does not compete with the same site on the enzyme C_0 as the substrate S. The inhibitor can form an inhibitor–enzyme complex C_2, as well as a substrate–inhibitor–enzyme complex C_3. The latter may yield the product P at a different rate from the enzyme–substrate complex C_1, or the complex may not

break down at all. In the latter case the inhibitory effect is achieved by tying up some of the enzyme so that it is not available for the conversion of substrate to product. Such inhibition is termed **pure noncompetitive** inhibition. Assume that such is the case and that the binding sites for the substrate and the inhibitor are completely independent of each other. The reactions are represented as follows:

$$C_0 + S \underset{k_{-1}}{\overset{k_{+1}}{\rightleftharpoons}} C_1 \overset{k+2}{\rightarrow} C_0 + P,$$

$$C_0 + I \underset{k_{-3}}{\overset{k_{+3}}{\rightleftharpoons}} C_2,$$

$$C_1 + I \underset{k_{-3}}{\overset{k_{+3}}{\rightleftharpoons}} C_3, \tag{2.120}$$

$$C_2 + S \underset{k_{-1}}{\overset{k_{+1}}{\rightleftharpoons}} C_3.$$

Draw the graph for the reactions in a steady state, and derive by graph-theoretic means the expression for the velocity of the reaction. Compare the result with that for competitive inhibition. Show schematically the effect on a Lineweaver–Burk plot of varying the initial inhibitor concentration i_0.

Suppose the complex C_3 can be converted to the product P at the same rate as the complex C_1. This case is called **apparent competitive** inhibition. Discuss the effect of such competition in relation to Lineweaver–Burk plots.

3

TRACERS IN PHYSIOLOGICAL SYSTEMS

A living organism is not isolated from its environment but is rather in continual interaction with it. First and foremost, it requires energy to maintain itself. Thus, it is continually obtaining energy from its environment and excreting waste products to the environment. The energy is in the form of radiation for plants, and oxidative nutrients for animals. These processes require the passage of matter across the boundary between the organism and its environment, as well as the passage of matter across boundaries contained within the organism. Because of this exchange, a living organism is said to be an **open system**. The interaction of a living organism with its environment can be characterized approximately as a steady state, in which there are constant interchanges of energy and matter with the environment.

At the same time, organisms are unstable in the sense that they produce large responses to relatively insignificant alterations in their environment, as, for example, the response of an owl to a field mouse observed in its peripheral vision. However, the higher multi-cellular animals, when viewed internally, are very stable. They attempt to maintain their processes and characteristics such as temperature and blood sugar level constant within narrow limits, despite small disturbances. The maintenance of the many internal steady-state processes that characterize the higher animals is called **homeostasis**.

3.1. COMPARTMENT SYSTEMS

The above paragraphs describe in a qualitative manner some salient features of living systems. We shall now look into the problem of describing some characteristics of such systems in quantitative terms. The system may be an entire organism, a collection of organisms, or more usually, one small aspect of a multi-cellular organism. In particular, we shall concern ourselves with the detection and determination of properties of physiological systems in a steady state by means of **tracers**. A tracer is a detectable quantity of matter such as a dye or a radioactive isotope. Because of the visual or radioactive property of the tracer, it is also called **labeled** material. The point of view adopted is to consider a biological system as a collection of pools or compartments.

A **compartment** is defined as a substance with a given size. The **size** is usually meant to be the mass of the substance. Sometimes, however, the size is determined by the volume occupied by the substance. In such a case, the term **volume of distribution** is used synomymously for size. The substance characteristic of one compartment is often the same as the substance characteristic of another compartment. In that case, the compartment is identified by means of its spatial assignment; for example, intracellular glucose and extracellular glucose may be viewed as two distinct compartments. On the contrary, two different substances may occupy the same physical volume, but be chemically identifiable. For example, free iodine in the blood may be one compartment, and serum bound iodine may be another. Furthermore, in one experiment, a tissue or collection of cells may be considered to be a compartment. In another experiment, a cell or a subcellular entity may be considered to be a compartment. The latter view can be justified by the fact that all the cells under consideration are more or less identical in behavior, and by the fact that the behavior of the cell with respect to one particular entity only is being considered. A compartment is further characterized by its having a **steady-state flux** of matter into and/or out of it.

We shall consider a **compartment system** consisting of two or more inter-connected compartments. By inter-connected we mean that between any two compartments, there exists a steady state of material flux. Each flux has its own channel associated with it, so that the flux from compartment i to compartment j is distinct from, and in general unequal to, the flux from compartment j to compartment i. It is further assumed that matter can enter a given compartment from the exterior at a steady rate. It is generally assumed that the net flux into any compartment equals the net flux out. This implies that no compartment is a **source** (for which the net outflux exceeds the net influx) or a **sink** (for which the net influx exceeds the net outflux). The exterior represents both an inexhaustible supply of raw material (source) and an infinite container for waste products (sink). The transfer of material from one compartment to another may be an actual transport of matter from one region to another, due to causes such as pressure gradients, diffusion, or osmosis. However, the compartment system may also represent a sequence of steady state chemical reactions of the type we considered in the previous chapter. In such a case, the reactants can all occupy the same physiological region, and the fluxes represent the transformations of substances from one chemical form to another.

We seek to determine the material fluxes among the compartments, and the size of each compartment. Usually, the number of compartments in the system is unknown and likewise needs to be determined. To determine these quantities, a known amount of labeled material is injected into one compartment. Consequently, the injection **perturbs** the system in that a small excess of material is suddenly introduced. However, the ratio of labeled to unlabeled material is assumed to be very small at all times, and the steady-state characteristics of the system are assumed to be unaltered. The labeled material itself can be followed in time to some extent, and its behavior represents a **transient** property of the system. It is this transient behavior that is studied in order to infer the steady-state characteristics.

It is further assumed that the material in a compartment consists of a **homogeneous** mixture of labeled and unlabeled elements. The implications of compartmental homogeneity are that (*a*) the behavior of the labeled material in a compartment is representative of the behavior of the unlabeled material in a compartment; (*b*) when labeled material enters a compartment, it is instantaneously "mixed," as if a blender were standing by for that purpose.

The abstract problem of a compartment system introduced in the above paragraphs can be perhaps understood best by the following hydrodynamic "bathtub model." Imagine a set of bathtubs or reservoirs, each containing a given characteristic amount of water. Assume that every pair of bathtubs is connected by two pipes, with water flowing through the pipes. The flux of water through any pipe is arbitrary, but fixed, so that in general more water flows in one direction than the other. Let each bathtub have a faucet connected to the exterior, turned on or off, a drain pipe going to the exterior, either open or closed, and a blender, which mixes the water thoroughly. The water fluxes are maintained in a steady state so that the level of water in every bathtub is not changing with time. Now assume that at $t = 0$, a known small amount of dye is injected into one of the bathtubs. The volume occupied by the dye is considered to be negligible. The blender insures that the dye concentration is homogeneous throughout the bathtub. Because of the interconnections the dye spreads throughout the system, and is ultimately cleared from the system because of the drain pipes. Suppose that the concentration of dye can be observed at regular time intervals in one or more of the bathtubs. The objective of the dye injection and the dye concentration observations is to determine the volume of water in each bathtub and the flux of water through each pipe, including the drain pipes.

We shall consider one-, two-, and three-compartment systems in some detail, as these are the compartment systems most commonly encountered.

3.2.   THE ONE-COMPARTMENT SYSTEM

The simplest example of a compartment system is that consisting of a single compartment. Consider a compartment whose size is denoted by the constant V. Let there be a steady flux out of the compartment of its characteristic substance at a rate K. Of course, this outward flux is balanced by an inward flux of the same magnitude. Suppose that at $t = 0$, a small amount of labeled material ρ_0 is introduced into the compartment. The amount of labeled material remaining in the compartment at any subsequent time t is denoted by $\rho(t)$. Because of the assumption of homogeneity, the density of labeled material $x(t)$ is constant throughout the compartment and satisfies the relation

$$\rho(t) = Vx(t). \tag{3.1}$$

When the material is labeled by a radioactive marker, the quantity ρ represents the total amount of radioactive material which is measured in units of radioactivity or simply **activity**, such as microcuries (μCi), disintegrations per second (dis/sec), or counts per minute (cpm). The **curie** is defined as the number of disintegrations per second occurring in 1 g of radium, or 1 Ci = 3.7×10^{10} dis/sec. The quantity x then represents the amount of radioactive material per unit mass in the compartment, and is measured in units of **specific activity**, or activity per unit mass, for example, microcuries per gram.

In time Δt, the change in the amount of tracer material $\Delta\rho(t)$ is proportional to Δt, the flux rate K, and the concentration $x(t)$, or,

$$\Delta\rho(t) = -Kx(t)\Delta t.$$

The minus sign expresses the fact that the tracer material in the compartment decreases with time because of the outward flux of unlabeled material. Dividing by Δt and taking the limit as $\Delta t \to 0$, we obtain the differential equation

$$\frac{d\rho}{dt} = -Kx. \tag{3.2}$$

We define the rate at which material leaves the compartment per unit amount of material in the compartment as

$$L = \frac{K}{V}. \tag{3.3}$$

L is called the **clearance rate** or **turnover rate** of the compartment. With this definition, equation (3.2) can be expressed as a differential equation for $\rho(t)$, by substituting for $x(t)$ on the right-hand side its value from equation (3.1), namely,

$$\frac{d\rho}{dt} = -L\rho. \tag{3.4}$$

Alternatively, we can use (3.1) to substitute for $\rho(t)$ on the left-hand side of (3.2) and obtain an equation for $x(t)$,

$$\frac{dx}{dt} = -Lx. \tag{3.5}$$

Either (3.4) or (3.5) is the fundamental differential equation representing a one-compartment system. The choice of equation depends on whether $\rho(t)$ or $x(t)$ is being observed. If the total activity in the compartment is being observed, then equation (3.4) would be utilized, and if the compartment is being sampled so that only the concentration of tracer material in the compartment is being observed, then equation (3.5) would be utilized. Because observations of the concentration are more usually made, we shall select equation (3.5) for study. We have of course solved this equation before, and its solution is

$$x(t) = x(0)e^{-Lt} = \frac{\rho_0}{V} e^{-Lt}. \tag{3.6}$$

In this solution, we have satisfied the initial condition with the aid of (3.1).

Suppose now that the concentration of labeled material in the compartment has been observed at various times t_j, where $j = 1, 2, \ldots, N$. Call the corresponding values of the concentration $x^*(t_j)$. The units of x^*, if the tracer material is a radioactive substance, will be disintegrations per minute per gram, or counts per second per milliliter, or something similar. We make a semilog plot of $x^*(t_j)$, or use the method of least squares, to fit the observations to the simple exponential expression

$$x^*(t) = Ae^{\lambda t}. \tag{3.7}$$

The best values of A and λ are determined from the observations. Here, λ will always turn out to be negative, because a positive value of λ would violate the law of conservation of mass by permitting the amount of labeled material in the compartment to grow in time without bound.

We now equate equation (3.6), the theoretical solution, to equation (3.7), the observed result. From this equation, we immediately infer the values of L and V to be

$$\begin{aligned} L &= -\lambda, \\ V &= \frac{\rho_0}{A}. \end{aligned} \tag{3.8}$$

Since we know the initial amount of injected tracer ρ_0, equations (3.8) determine the compartment size V and turnover rate L. These parameters characterize the steady-state behavior of the compartment system. The values were determined by means of an investigation of the transient response of the system to a small perturbation, namely, the injection of some tracer material. The result (3.8) completes the solution to the problem.

The units of L will be $(\text{time})^{-1}$. The units of V will be either mass or volume, depending on the units of ρ_0 and A. Note that disintegrations or counts do not enter into either of these expressions, as these units of detection cancel themselves in the second expression given above.

3.3. INDICATOR–DILUTION THEORY

The method of measurement of the volume of human blood or mean blood flow rate through the cardiovascular system or a segment of it is usually called the **indicator–dilution method** or the **Stewart–Hamilton method**, although the origins of the method antedate the work of Stewart (1921) and Hamilton (Kinsman, Moore, and Hamilton, 1929; Moore, Kinsman, Hamilton, and Spurling, 1929). In hydraulic engineering, the method for determining the volume of water in a conduit, which is an analogous problem, is called Allen's method. The simplest theoretical formulation of this method is based on a one-compartment system.

Suppose an amount of dye (indicator) ρ_0 is injected into the blood stream on the venous side of the left ventricle at a given instant of time. The ventricle may be thought of as a reservoir of constant level with a constant rate of influx and efflux. Obviously, this view neglects the periodic variations in blood flow due to the beating of the heart. The dye is selected so that it binds to albumin or other serum proteins. As a result, a negligible amount of it passes through the capillary walls and out of the vascular system during the time of blood flow observations. Blood in the veins will, after a short while, enter the left ventricle and be pumped from there into the aorta, the major artery of the vascular system. Imagine that we can make measurements of the dye concentration in the aorta as a function of the time. A typical observed curve is shown in Figure 3.1.

Let $c(t)$ denote the concentration of dye in the left ventricle, V the mean blood volume in the left ventricle, and K the steady flux of blood through the ventricle, or **cardiac output**. We assume that all the dye leaving the left ventricle must pass by the point of observation in the aorta that has been selected. This assumption neglects the effect of the presence of the coronary artery and other upstream arterial branches through which dye can escape before reaching the point of observation. Assume that all the dye enters the ventricle instantaneously at a time t_0 following its injection into the vein at

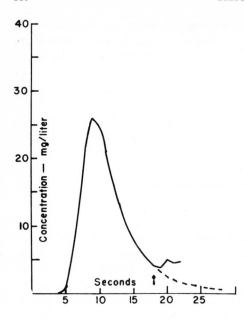

Figure 3.1. Concentration of indicator is shown schematically as a function of time after it is injected into a systemic vein at time zero and sampled subsequently from a systemic artery. The arrow indicates the onset of recirculation. From Meier and Zierler (1954), with permission.

time zero. Then the amount of dye initially injected is $\rho_0 = Vc(t_0)$. Assume further that the time it takes the dye to reach the point of observation after leaving the left ventricle is negligibly small. The equation representing the dye concentration in the ventricle is assumed to be the following:

$$\frac{d}{dt}(cV) = -Kc, \; t > t_0, \tag{3.9}$$

which is just the equation for a one-compartment system. The solution is

$$c(t) = \frac{\rho_0}{V} e^{-(K/V)(t-t_0)}, \; t > t_0. \tag{3.10}$$

The actual experimental curve, which we call $c^*(t)$, displays a time delay of about 5 sec representing the time it takes for the injected dye to reach the heart, as well as a rapid but not instantaneous increase in concentration to a maximum value. However, the injection into the heart is not instantaneous, for a number of reasons such as the diffusion of dye, the finite time interval necessary for injection, the precise mechanism of convection of the dye from the injection site to the heart, etc. Nevertheless, if equation (3.9) above represents our system, the tail of the curve in the figure should be representable as an exponential, and this is found to be the case. At the point of the curve marked by an arrow, the curve deviates from exponential form.

The reason for this is the recirculation of dye material, which is not represented in the model: The dye entering the aorta ultimately is distributed to the venous side of the circulatory system and reenters the left ventricle.

The result of making a semilog plot of the tail of the curve $c^*(t)$ (disregarding the recirculation portion) is the expression

$$c^*(t) = Ae^{\lambda t}, \tag{3.11}$$

where λ is a negative quantity of dimension $(\text{time})^{-1}$. However, we know this representation is only appropriate beyond a certain time t_0, whose value we do not know yet. Nevertheless, we can compare this expression with the theoretical solution, and obtain the relations

$$\frac{K}{V} = \lambda,$$
$$\frac{\rho_0}{V} = Ae^{\lambda t_0}. \tag{3.12}$$

Because we have three unknowns, K, V, and t_0, and two equations, an additional relation is needed to be able to solve for the unknowns. It is obtained by determining the area under the curve $c^*(t)$, with the portion of the curve representing recirculation replaced by the extrapolated value of the curve determined from equation (3.11). This extrapolation is done so as to obtain the result of "first passage" only of the dye material. The area under this extrapolated curve (see Figure 3.1) can be determined by means of a numerical integration procedure, which replaces the integral by a sum. We shall not explain such a procedure, which depends in essence on the fundamental definition of an integral, and assume that the integration is accomplished. Let us denote the area so obtained by the symbol S:

$$S = \int_0^\infty c^*(t)dt. \tag{3.13}$$

By integrating equation (3.10), we find that the theoretical value of this area is

$$S = \int_0^\infty c(t)dt = \frac{\rho_0}{V} \int_{t_0}^\infty e^{-(K/V)(t-t_0)}\, dt = \frac{\rho_0}{V} \int_0^\infty e^{-(K/V)\tau}\, d\tau. \tag{3.14}$$

The last integral is obtained from the second integral by changing the variable of integration from t to $\tau = t - t_0$. By integrating the last expression and combining the result with Equation (3.13), we obtain

$$K = \frac{\rho_0}{S} = \frac{\rho_0}{\displaystyle\int_0^\infty c^*(t)dt}. \tag{3.15}$$

The equality of K with the last expression above is called **Hamilton's formula**. By making use of this result in (3.12), we infer that

$$V = -\frac{\rho_0}{\lambda S},$$

$$t_0 = -\frac{1}{\lambda} \log \frac{A}{\lambda S}.$$

(3.16)

It is seen that the constant A only enters into the determination of t_0, which is the mean delay time for the tracer to enter the compartment. The parameters V and K, on the other hand, depend only on the decay rate λ and the area S under the concentration curve.

Figure 3.2 represents the theoretical form of the curve based on equation

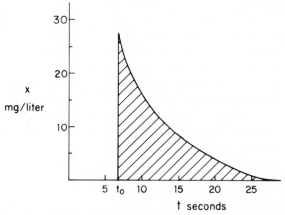

Figure 3.2.   The theoretical curve of dye concentration x as a function of the time, given by equation (3.10), representing an instantaneous injection of dye at time t_0.

(3.10) for an instantaneous injection of dye into the compartment at time t_0. The time t_0 of the theoretical curve has been adjusted to satisfy equation (3.16), so that the areas under the theoretical and experimental curves are equal. The fact that the experimental curve is more rounded initially and does not rise abruptly to its maximum value is ascribed, as indicated previously, to the lack of instantaneity of dye injection into the ventricle. This view can be supported by more sophisticated models of the injection process which are not compartmental in form (see Chapter 5, problem 15). A slightly more sophisticated model which is compartmental in nature is found in problem 4 at the end of this chapter.

The mean lifetime or "**transit time**" $\bar{t}$ of indicator material in the compartment is defined as

$$\bar{t} = \frac{\int_0^\infty tc(t)dt}{\int_0^\infty c(t)dt}. \tag{3.17}$$

Given a density function such as $c(t)$, the above quantity $\bar{t}$ is a general definition for the mean value of t associated with it. If t is thought of as the age of the dye molecules in the compartment, then $c(t)$ can be viewed as representing the fraction of the original number of dye molecules surviving in the compartment in the age interval t to $t + dt$. The function $c(t)$ could be called an age density function, and $\bar{t}$ would then represent the average age of the population of dye molecules in the compartment. If $c(t)$ is given by (3.10), then inserting it into (3.17) yields the expression

$$\bar{t} = \frac{\int_{t_0}^\infty te^{-(K/V)(t-t_0)}\,dt}{\int_{t_0}^\infty e^{-(K/V)(t-t_0)}\,dt} = \frac{\int_0^\infty (t_0 + \tau)e^{-(K/V)\tau}\,d\tau}{\int_0^\infty e^{-(K/V)\tau}\,d\tau}.$$

After utilizing integration by parts to evaluate the numerator, we obtain the result

$$\bar{t} = t_0 + \frac{V}{K}. \tag{3.18}$$

In practice, t_0 is often assumed to be the time interval from the origin to the beginning of the nonzero portion of the curve $c^*(t)$. Then equation (3.18) can be utilized instead of equation (3.12) to determine V, that is,

$$V = K(\bar{t} - t_0), \tag{3.19}$$

with K given by (3.15). This practice replaces the curve-fitting procedure to determine λ by an integration procedure to determine $\bar{t}$.

When $t_0 = 0$, comparison of equations (3.12) and (3.18) shows that $\bar{t} = -V\lambda$. This equality indicates that an integration method can be utilized to determine decay constants for exponential curves as an alternative to a curve-fitting method.

3.4. CONTINUOUS INFUSION

A frequently used method of studying compartment systems is the **continuous infusion** of labeled substance, rather than an instantaneous injection, which commences at time zero. As an example, we shall consider the glucose tolerance test. Such tests are routinely given by oral administration of sugar, in order to determine carbohydrate metabolism levels in man. In order to avoid variations in results due to variations in the rate of intestinal absorption of sugar, glucose can be administered intravenously at a continuous rate. The blood sugar concentration level is determined from samples of blood taken at regular intervals subsequent to the onset of glucose infusion. A typical curve of glucose concentration in the blood as a function of the time for a normal human (one not suffering from diabetes or other diseases known to affect carbohydrate metabolism) is shown in Figure 3.3 (Jokipii and Turpeinen, 1954).

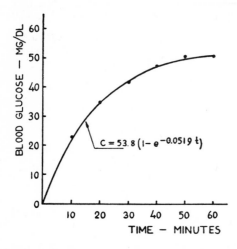

Figure 3.3. Average blood glucose concentration in 24 normal subjects during the course of continuous infusion. Mean blood glucose level before infusion = 84.9 mg/dl (zero level of ordinate). Mean continuous infusion rate $J = 297$ mg/min. From Jokipii and Turpeinen (1954), with permission.

In the figure, the zero level of the glucose concentration refers to the mean level of blood glucose concentration in the fasting state, which was 84.9 mg/dl (dl = deciliter). Consequently, the glucose concentration shown represents the mean increase of glucose concentration above the aforementioned base value. In this experiment, the continuous infusion rate of glucose into the blood was

$$J = 297 \text{ mg/min.}$$

Let us denote the observed excess glucose concentration values by $x^*(t)$. The curve is seen to approach an asymptotic value A_0, which is found to be

$$A_0 = 53.8 \text{ mg/dl.}$$

When the difference between A_0 and $x^*(t)$ is plotted on semilog paper, as a function of the time, the resulting curve is found to be a straight line, from which the decay rate λ is inferred to be

$$\lambda = -0.0519/\text{min.}$$

Hence, $x^*(t)$ can be represented by the expression

$$x^*(t) = A_0 - A_0 e^{\lambda t}. \tag{3.20}$$

This result is suggestive of a compartmental model. However, the experimental conditions require that there be a continuous infusion of glucose, rather than an initial injection of glucose. Let V be the volume of distribution of glucose in the blood. At first thought, it might be expected that V should represented the blood volume, but we should more properly let the experimental facts determine whether such an identification is correct. Then, the differential equation governing the temporal behavior of the blood glucose concentration $x(t)$ is assumed to be

$$\frac{d}{dt}(Vx) = -LVx + J, \tag{3.21}$$

where L represents the fractional rate at which glucose leaves, or is cleared from, the blood. Note that Vx is the actual amount of glucose in the blood in excess of the normal amount present in the fasting state. This equation is not exactly in the form of the fundamental compartment equation (3.5), but it can be placed in that form if we introduce $\xi(t)$ representing the blood glucose concentration in excess of the steady-state value $\bar{x}$. The latter is determined by setting $d(Vx)/dt = 0$ in equation (3.21). Thus,

$$\bar{x} = \frac{J}{LV},$$
$$\xi(t) = \bar{x} - x(t). \tag{3.22}$$

From (3.21) and (3.22), the equation for $\xi(t)$ is

$$\frac{d\xi}{dt} = -L\xi, \tag{3.23}$$

which is the equation for a one-compartment system. The solution of it that satisfies the initial condition $x(0) = 0$ is

$$\xi(t) = \bar{x}e^{-Lt}. \tag{3.24}$$

By examining (3.20) as $t \to \infty$, we identify the steady-state value of $x(t)$ as

$$\bar{x} = A_0. \tag{3.25}$$

Therefore, the experimentally determined representation of $\xi(t)$, which we denote as $\xi^*(t)$, is expressible from (3.20) and (3.22) as

$$\xi^*(t) = A_0 e^{\lambda t}. \tag{3.26}$$

By comparing (3.24) with (3.26), we find that

$$L = -\lambda. \tag{3.27}$$

Next, by comparing (3.22) and (3.25), we infer that

$$V = \frac{J}{A_0 L}.$$ (3.28)

By inserting the numerical values of λ, J, and A_0, we find that

$$L = 0.0519 \text{ per minute},$$
$$V = 10.6 \text{ liter}.$$

The volume of distribution V that is obtained is about twice the mean blood volume in an adult, which is approximately 5 liter. This difference suggests that V represents both the actual blood volume and an extravascular volume component.

Although the regulation of glucose levels in the blood is without doubt a complex homeostatic process, it must be considered remarkable that the observed concentration levels are fitted by such a simple model. It is noteworthy that this is so even though the amount injected leads to a new steady-state level of blood glucose which is more than 50% (more precisely, $53.8/84.9 = 63.4\%$) greater than the normal level of 84.9 mg/dl.

The complementary experiment of stopping the infusion and allowing the system to decay back to its equilibrium state has also been performed. It is generally found that the phenomenon of "overshoot" is observed, the blood glucose level after a time decaying to a value which is less than that of the normal fasting level. Such a result is clear evidence that a simple linear one-compartment model is not adequate to explain the decay. However, it is known that the hormone **insulin**, manufactured in the pancreas and released to the blood, regulates sugar metabolism. More complicated models of glucose metabolism have been proposed which utilize insulin as an additional variable. These models are essentially compartmental in nature, and they do predict an oscillatory decay to equilibrium of an excess glucose injection. [See, for example, Ackerman, Rosevear, and McGuckin (1964).]

3.5. THE TWO-COMPARTMENT SYSTEM

We shall now consider the solution to a two-compartment system of equations. Let the two compartments be called 1 and 2, and let the amount of tracer in each compartment at time t be $\rho_1(t)$ and $\rho_2(t)$, respectively. The rate equations of the compartment system governing the dependence of ρ_1 and ρ_2 on the time are assumed to be the following,

$$\frac{d\rho_1}{dt} = -K_{11}x_1 + K_{12}x_2,$$
$$\frac{d\rho_2}{dt} = K_{21}x_1 - K_{22}x_2,$$ (3.29)

where $x_1(t)$ and $x_2(t)$ are the concentrations of tracer in the two compartments 1 and 2, respectively. They are defined by the equations

$$\rho_1 = V_1 x_1,$$
$$\rho_2 = V_2 x_2, \tag{3.30}$$

where V_1 and V_2 are the compartment sizes of compartments 1 and 2, respectively. In (3.29), K_{11} represents the rate at which unlabeled matter leaves compartment 1 in the steady state, and K_{22} represents the rate at which unlabeled matter leaves compartment 2 in the steady state. The quantity K_{12} represents the rate at which matter is transferred from compartment 2 to compartment 1. Conversely, K_{21} represents the rate at which matter is transferred from compartment 1 to compartment 2. According to these definitions, the K's are all nonnegative quantities. Furthermore, because of the possibility that matter can leave either compartment to go to the exterior, it must be the case that

$$K_{11} \geqq K_{21},$$
$$K_{22} \geqq K_{12}.$$

We shall denote the rate at which matter leaves compartment 1 or 2 to go to the exterior by K_{01} or K_{02}, respectively. Then, because as much matter flows into a compartment as flows out,

$$K_{11} = K_{01} + K_{21},$$
$$K_{22} = K_{02} + K_{12}. \tag{3.31}$$

The dimensions of the K's are mass (or volume) $\cdot$ time^{-1}.

We shall now express the rate equations in terms of the dependent variables $\rho_1(t)$ and $\rho_2(t)$ by eliminating x_1 and x_2 from the right-hand sides of equations (3.29). To this end, define

$$L_{ij} \equiv \frac{K_{ij}}{V_j}, \qquad i, j = 1, 2. \tag{3.32}$$

Then L_{ii} is the rate at which matter leaves compartment i per unit amount of mass (or volume) in compartment i, $i = 1, 2$. L_{ii} is called the **total turnover rate** of compartment i. Similarly, L_{ij} for $i \neq j$ is the rate at which matter leaves compartment j to go to compartment i, per unit amount of matter in compartment j. L_{ij} is called the **fractional turnover rate** of compartment j with respect to i. Similarly, L_{0j} is called the **fractional excretion rate** from compartment j, $j = 1, 2$. It follows from equations (3.31) that the L_{ij} satisfy similar equations to them, namely,

$$L_{11} = L_{01} + L_{21},$$
$$L_{22} = L_{02} + L_{12}. \tag{3.33}$$

With the aid of equations (3.30) and (3.32), the rate equations (3.29) for ρ_1 and ρ_2 become

$$\frac{d\rho_1}{dt} = -L_{11}\rho_1 + L_{12}\rho_2,$$

$$\frac{d\rho_2}{dt} = L_{21}\rho_1 - L_{22}\rho_2. \tag{3.34}$$

If ρ_1 and ρ_2 are the observed variables, then these are the equations to be solved. If however, x_1 or x_2 is to be observed, it is better to make x_1 and x_2 the dependent variables. The above two equations become, after substituting equations (3.30) and dividing by V_1 and V_2, respectively,

$$\frac{dx_1}{dt} = -L_{11}x_1 + M_{12}x_2,$$

$$\frac{dx_2}{dt} = M_{21}x_1 - L_{22}x_2, \tag{3.35}$$

where M_{12} and M_{21} are defined as

$$M_{12} = V_1^{-1}L_{12}V_2,$$

$$M_{21} = V_2^{-1}L_{21}V_1. \tag{3.36}$$

Unlike the one-compartment model, the coefficients of x_1 and x_2 on the right-hand side of equations (3.35) are not the same as the coefficients of ρ_1 and ρ_2 on the right-hand side of equations (3.34). As a result, if concentrations in compartments are going to be measured, it is in general a much more difficult task to infer the values of L_{ij} than it would be if the total amounts of tracer material in the compartments of the system were observed. Furthermore, the quantities M_{12} and M_{21} do not have a useful biophysical significance as do the fractional clearance rates. Nevertheless, because observations of compartmental concentrations are most commonly encountered, we shall take equations (3.35) to be the fundamental compartmental equations for a two-compartment system.

Let us attempt a solution of equations (3.35) in the form

$$x_1(t) = A_1 e^{\lambda t},$$

$$x_2(t) = A_2 e^{\lambda t}, \tag{3.37}$$

where the same constant λ appears in both x_1 and x_2. Substituting (3.37) into (3.35) leads to the equations

$$0 = [-\lambda A_1 - L_{11}A_1 + M_{12}A_2]e^{\lambda t},$$

$$0 = [-\lambda A_2 + M_{21}A_1 - L_{22}A_2]e^{\lambda t}.$$

These equations must be true whatever t is. Because $e^{\lambda t}$ is never zero for any value of t, the coefficient of $e^{\lambda t}$ in each of these equations must vanish, or,

$$-(L_{11} + \lambda)A_1 + M_{12}A_2 = 0,$$
$$M_{21}A_1 - (L_{22} + \lambda)A_2 = 0.$$

These are two simultaneous homogeneous linear equations for the constants A_1 and A_2. The solution $A_1 = A_2 = 0$ is called the **trivial** solution and is of no interest. A **nontrivial** solution exists, and is given by the expression

$$\frac{A_1}{A_2} = \frac{M_{12}}{L_{11} + \lambda} = \frac{L_{22} + \lambda}{M_{21}}. \tag{3.38}$$

The second equality is a necessary condition on the constant λ in order for this nontrivial solution to exist. After cross-multiplication, equation (3.38) can be expressed as a quadratic equation in λ,

$$\lambda^2 + (L_{11} + L_{22})\lambda + L_{11}L_{22} - M_{12}M_{21} = 0. \tag{3.39}$$

We could readily find the two solutions λ_1 and λ_2 to this equation by means of the quadratic formula. This formula would express λ_1 and λ_2 in terms of the coefficients L_{11}, L_{22}, M_{12}, and M_{21}. However, our purpose is to find these coefficients, given experimentally determined values of λ_1 and λ_2. Therefore, it is more desirable to express the coefficients in terms of the quantities λ_1 and λ_2. To this end, we note that equation (3.39) is expressible in factored form as

$$(\lambda - \lambda_1)(\lambda - \lambda_2) = \lambda^2 - (\lambda_1 + \lambda_2)\lambda + \lambda_1\lambda_2 = 0,$$

if λ_1 and λ_2 are the roots of this equation. By comparing the last expression with the previous one, and equating the coefficients of the powers of λ, we see that

$$L_{11} + L_{22} = -\lambda_1 - \lambda_2, \tag{3.40a}$$
$$L_{11}L_{22} - M_{12}M_{21} = \lambda_1\lambda_2. \tag{3.40b}$$

Thus, there are two equations relating the coefficients appearing in equations (3.35) and λ_1 and λ_2.

For each of these two permissible values of λ, there is a corresponding value of the amplitude ratio A_1/A_2, given by equation (3.38). In other words, we have found two solutions to equations (3.35), namely,

$$\left.\begin{array}{r} x_1 = A_{11}e^{\lambda_1 t} \\ x_2 = A_{21}e^{\lambda_1 t} \end{array}\right\}, \text{ and } \left.\begin{array}{r} x_1 = A_{12}e^{\lambda_2 t} \\ x_2 = A_{22}e^{\lambda_2 t} \end{array}\right\}. \tag{3.41}$$

Here, the ratios A_{11}/A_{21} and A_{12}/A_{22} are determined from equation (3.38) with $\lambda = \lambda_1$, and $\lambda = \lambda_2$, respectively. Because the compartment equations are linear, the sum of these two solutions is also a solution. Hence, the general solution of the compartment equations is

$$x_1(t) = A_{11}e^{\lambda_1 t} + A_{12}e^{\lambda_2 t},$$
$$x_2(t) = A_{21}e^{\lambda_1 t} + A_{22}e^{\lambda_2 t}. \tag{3.42}$$

Only the ratio of the amplitudes are determined by (3.38), so that there is still some arbitrariness in the amplitudes. This arbitrariness is removed by the initial condition which prescribes the concentrations of x_1 and x_2 at $t = 0$. From equations (3.42), it follows that

$$x_1(0) = A_{11} + A_{12},$$
$$x_2(0) = A_{21} + A_{22}. \tag{3.43}$$

Equations (3.42) together with the supplementary equations (3.38), (3.40), and (3.43), constitute the solution to the system of compartment equations plus the prescribed initial condition.

Suppose now that the concentration of labeled material in compartment 1 has been observed at various times. We denote these observations by $x_1^*(t_j)$, $j = 1, 2, \ldots, N$, where t_j represents the observation time. According to (3.42), x_1^* is representable as a sum of two exponentials. We must therefore convert the data points $x_1^*(t_j)$ so that they have a representation as a sum of two exponentials. We shall set aside the consideration of how to make this conversion until Section 3.7, and assume that it has been done. Similarly, we will assume that observations have been made of compartment 2, and that these observations are representable likewise as a sum of two exponentials. In other words, we assume that the constants A_{11}, A_{12}, A_{21}, A_{22}, λ_1, and λ_2 of equations (3.42) are all determined from observation. From these experimentally determined quantities, we wish to infer the properties of the system under investigation. Such a task is inverse to the derivation of the solution we have described above. Instead of deducing the solution from the model, we wish to deduce from the solution the values of the parameters of the model, namely, L_{11}, L_{22}, L_{12}, L_{21}, V_1, and V_2.

A systematic procedure for obtaining all the relations between the parameters and the constants is to substitute the solution (3.42) into the compartment equations (3.35), and to equate the coefficients of $e^{\lambda_1 t}$ and $e^{\lambda_2 t}$. Thus, we obtain

$$-L_{11}A_{11} + M_{12}A_{21} = A_{11}\lambda_1, \tag{3.44a}$$
$$-L_{11}A_{12} + M_{12}A_{22} = A_{12}\lambda_2, \tag{3.44b}$$
$$M_{21}A_{11} - L_{22}A_{21} = A_{21}\lambda_1, \tag{3.44c}$$
$$M_{21}A_{12} - L_{22}A_{22} = A_{22}\lambda_2. \tag{3.44d}$$

These linear algebraic equations are readily solved in pairs, and yield the result,

$$L_{11} = \frac{-A_{11}A_{22}\lambda_1 + A_{12}A_{21}\lambda_2}{\Delta},$$

$$M_{12} = \frac{A_{11}A_{12}(\lambda_2 - \lambda_1)}{\Delta},$$

$$L_{22} = \frac{-A_{11}A_{22}\lambda_2 + A_{12}A_{21}\lambda_1}{\Delta}, \qquad (3.45)$$

$$M_{21} = \frac{A_{21}A_{22}(\lambda_1 - \lambda_2)}{\Delta},$$

$$\Delta = A_{11}A_{22} - A_{12}A_{21}.$$

We see from these equations that complete knowledge of the solution to our fundamental system of compartment equations leads us to the determination of L_{11}, L_{22}, M_{12}, M_{21}. By contrast, if the corresponding solution to the compartment equations in the form of equations (3.34) were known, we would determine the fractional clearance rates as well as the total clearance rates.

If independent knowledge of the compartment sizes exists, then the fractional clearance rates can be deduced from the knowledge of M_{12} and M_{21}. For example, suppose the absolute amount of tracer material injected into each compartment initially is known. Call these amounts ρ_{10} and ρ_{20}. Then from (3.30) and (3.43),

$$V_1 = \frac{\rho_{10}}{x_1(0)} = \frac{\rho_{10}}{A_{11} + A_{12}}, \qquad (3.46a)$$

$$V_2 = \frac{\rho_{20}}{x_2(0)} = \frac{\rho_{20}}{A_{21} + A_{22}}. \qquad (3.46b)$$

These equations, combined with (3.45) and (3.36), constitute the complete determination of the steady state parameters of the two-compartment system.

Usually it is the case that injection of tracer material is made initially into one compartment only, say compartment 1. Then equation (3.46a) remains valid and determines V_1. The attempt to make use of equation (3.46b) leads to the indeterminate form 0/0, and leaves V_2 unknown. In such a case, we cannot in fact determine V_2.

Suppose now that only one compartment is accessible to observation. It would appear that we could infer considerably less about the properties of the system, but the problem is not as bad as it appears. For definiteness, suppose compartment 1 is injected initially with an amount of tracer material

ρ_{10}, and that only compartment 1 is observed subsequently. We shall assume that the observations are fitted by the functional form shown in the first of equations (3.42), so that the known constants are ρ_{10}, A_{11}, A_{12}, λ_1, and λ_2.

By adding the first two equations in (3.44), we obtain

$$-L_{11}(A_{11} + A_{12}) + M_{12}(A_{21} + A_{22}) = A_{11}\lambda_1 + A_{12}\lambda_2. \quad (3.47)$$

However, because no tracer material was injected initially into compartment 2, $x_2(0) = 0$, and therefore, from equations (3.43), we see that the coefficient of M_{12} above vanishes. Hence, from equation (3.47) we determine the value of L_{11} to be

$$L_{11} = -\frac{A_{11}\lambda_1 + A_{12}\lambda_2}{A_{11} + A_{12}}. \quad (3.48)$$

If we combine this result with (3.40a), we deduce also the value of L_{22} to be

$$L_{22} = -\frac{A_{11}\lambda_2 + A_{12}\lambda_1}{A_{11} + A_{12}}. \quad (3.49)$$

By combining equations (3.48) and (3.49) with equation (3.40b), we infer the value of the product $M_{12}M_{21}$. Equation (3.46a) tells us the value of V_1.

To summarize the case when only compartment 1 is injected and observed, the equations which are useful are (3.40), (3.46a), and (3.48). These relations determine the parameters L_{11}, L_{22}, and V_1. The quantity V_2 is unknown, and so are L_{12} and L_{21}, but the product $L_{12}L_{21}$ is known from equation (3.40b). This is so because $M_{12}M_{21} = L_{12}L_{21}$, according to (3.36).

3.6. LEAKY COMPARTMENTS AND CLOSED SYSTEMS

When matter can leave a compartment j to go to the exterior, we have seen that $K_{0j} > 0$ and $L_{0j} > 0$. If this is so, compartment j is said to be **leaky**. If $L_{0j} = 0$, then the compartment is said to be **leakproof**. If every compartment of a system is leakproof, then the compartment system is said to be a **closed** system. By conservation of mass, a closed system cannot have any material influx from the exterior. Therefore, it is isolated from its environment. A compartment system that is not closed is said to be an **open** system.

It is often the case that, from other biological knowledge of the system under study, we know that a particular· compartment is leakproof, or a particular compartment system is closed, or that the flux in a given direction between a pair of compartments cannot exist. Such knowledge, when expressed in quantitative terms, yields equations of constraint that the parameters of the compartment system must satisfy. This additional knowledge can be sufficient to permit us to uniquely determine the parameters of the system, even when the complete solution of the compartment equations has not been experimentally determined.

In this connection, we remark that the representation of compartment models by means of graphs will prove to be very useful for the purpose of visualizing the compartment system under consideration. Thus, to every compartment we assign a node with a sequentially chosen numerical label. To every fractional clearance rate L_{ij} representing the fractional transfer of matter from compartment j to compartment i, we assign the directed line segment from the node j to the node i. In addition, we assign the directed line segment running from compartment j to the exterior to represent the excretion rate L_{0j}. The omission of a directed line segment from a graph is to be interpreted as meaning that the given clearance rate it represents is null. Consequently, there is a unique one-to-one relationship between the clearance rates and the graph. It is to be kept in mind that the graph indicates the excretion rates to the exterior, so that in order to infer the element L_{jj}, representing the total clearance rate, from the graph, use must be made of equation (3.33) or its equivalent, for a many-compartment system. An example of a graph for a three-compartment system is given in Figure 3.4.

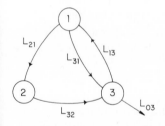

Figure 3.4. Graph for a three-compartment system for which $L_{12} = L_{23} = 0$, and for which compartments 1 and 2 are leakproof, $L_{01} = L_{02} = 0$. The branches corresponding to such null clearance rates are omitted from the graph.

As an example of the application of constraints, we shall return to the two-compartment system considered at the end of the previous section, for which tracer material is injected into a given compartment (called compartment 1), and that same compartment is observed. Suppose we believe the compartment system is closed. Then the tracer material could never escape the system, and the amount of tracer in a given compartment would eventually settle down to some stationary value. Consequently, $x_1(t)$ would approach a non-zero constant value as $t \to \infty$. If it does not, we know that the system is open, and that at least one compartment is leaky.

Suppose instead that compartment 2 is known to be leakproof. Such a constraint could arise, for example, if compartment 1 represents blood plasma, and compartment 2 represents tissue. It is frequently the case for physiological systems that the only exit possible of a given chemical substance from tissue is via the plasma. The graph for such a compartment system is shown in Figure 3.5. Because compartment 2 is leakproof, $L_{02} = 0$,

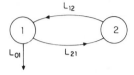

Figure 3.5. Graph for a two-compartment system for which compartment 2 is leakproof, so that $L_{02} = 0$. In the graph, the branch leaving compartment 2 and going to the exterior is omitted.

and, therefore, from equation (3.33),

$$L_{22} = L_{12}. \tag{3.50}$$

For a two-compartment system, a concomitant inference that frequently can be made about leakproof compartments, although it is not a necessary conclusion, is that there is no influx of unlabeled material to it from the exterior. In the example cited above, it is almost always the case that metabolites or other chemical substances can only get to the tissue via the plasma. In the bathtub model, a leakproof compartment implies only that there is no drainpipe, but it is often the case for biological systems that the faucet is also turned off for a leakproof compartment. If that is so, then the net material inflow from other compartments equals the net material outflow to other compartments. This property is easily expressed in terms of the coefficients K_{ij} that were introduced previously. For example, for a two-compartment system for which compartment 2 is leakproof and has no influx from the exterior, we have

$$K_{21} = K_{12}.$$

Here K_{21} represents the influx into compartment 2 coming from compartment 1, and K_{12} represents the efflux from compartment 2 to compartment 1. The equivalent equation expressed in terms of the L_{ij} is

$$L_{21}V_1 = L_{12}V_2. \tag{3.51}$$

Equations (3.50) and (3.51) can be utilized in conjunction with equations (3.40), (3.46a), and (3.48). These six equations uniquely determine all the clearance rates and compartment sizes. Hence, even though only one compartment of the system is observed, we are able to determine all the steady state parameters of the system.

3.7. THE METHOD OF EXPONENTIAL PEELING

We now turn to the problem of fitting observed data points to a special functional form—a sum of exponentials. Suppose that the concentration

$x^*(t)$ of a labeled substance in a compartment has been measured at various discrete values of the time t, so that a table of values of x^* exists, for the N time values $t_1, t_2, \ldots, t_N$. Let us plot on semilog paper the quantity log $x^*(t)$ at the observation times t_j, where only the dimensionless values of $x^*(t)$ are utilized in taking logarithms. We presume that $x^*(t)$ is representable by the functional form of either of equations (3.42):

$$x^*(t) = A_{11}e^{\lambda_1 t} + A_{12}e^{\lambda_2 t}.$$

If it is not, then a compartmental model of the system is not applicable, and we should better spend our time thinking of an alternative model.

Without loss of generality, assume the λ's are ordered, so that $\lambda_1 > \lambda_2$. Remember that, for compartmental systems, the λ's are never positive, so that $\lambda_1 > \lambda_2$ implies that $|\lambda_2| > |\lambda_1|$. Unless $x^*(t)$ consists of exactly one exponential, the semilog plot of $x^*(t)$ will not yield a straight line. Rather, it will be of the form shown by the solid circles in Figure 3.6. The feature that immediately strikes our attention in the figure is that the curve does become a straight line for large values of the time. The reason for this is

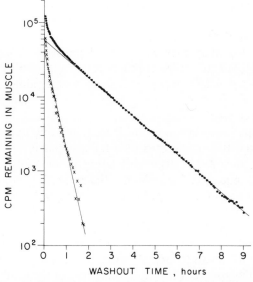

Figure 3.6. Semilog plot illustrating the method of exponential peeling for a washing-out experiment (see Section 3.9). The solid circles represent the radioactivity of sodium remaining in muscle tissue treated with a 10^{-5} M solution of ouabain, a drug known to affect sodium and potassium transport in cells. The crosses represent the experimental observations after subtraction of the slowest exponential term, as explained in the text. The two straight lines represent the two exponentials obtained to fit the data. From Rogus and Zierler (1973), with permission.

that only one exponential term in (3.42) survives for large values of the time, the one with the largest λ_j, $e^{\lambda_1 t}$. In mathematical language, we say $x^*(t)$ **behaves like** or **is asymptotic to** $A_{11}e^{\lambda_1 t}$ for t large. This is written as

$$x^*(t) \sim A_{11}e^{\lambda_1 t}, \qquad t \gg 1. \tag{3.52}$$

Therefore,

$$\log x^*(t) \sim \log A_{11} + \lambda_1 t, \qquad t \gg 1. \tag{3.53}$$

This straight line is the one passing through the solid circles in the figure. From it, we can determine the slope λ_1 and the amplitude A_{11}. Because of experimental errors, it is usually desirable to utilize the method of least squares to determine the best straight line that fits the "tail" of the curve, as discussed in Chapter 1. In doing so, a certain arbitrariness exists in determining where the tail of the curve begins.

Having determined A_{11} and λ_1, we can now form the function $x^{(1)}(t) \equiv x^*(t) - A_{11}e^{\lambda_1 t}$. The experimentally determined values of $x^{(1)}(t)$ are shown by crosses in Figure 3.6. This function too is an exponential, according to (3.42). We then repeat the above procedure, obtaining in this manner A_{12} and λ_2, from the straight line passing through the crosses in the figure. The above procedure constitutes the **method of exponential peeling** (see Smith and Morales, 1944; Perl, 1960; Van Liew, 1967).

It should be clear that the same method can be used, regardless of the number of exponential terms needed to represent $x^*(t)$. Thus, in determining the number of exponential terms, the method determines the number of compartments in the model: There must be one compartment for each exponential decay term.

There are statistical procedures for determining the best values of n, A_{ij}, and λ_j associated with given observations. It is also possible to fit the observed data directly to the sum of exponentials by the method of least squares. In that case, the equations determining the A_{ij} and λ_j are not linear. An interesting novel suggestion recently made (Dyson and Isenberg, 1971) is to replace the data by moments and fit it with moments of an exponential sum.

As a matter of practice, because of experimental errors associated with the determination of $x^*(t)$, each exponential term requires at least several points to determine its amplitude and decay constant. In general, the more points utilized, the greater will be the confidence in the result. Moreover, in the method of exponential peeling, the data points utilized for one exponential term are discarded for the determination of the remaining exponential terms. Thus, there is usually a practical limitation to the number of exponential terms that is detectable.

3.8.   CREATININE CLEARANCE A TWO-COMPARTMENT SYSTEM

Creatinine is a nitrogenous waste product of the metabolism of three amino acids: glycine, methionine, and arginine. It is a fairly constant component of urine, and there is evidence that the amount excreted daily is proportional to the amount of muscle tissue in the body. In humans, the amount excreted is of the order of 1 mg/hr/kg, meaning 1 mg/hr for each kilogram of body weight. We shall describe and analyze an experiment to determine some quantitative features of the elimination of creatinine from the tissues.

Consider a single injection of creatinine into the vein of a dog for which the level of creatinine in the blood is null. From subsequent blood samples, the concentration of creatinine in the plasma can be determined as a function of the time. Such an experiment was carried out in ten dogs (Sapirstein, Vidt, Mandel, and Hanusek, 1955). A typical result for the concentration $x_1(t)$ of creatinine in the plasma, after fitting the curve to a sum of exponentials, was the following:

$$
\begin{aligned}
x_1(t) &= A_{11}e^{\lambda_1 t} + A_{12}e^{\lambda_2 t}, \\
A_{11} &= 0.188 \text{ mg/ml}, \ \lambda_1 = -0.0161/\text{min}, \\
A_{12} &= 0.321 \text{ mg/ml}, \ \lambda_2 = -0.1105/\text{min}.
\end{aligned} \tag{3.54}
$$

The initial amount of creatinine injected at $t = 0$ was

$$
\rho_{10} = 2 \text{ g}.
$$

Because $x_1(t)$ is described as a sum of two exponentials, the number of compartments is identified as two. One compartment is obviously taken to be the plasma and designated as compartment 1. The other compartment is presumed to be extravascular and is simply referred to as the tissue compartment, designated by the subscript 2.

Now let us examine the compartment model more closely. The tissue compartment is assumed to be leakproof, because it only exchanges material with the blood. The blood plasma compartment, on the other hand, is presumably leaky because creatinine can be excreted via the kidney. The graph representing the compartment system is the one shown in Figure 3.5. It is an example of an **almost closed compartment system**: one for which all compartments are leakproof, save one.

In this system, the tissue compartment is actually a source of creatinine, because of normal muscle metabolism. However, the level of creatinine in the plasma resulting from this production is usually not observable, and we shall take this production to be nil. Thus, we assume compartment 2 is leakproof, and does not produce creatinine or receive it from the exterior of the dog.

We now substitute the empirically determined values of the constants given in (3.54) into equations (3.48), (3.49), (3.50), and (3.40b) (replacing $M_{12}M_{21}$ there by $L_{12}L_{21}$). We find that the clearance rates are determined to be

$$L_{11} = \frac{A_{11}\lambda_1 + A_{12}\lambda_2}{A_{11} + A_{12}} = 0.0786/\text{min},$$

$$L_{12} = L_{22} = 0.0479/\text{min},$$

$$L_{22} = \frac{A_{11}\lambda_2 + A_{12}\lambda_1}{A_{11} + A_{12}} = 0.0479/\text{min}, \qquad (3.55)$$

$$L_{21} = \frac{L_{11}L_{22} - \lambda_1\lambda_2}{L_{12}} = 0.0391/\text{min}.$$

Because the amount of creatinine injected into compartment 1 ρ_{10} is known, the size of the plasma compartment is determined from (3.46a) to be

$$V_1 = \frac{\rho_{10}}{A_{11} + A_{12}} = 3.93 \text{ liter}. \qquad (3.56)$$

Independent corroboration of the excretion rate of creatinine from the plasma, $K_{01} = V_1 L_{01}$, was obtained by comparing it to the elimination rate of **mannitol**, a common sugar, from the plasma. It was found, as is plausible, that these two clearance rates were approximately equal.

Finally, using equation (3.51) and the derived values of L_{21}, V_1, and L_{12}, the value of V_2 is determined to be

$$V_2 = V_1 \frac{L_{21}}{L_{12}} = 3.21 \text{ liter}. \qquad (3.57)$$

3.9. THE "SOAKING OUT" EXPERIMENT

In a "**soaking out**" or "**washing out**" experiment (Huxley, 1960), a piece of tissue (a nerve or a muscle, say) is left in a solution containing a radioactive ionic species for a time sufficiently long for equilibrium to be reached. At $t = 0$, the tissue is transferred to a large volume of unlabeled solution. The total activity of the tissue is then followed as a function of time. It is found to consist of a sum of two exponentials. What can be inferred about the tissue, viewed as a compartmental system?

Because the activity consists of two exponentials, we infer that the tissue is representable as a two-compartment system. The obvious candidates for the compartments are the tissue and the bathing fluid, but it is apparent that because the bathing fluid is so large, it acts essentially as a sink and makes a negligible contribution to the activity in the tissue. Therefore, the usual assumption made is that the two compartments are inside the tissue, and represent the cells themselves and the extracellular fluid in the tissue.

Actually, the experiment consists of two parts: a "soaking in" part during which a third compartment, the bathing solution, is initially labeled, while the cellular compartments are unlabeled. This establishes the initial conditions for the second part, a "soaking out": The bathing solution is initially an unlabeled solution, while the cellular compartments are labeled. We shall first examine the "soaking in" part of the experiment. From the knowledge of the structure of tissue, this system is assumed to be a closed three-compartment system, as shown in Figure 3.7. By anology with equations (3.29)

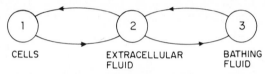

CELLS   EXTRACELLULAR   BATHING
                 FLUID   FLUID

Figure 3.7. Closed catenary compartment model for a "soaking in" experiment. Initially, the bathing fluid is labeled, and the cells and extracellular fluid are unlabeled.

for a two-compartment system, we can readily write down the following compartment equations for this three-compartment system,

$$\frac{d\rho_1}{dt} = -K_{11}x_1 + K_{12}x_2, \tag{3.58a}$$

$$\frac{d\rho_2}{dt} = K_{21}x_1 - K_{22}x_2 + K_{23}x_3, \tag{3.58b}$$

$$\frac{d\rho_3}{dt} = K_{32}x_2 - K_{33}x_3. \tag{3.58c}$$

What are the equilibrium values $\bar{x}_i$, $i = 1, 2, 3$, of the concentrations in the compartments, attained after a sufficiently long period of soaking in? These are determined by setting the time derivatives in (3.58) equal to zero, which yields

$$-K_{11}\bar{x}_1 + K_{12}\bar{x}_2 = 0,$$
$$K_{21}\bar{x}_1 - K_{22}\bar{x}_2 + K_{23}\bar{x}_3 = 0, \tag{3.59}$$
$$K_{32}\bar{x}_2 - K_{33}\bar{x}_3 = 0.$$

In order to solve these equations for the equilibrium values $\bar{x}_i$, we shall first consider some relationships that exist among the K_{ij}, independently of any labeling experiment. If the system were not closed, the conservation of the steady-state flux of unlabeled material into and out of each compartment would be expressed as follows:

$$K_{11} = K_{12} + J_1,$$
$$K_{22} = K_{21} + K_{23} + J_2, \qquad (3.60)$$
$$K_{33} = K_{32} + J_3.$$

In these equations, J_i, $i = 1, 2, 3$, represents the material flux into the ith compartment from the exterior. The left-hand side of any of the above equations represents the net efflux rate, while the right-hand side represents net influx rate from the neighboring compartments and the exterior, for a given compartment. Because the compartment system is closed, there is no net flux from the system to the outside. Therefore, there is no net flux into the system from the outside, or $J_1 = J_2 = J_3 = 0$. Hence, for a closed three-compartment system, the conservation of flux into and out of each compartment is expressed as

$$-K_{11} + K_{12} = 0,$$
$$K_{21} - K_{22} + K_{23} = 0, \qquad (3.61)$$
$$K_{32} - K_{33} = 0.$$

A comparison of equations (3.59) and (3.61) shows that they are compatible provided $\bar{x}_1 = \bar{x}_2 = \bar{x}_3$ in (3.59). In other words, the equilibrium in the various compartments of a closed compartment system is attained when the concentrations (or specific activity, in the case of radioactivity labeled compartments) are equal, a result that is perhaps intuitively obvious. This result is independent of the number of compartments in the system.

The soaking-in part of the experiment determines the initial values of the concentrations in the soaking-out part as $x_1(0) = x_2(0)$. Also, because the steady-state bathing solution is replaced with a fresh solution initially, $x_3(0) = 0$. We have seen that, for a one-compartment system, the tracer concentration could be represented by a single exponential, and that for a two-compartment system, two exponential terms are required. Our intuition would lead us to believe that the total activity of the tissue in the soaking-out part of the experiment should be expressible as a sum of three exponentials, because the system is a three-compartment system. The fact that it is observed to be a sum of two exponentials is interpreted to mean that the third exponential has too small an amplitude to be detectable.

The reason for the lack of detectability of a third exponential term can be ascribed to the fact that the size of the bathing solution compartment is very large compared to the size of either of the other two compartments. To see this in a more quantitative fashion, we examine the system represented

by equations (3.58). Consider first equation (3.58c). By expressing the coefficients in terms of clearance rates, and setting $\rho_3 = V_3 x_3$, the solution to
this equation subject to the initial condition $x_3(0) = 0$ can be expressed
quite generally with the aid of equation (1.21) as

$$x_3(t) = \frac{V_2}{V_3} L_{32} e^{-L_{33}t} \int_0^t e^{L_{33}\tau} x_2(\tau) d\tau. \tag{3.62}$$

From this equation we see that $x_3(t)$ is a small quantity because $V_3 \gg V_2$.
Consequently, to good approximation we can neglect the bathing solution
compartment and represent the closed three-compartment system as a leaky
two-compartment system, as shown in Figure 3.8. We see from these considerations that the action of permitting a compartment size to become very

Figure 3.8. Effective two-compartment model for a soaking-out
experiment in which the volume of the bathing fluid is assumed to
be very large compared to the volume of the tissue.

large has the same effect on a compartmental system that the action of
saturation does in a chemical reaction system (see Section 2.1): It eliminates
an independent variable from the system.

In a soaking-out experiment, observations of the total activity of the
tissue are commonly made. From these observations, we determine by a
curve-fitting procedure the quantity

$$\rho_1(t) + \rho_2(t) = Ae^{\lambda_1 t} + Be^{\lambda_2 t}, \tag{3.63}$$

where A, B, λ_1, and λ_2 are known constants. In particular, we note that

$$\rho_1(0) + \rho_2(0) = A + B. \tag{3.64}$$

Because $x_1(0) = x_2(0)$, the initial amount of the radioactive material in the
cell compartment is related to the initial amount of the radioactive material
in the extracellular compartment by means of the equation

$$\rho_1(0) = \frac{V_1}{V_2} \rho_2(0). \tag{3.65}$$

The value of this initial amount is otherwise unknown. The foregoing relation
reflects the fact that the amount of labeled material in a compartment is
proportional to the size of the compartment, when the specific activity is
constant throughout the system.

What makes this problem different from the problem of creatinine clearance discussed in the previous section is that the initial condition is different,

and that the quantity observed is the total amount of labeled material in the system, rather than the concentration of labeled material in a single compartment. Nevertheless, for this problem too, the information inferred from the experiment is sufficient to determine all the steady-state parameters of the system. To show this, let us write down the compartmental equations for this soaking-out part of the experiment. They are most conveniently expressed in terms of the variables ρ_1 and ρ_2. Thus, equations (3.58) with $x_3 = \rho_3 = 0$ are the same as equations (3.34),

$$
\begin{aligned}
\frac{d\rho_1}{dt} &= -L_{11}\rho_1 + L_{12}\rho_2, \\
\frac{d\rho_2}{dt} &= L_{21}\rho_1 - L_{22}\rho_2.
\end{aligned}
\tag{3.66}
$$

In these equations we recognize that compartment 1 is leakproof, so that $L_{01} = 0$, and $L_{21} = L_{11}$. However, $L_{02} \neq 0$. The solution to equations (3.66) is

$$
\begin{aligned}
\rho_1 &= B_{11}e^{\lambda_1 t} + B_{12}e^{\lambda_2 t}, \\
\rho_2 &= B_{21}e^{\lambda_1 t} + B_{22}e^{\lambda_2 t}.
\end{aligned}
\tag{3.67}
$$

Then, substituting (3.67) into (3.63) and equating to zero each of the coefficients of $e^{\lambda_1 t}$ and $e^{\lambda_2 t}$ leads to the relations

$$
\begin{aligned}
B_{11} + B_{21} &= A, \\
B_{12} + B_{22} &= B.
\end{aligned}
\tag{3.68}
$$

By combining (3.64) with (3.65), we see that the amount of labeled material in a given compartment is directly proportional to the relative size of the compartment, namely,

$$
\begin{aligned}
\rho_1(0) &= \frac{V_1}{V_1 + V_2}(A + B), \\
\rho_2(0) &= \frac{V_2}{V_1 + V_2}(A + B).
\end{aligned}
\tag{3.69}
$$

In the same manner that equations (3.44) were derived for the case when the compartment equations were given in terms of the concentrations [compare equations (3.66) and (3.67) to equations (3.35) and (3.42)], it follows that equations (3.44) remain true if M_{21}, M_{12}, and A_{ij} are replaced by L_{21}, L_{12}, and B_{ij}, respectively. Therefore, by analogy to the derivation of equation (3.47), we infer that

$$
-L_{11}\rho_1(0) + L_{12}\rho_2(0) = B_{11}\lambda_1 + B_{12}\lambda_2.
\tag{3.70}
$$

Because there is no external influx into the cellular compartment, it follows from the first equation in (3.61) that

$$-L_{11}V_1 + L_{12}V_2 = 0. \tag{3.71}$$

However, the external flux into the extracellular fluid compartment $J_2 \neq 0$, because of the presence of the bathing solution. By substituting equations (3.65) and (3.71) into (3.70), we see that the left-hand side of equation (3.70) is zero, and therefore so is the right-hand side, or,

$$-L_{11}\rho_1(0) + L_{12}\rho_2(0) = B_{11}\lambda_1 + B_{12}\lambda_2 = 0. \tag{3.72}$$

By a derivation similar to that of equation (3.70), it follows from equations (3.44c) and (3.44d) after replacing M_{21} by L_{21} and A_{ij} by B_{ij}, that

$$L_{21}\rho_1(0) - L_{22}\rho_2(0) = B_{21}\lambda_1 + B_{22}\lambda_2.$$

Now we eliminate B_{21} and B_{22} from the above equation by means of (3.68). Then with the aid of equation (3.72), we find that

$$L_{21}\rho_1(0) - L_{22}\rho_2(0) = \lambda_1 A + \lambda_2 B. \tag{3.73}$$

In addition, we utilize equations (3.40) with $M_{12}M_{21}$ replaced by $L_{12}L_{21}$, which are

$$L_{11} + L_{22} = \lambda_1 + \lambda_2,$$
$$L_{11}L_{22} - L_{12}L_{22} = \lambda_1\lambda_2. \tag{3.74}$$

Because $L_{01} = 0$, we have the following additional relation, which follows from the defining equation (3.33),

$$L_{11} = L_{21}. \tag{3.75}$$

Equations (3.64) and (3.72)–(3.75) constitute six equations for the four unknown clearance rates and the quantities $\rho_1(0)$ and $\rho_2(0)$. Their solution yields the following results:

$$L_{11} = L_{21} = -\frac{\lambda_1\lambda_2(\lambda_1 A + \lambda_2 B)}{\lambda_1^2 A + \lambda_2^2 B},$$

$$L_{12} = -\frac{\lambda_1\lambda_2(\lambda_1 - \lambda_2)^2 AB}{(\lambda_1^2 A + \lambda_2^2 B)(\lambda_1 A + \lambda_2 B)},$$

$$L_{22} = \frac{\lambda_1^3 A + \lambda_2^3 B}{\lambda_1^2 A + \lambda_2^2 B},$$

$$L_{02} = -\frac{\lambda_1^2 A + \lambda_2^2 B}{\lambda_1 A + \lambda_2 B}, \tag{3.76}$$

$$\rho_1(0) = \frac{(\lambda_1 - \lambda_2)^2 AB}{(\lambda_1^2 A + \lambda_2^2 B)},$$

$$\rho_2(0) = \frac{(\lambda_1 A + \lambda_2 B)^2}{(\lambda_1^2 A + \lambda_2^2 B)}.$$

From (3.71) and the first two equations above, we determine the compartment size ratio to be

$$\frac{V_1}{V_2} = \frac{(\lambda_1 - \lambda_2)^2 AB}{(\lambda_1 A + \lambda_2 B)^2}. \tag{3.77}$$

By observing the specific activity at equilibrium of the bathing solution in the soaking-in experiment, we determine both $x_1(0)$ and $x_2(0)$. From these and the knowledge of $\rho_1(0)$ and $\rho_2(0)$ from (3.76), the individual compartment sizes V_1 and V_2 can also be inferred.

Quite different conclusions would be reached if one were, somewhat naively, to assume that the tissue system is represented by two compartments, both exchanging material with the bathing solution, but otherwise disconnected. Then $L_{12} = L_{21} = 0$, but $L_{01} \neq 0$. The solution to the compartment equations becomes, instead of (3.67),

$$\begin{aligned}
\rho_1(t) &= Ae^{\lambda_1 t}, \\
\rho_2(t) &= Be^{\lambda_2 t}.
\end{aligned} \tag{3.78}$$

Now

$$\begin{aligned}
\rho_1(0) &= A, \, L_{11} = -\lambda_1, \\
\rho_2(0) &= B, \, L_{22} = -\lambda_2.
\end{aligned} \tag{3.79}$$

Because it is still true that $x_1(0) = x_2(0)$, it follows from (3.30) and (3.79) that

$$\frac{V_1}{V_2} = \frac{A}{B}. \tag{3.80}$$

The results (3.79) and (3.80) are quite at variance with the corresponding inferences for the model we discussed for which L_{12} and L_{21} were not zero. For example, if $A/B = 1$, and $\lambda_1/\lambda_2 = 3$, then according to (3.77), $V_1/V_2 = 0.25$, which is a much smaller estimate of V_1/V_2 than the value 1 given by (3.80). Conversely, with $A/B = 1/2$, and $\lambda_1/\lambda_2 = 6$, equation (3.77) yields $V_1/V_2 = 0.78$, which is much larger than the value of $1/2$ given by (3.80).

3.10. THE THREE-COMPARTMENT CATENARY SYSTEM

In many physiological and/or pharmacological studies of mammals, similar compartmental models appear, consisting of cells, interstitial fluid, and blood plasma. Largely because of the kidney, the blood plasma is a leaky compartment in that material can leave it to go to the exterior. Consequently, a model of general interest is the slightly leaky or almost closed compartment system that is represented by the graph in Figure 3.9, with the number of compartments equal to three. Because of the chain-like appearance of the

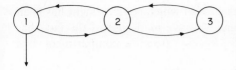

Figure 3.9. An almost closed three-compartment catenary system of general physiological interest. A typical compartmental identification is the following: (1), blood plasma; (2), interstitial fluid; (3), cells.

graph, such a compartmental model is called a **catenary system**. It is defined as a system that consists of serially connected compartments. For it, L_{ii}, $L_{i,i\pm 1} \neq 0$ for all permissible values of i, while all other $L_{ij} = 0$. In practice, an initial injection of labeled material into the blood would not be instantaneously mixed, but usually, the mixing time is small compared to the mean time spent in the blood by the labeled material. Other examples of such a catenary system are the following: three compartments representing extracellular space, the cytoplasm, and a particular intracellular component such as the nucleus; a sequence of biochemical reactions.

We suppose that initially, a given amount of labeled material $\rho_1(0)$ is injected into compartment 1. Subsequently, the concentration in this compartment $x_1(t)$ is determined as a function of time and found to be representable as a sum of three exponentials, namely,

$$x_1(t) = A_{11}e^{\lambda_1 t} + A_{12}e^{\lambda_2 t} + A_{13}e^{\lambda_3 t}. \tag{3.81}$$

Here, A_{11}, A_{12}, A_{13}, λ_1, λ_2, and λ_3 are known quantities. We desire to make all the quantitative inferences possible based on this information. The system of compartment equations is the same as that given by equations (3.58). In terms of the concentrations and the turnover rates, these become

$$\frac{dx_1}{dt} = -L_{11}x_1 + M_{12}x_2,$$

$$\frac{dx_2}{dt} = M_{21}x_1 - L_{22}x_2 + M_{23}x_3, \tag{3.82}$$

$$\frac{dx_3}{dt} = M_{32}x_2 - L_{33}x_3,$$

where M_{12} and M_{21} are defined by equations (3.36), and

$$\begin{aligned} M_{23} &= V_2^{-1}L_{23}V_3, \\ M_{32} &= V_3^{-1}L_{32}V_2. \end{aligned} \tag{3.83}$$

If we analyze equations (3.82) and their solution as we did the two-compartment system in Section 3.5, we can derive the following results,

$$L_{11} + L_{22} + L_{33} = -\lambda_1 - \lambda_2 - \lambda_3,$$

$$L_{11}L_{22} + L_{11}L_{33} + L_{22}L_{33} - L_{12}L_{21} - L_{23}L_{32} = \lambda_1\lambda_2 + \lambda_1\lambda_3 + \lambda_2\lambda_3, \tag{3.84}$$

$$L_{11}L_{22}L_{33} - L_{12}L_{21}L_{33} - L_{23}L_{32}L_{11} = -\lambda_1\lambda_2\lambda_3.$$

These three equations hold true regardless of which compartment has been observed. In deriving these equations, we have made use of the fact that $M_{12}M_{21} = L_{12}L_{21}$, and $M_{23}M_{32} = L_{23}L_{32}$. Because compartment 1 is observed, it can also be shown that

$$L_{11} = -\frac{A_{11}\lambda_1 + A_{12}\lambda_2 + A_{13}\lambda_3}{x_1(0)}, \tag{3.85a}$$

$$L_{11}^2 + L_{12}L_{21} = \frac{A_{11}\lambda_1^2 + A_{12}\lambda_2^2 + A_{13}\lambda_3^2}{x_1(0)}, \tag{3.85b}$$

$$x_1(0) = A_{11} + A_{12} + A_{13}. \tag{3.85c}$$

Because compartments 2 and 3 are leakproof, L_{02} and L_{03} are null, and it follows from the extension of equations (3.33) to three compartments that

$$\begin{aligned} L_{22} &= L_{12} + L_{32}, \\ L_{33} &= L_{23}. \end{aligned} \tag{3.86}$$

The total turnover rate of compartment 1 is given by equation (3.85a). The remaining equations in (3.84)–(3.86) constitute six equations for the remaining six unknown turnover rates L_{ij}. Their solution is found to be as follows:

$$\begin{aligned} L_{23} = L_{33} &= \frac{L_{11}(s_2 + b - L_{11}s_1) - s_3}{b - L_{11}^2}, \\[2mm] L_{22} &= s_1 - L_{11} - L_{33}, \\[2mm] L_{12} &= \frac{L_{11}s_1 - s_2 - b}{L_{33}}, \\[2mm] L_{32} &= L_{22} - L_{12}, \\[2mm] L_{21} &= \frac{b - L_{11}^2}{L_{12}}, \end{aligned} \tag{3.87}$$

where the constants s_1, s_2, s_3, and b are defined as

$$\begin{aligned} s_1 &= -\lambda_1 - \lambda_2 - \lambda_3, \\ s_2 &= \lambda_1\lambda_2 + \lambda_1\lambda_3 + \lambda_2\lambda_3, \\ s_3 &= -\lambda_1\lambda_2\lambda_3, \\ b &= \frac{A_{11}\lambda_1^2 + A_{12}\lambda_2^2 + A_{13}\lambda_3^2}{x_1(0)}. \end{aligned}$$

The compartment sizes can also be determined. Because $\rho_1(0)$ and $x_1(0)$ are both known, we determine the compartment size V_1 as

$$V_1 = \frac{\rho_1(0)}{x_1(0)}. \tag{3.88}$$

Let us suppose that influx of unlabeled material from the exterior is received only by compartment 1. Then $J_2 = J_3 = 0$, and it follows as in equation (3.61) that

$$L_{21}V_1 - L_{22}V_2 + L_{23}V_3 = 0,$$
$$L_{32}V_2 - L_{33}V_3 = 0. \tag{3.89}$$

These two equations determine V_2/V_1 and V_3/V_1. With V_1 known from (3.88), V_2 and V_3 are also determined.

We have shown above that for an almost closed three-compartment catenary system, under reasonable assumptions about the mode of observation and the nature of the system, the clearance rates and the compartment sizes are uniquely determined even though only one compartment has been observed (Rubinow, 1973). Actually, it makes a difference which compartment has been observed. In this example, the observed compartment was the same as the injected compartment. In general, the more distant the observed compartment is from the injected compartment, the less one can infer about the parameters of the system.

It can be shown that for an almost closed n-compartment catenary system, with initial injection and subsequent observation of compartment 1, and similar assumptions about the means of external influx of unlabeled material, similar inferences can be made: All the clearance rates and compartment sizes can be determined. Because of the algebraic nature of the equations to be solved, it may happen that there is more than one solution to the equations. In such a case, there are more than one compartment models which are compatible with the observations. This multiplicity is never greater than $n - 1$.

We shall illustrate the fact that it is most informative to observe the compartment in which the injection of labeled material has been made. Suppose compartment 2 had been observed in the above example, instead of compartment 1. This means that $x_2(t)$ is represented as a sum of exponentials as in (3.81), except that A_{21}, A_{22}, and A_{23} replace A_{11}, A_{22}, and A_{13} as known quantities. Equations (3.84) and (3.86) are unaffected by these considerations. However, equations (3.85) are replaced by the relations

$$L_{21} = V_2 \frac{A_{21}\lambda_1 + A_{22}\lambda_2 + A_{23}\lambda_3}{\rho_1(0)},$$
$$-L_{21}(L_{11} + L_{22}) + L_{23}L_{31} = V_2 \frac{A_{21}\lambda_1^2 + A_{22}\lambda_2^2 + A_{23}\lambda_3^2}{\rho_1(0)}. \tag{3.90}$$

Because of the appearance of V_2 in these equations, we cannot now determine the turnover rates as we could for the case when compartment 1 was observed. Nor can we determine the compartment size V_1. We can consider V_2 an unknown parameter in the above equations. We can then solve for all the turnover rates as we did previously in equations (3.87). The result is that we have a one-parameter set of clearance rates, compatible with the observations.

Even less information is obtained if we observe compartment 3, which is further away from the injected compartment. Then instead of equations (3.85) or (3.87), we obtain the relations

$$L_{31} = V_3 \frac{A_{31}\lambda_1 + A_{32}\lambda_2 + A_{33}\lambda_3}{\rho_1(0)},$$

$$-L_{31}(L_{11} + L_{33}) + L_{32}L_{21} = V_3 \frac{A_{31}\lambda_1^2 + A_{32}\lambda_2^2 + A_{33}\lambda_3^2}{\rho_1(0)}, \quad (3.91)$$

where all the constants on the right-hand sides of these equations except V_3 are known. However, because of the known catenary structure of the compartment system, L_{31} must be zero. Therefore, the first equation above tells us nothing new, and the equation merely corroborates our knowledge of the compartment structure (the right-hand side of this equation turns out to be zero). Hence, we have one less equation to aid us in determining the turnover rates and the compartment sizes, in comparison with the previous cases.

Inferences similar to those made above for an almost closed catenary system can be made for an almost closed mammillary compartment system. A **mammillary compartment system** is one for which the graph has a star-shaped appearance. It has a central compartment, call it 1, connected to $n - 1$ other compartments, which are otherwise disconnected. For it, L_{ii}, L_{i1}, $L_{1i} \neq 0$ for all i, while the remaining $L_{ij} = 0$. A graph of an almost closed four-compartment mammillary system is shown in Figure 3.10.

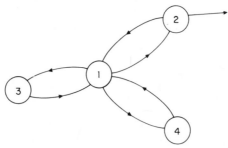

Figure 3.10.   Illustrative graph of an almost closed four-compartment mammillary compartment system.

3.11*.   THE n-COMPARTMENT SYSTEM

Here we wish to generalize our considerations of the previous sections to an n-compartment system. In particular, we shall express the turnover rates and the compartment sizes solely in terms of constants derived from observation of one or more compartments of the system (Berman and Schoenfeld, 1956; Rubinow and Winzer, 1971). The generalization is achieved most readily with the aid of matrix algebra, and an acquaintance with this subject is presumed in what follows (see Appendix B). For the sake of completeness of presentation, some definitions will be repeated.

The fundamental rate equations governing the behavior of labeled material in the compartment system are assumed to be the following:

$$\frac{d\rho}{dt} = L\rho. \tag{3.92}$$

These are called compartment equations. Here $\rho = \rho(t)$ is a time-dependent n-dimensional column vector $\rho = \{\rho_i\}$, where ρ_i represents the amount of labeled material in the ith compartment at time t. The matrix L is defined to be

$$L = \{L_{ij}\} = \begin{bmatrix} L_{11}L_{12}L_{13} \cdots L_{1n} \\ L_{21}L_{22}L_{23} \cdots L_{2n} \\ L_{31}L_{32}L_{33} \cdots L_{3n} \\ \cdots\cdots\cdots\cdots\cdots \\ L_{n1}L_{n2}L_{n3} \cdots L_{nn} \end{bmatrix} \tag{3.93}$$

L is an $n \times n$ matrix and therefore square. As previously defined, L_{ij} for $i \neq j$ is the fractional turnover rate representing the material (labeled plus unlabeled) transport rate from compartment j to compartment i per unit amount of material in the jth compartment. The quantity $(-L_{ii})$ is the total turnover rate of the ith compartment. Note that, for convenience of matrix notation, we have changed the sign convention of L_{ii} from that used in the previous sections. L_{ii} is here the negative of what it was before.

Because of their physical significance, the coefficients L_{ij} must satisfy the conditions

$$L_{ij} \geqq 0, \qquad i,j = 1, 2, \ldots, n; i \neq j, \tag{3.94a}$$

$$\sum_{i=1}^{n} L_{ij} \leqq 0, \qquad j = 1, 2, \ldots, n. \tag{3.94b}$$

The **fractional excretion rate** is denoted by L_{0j} for the jth compartment, and is defined as

$$L_{0j} = -\sum_{i=1}^{n} L_{ij}, \qquad j = 1, 2, \ldots, n. \tag{3.95}$$

From equation (3.94b) it follows that $L_{0j} \geqq 0$. L_{0j} represents the fractional rate at which material is lost from compartment j to the exterior, but not to the other compartments. If we combine equation (3.95) with (3.94b), the latter equation can be written as

$$\sum_{i=0}^{n} L_{ij} = 0, \qquad j = 1, 2, \ldots, n. \tag{3.96}$$

The concentration of labeled material x_i in compartment i is introduced by means of the relation

$$\rho_i = V_i x_i$$

or   (3.97)

$$\rho = Vx.$$

In the last equation, $x = x(t)$ is an n-dimensional column vector $x = \{x_i\}$, and V is a diagonal matrix with nonzero diagonal elements $V_{ii} = V_i$, where V_i is the size of the ith compartment. Because V is a **constant matrix** (all its elements are constants), $d\rho/dt = V\, dx/dt$. Also, V has an inverse V^{-1} that is diagonal, with elements $(V^{-1})_{ii} = V_i^{-1}$, $i = 1, 2, \ldots, n$. Consequently, the equation satisfied by the vector x is

$$\frac{dx}{dt} = Mx, \tag{3.98}$$

where the matrix M is defined as

$$\begin{aligned} M &\equiv V^{-1}LV, \\ M_{ij} &= V_i^{-1} L_{ij} V_j. \end{aligned} \tag{3.99}$$

Note that the diagonal elements of M are the same as those of L, or $M_{ii} = L_{ii}$. The matrix M is introduced for mathematical convenience in solving the compartment equations, and its off-diagonal elements have no particularly useful biological significance.

There exists one further constraint on the elements of L and V, which is a consequence of the steady-state nature of the compartment system and the fact that whatever unlabeled material goes into a given compartment must come out of it. Let the material flux of unlabeled material from the exterior into compartment i be denoted by J_i. Then the steady-state behavior of the unlabeled material fluxes is represented mathematically by the equations

$$J_i + \sum_{j=1}^{n} L_{ij} V_j = 0, \qquad i = 1, 2, \ldots, n. \tag{3.100}$$

Here, the inward flux is $J_i + \sum_{j=1}^{n}{}' L_{ji} V_i$ and the outward flux is $-L_{ii} V_i =$

$\sum\limits_{j=0}^{n}{}' L_{ji}V_i$. We take note of the fact that for a closed system, $J_i = 0$ for all i, by definition.

Let us attempt a solution of equation (3.98) of the form

$$x = ae^{\lambda t}, \tag{3.101}$$

where a is a constant vector. If we substitute this expression for x into (3.98), we find that

$$Ma = \lambda a, \tag{3.102}$$

where we have canceled the factor $e^{\lambda t}$. The above equation is a set of n linear homogeneous equations for the components of the vector a. The solution to this equation, that is to say, the determination of a and λ for a given matrix M is called an **eigenvalue problem**. The value of λ that permits a nontrivial solution of (3.102) is called an **eigenvalue** or **characteristic value** of the matrix M, and the associated vector a is called an **eigenvector** or **characteristic vector**. Equation (3.102) can also be written as

$$(M - \lambda I)a = 0. \tag{3.103}$$

Here I is the unit matrix which has unit nonzero diagonal elements $I_{ii} = 1$, while $I_{ij} = 0$ for $i \neq j$.

The necessary and sufficient condition for a nontrivial solution of (3.103) is the vanishing of the determinant of the coefficient matrix of a, which is called the **secular determinant**,

$$|M - \lambda I| = \begin{vmatrix} M_{11} - \lambda & M_{12} & M_{13} & \dots & M_{1n} \\ M_{21} & M_{22} - \lambda & M_{23} & \dots & M_{2n} \\ M_{31} & M_{32} & M_{33} - \lambda & \dots & M_{3n} \\ \dots\dots\dots\dots\dots\dots\dots\dots\dots\dots\dots\dots \\ M_{n1} & M_{n2} & M_{n3} & \dots & M_{nn} - \lambda \end{vmatrix} = 0. \tag{3.104}$$

This equation is an nth degree polynomial in λ and has n roots. It is called the **characteristic equation** of the matrix M, and the roots of the polynomial are called **characteristic roots**. Under the conditions that the elements of L must satisfy equations (3.94), it is possible to show mathematically that the values of λ are restricted to being either real and negative, zero, or complex with negative real parts. The fact that the real part of λ can never be positive insures the stability of the underlying steady-state process. (The occurrence of complex roots implies that oscillations occur in the system.)

We shall assume the n eigenvectors λ_i are all distinct. We shall not actually find the eigenvectors associated with each eigenvector, but rather assume that they have been found, as they were for the two-compartment system previously discussed. We write the solution for the generic component x_i of the vector x as

$$x_i = A_{i1}e^{\lambda_1 t} + A_{i2}e^{\lambda_2 t} + A_{i3}e^{\lambda_3 t} + \cdots + A_{in}e^{\lambda_n t} = \sum_{j=1}^{n} A_{ij}e^{\lambda_j t},$$

$$i = 1, 2, \ldots, n.$$

(3.105)

where the A_{ij} are constants. (If the λ_i are not distinct, then powers of t appear in some of the A_{ij}.) In order to write the solution x conveniently in vector notation, we introduce the vector e defined as

$$e = e(t) = \{e^{\lambda_i t}\},$$

(3.106)

and the constant matrix A of order n,

$$A = \{A_{ij}\}.$$

(3.107)

Then the solution x is expressible as

$$x = Ae.$$

(3.108)

To show the relationship between A and M, we substitute (3.108) into (3.98). We observe that

$$\frac{de}{dt} = \Lambda e,$$

(3.109)

where $\Lambda = \{\Lambda_{ij}\}$ is a diagonal matrix with diagonal elements λ_i, namely,

$$\Lambda_{ij} = 0, \quad i \neq j,$$
$$\Lambda_{ii} = \lambda_i, \quad i = 1, 2, \ldots, n.$$

(3.110)

From (3.108), (3.109), and (3.98), it follows that

$$MA = A\Lambda.$$

(3.111)

If, now, we postmultiply (multiply on the right) both sides of equation (3.111) by A^{-1} (it can be shown that A is always nonsingular so that the matrix A^{-1} exists), the left-hand side equals $MAA^{-1} = MI = M$. Therefore, the matrix M is completely determined by the matrices A and Λ as

$$M = A\Lambda A^{-1}.$$

(3.112)

However, we emphasize that it is only the matrix M that is determined in this way. Unless we have other independent knowledge of V or L, we cannot unravel the matrices V and L from M.

We shall assume that at least one but not all of the compartments constituting the system have been observed, and that each observed concentration $x_i^*(t)$ has been converted into a sum of exponentials as discussed in

Section 3.7. If only some of the compartments have been observed, then only some elements of A are known, and (3.112) cannot be used to find M. In this case we ask: What are the equations that relate the elements of M to the known elements of A and to Λ (all of whose elements are known)?

To answer this question, we first note that the matrix A is subject to the initial condition

$$x(0) = Ae(0), \tag{3.113}$$

which can be expressed alternatively as

$$A^{-1}x(0) = e(0). \tag{3.114}$$

This follows directly by premultiplying equation (3.113) by A^{-1}. If now we postmultiply equation (3.112) by $x(0)$ and use (3.114), we obtain the relation

$$Mx(0) = A\Lambda e(0). \tag{3.115}$$

Both sides of this equation are n-dimensional vectors. Therefore, in component form, it is equivalent to n equations. However, there is only one row of A that is known for each compartment that is observed. Therefore, if only one compartment has been observed, only one number is determined on the right-hand side of (3.115). Hence for each observed compartment, (3.115) yields precisely one useful relation between the M_{ij} and the given data.

A second useful relation is obtained if we square both sides of equation (3.112). Then $M^2 = A\Lambda A^{-1}A\Lambda A^{-1} = A\Lambda I\Lambda A^{-1} = A\Lambda^2 A^{-1}$. This multiplication can be repeated, so that in general

$$M^p = A\Lambda^p A^{-1}, \qquad p = 1, 2, 3, \ldots. \tag{3.116}$$

If we now postmultiply by $x(0)$ and again use (3.114), there results

$$M^p x(0) = A\Lambda^p e(0), \qquad p = 1, 2, \ldots, n - 1. \tag{3.117}$$

If we set $p = 1$ above, we again obtain (3.115). Furthermore, we note that because Λ is a diagonal matrix, so is Λ^p, with diagonal elements $(\Lambda^p)_{ii} = \lambda_i^p$. We terminated the allowed values of p at $n - 1$ because choosing $p \geqq n$ does not provide any additional new relations between M and the known A_{ij} and Λ†. Each value of p in (3.117) provides us with one independent relation for the elements of M. If only the ith row of A is known, then the ith element of the right-hand side of (3.117) is known. Therefore if we consider the ith component of (3.117) for each value of p, we get $n - 1$ equations for the elements of M. Each compartment that is observed yields $n - 1$ such equations.

---

† The mathematical reason for this, which is beyond our scope to explain here, is a consequence of the Cayley–Hamilton theorem.

Given a square matrix of order n, there are n so-called **invariants** S_j, $j = 1, 2, \ldots, n$, associated with it. These invariants are readily expressed in terms of the eigenvalues λ_j that can be observed, and also in terms of the M_{ij} that are unknown. We shall consider them to be conditions that must be satisfied by the unknown elements of M. They are

$$S_1 \equiv \sum_{i=1}^{n} M_{ii} = \sum_{j=1}^{n} \lambda_j,$$

$$S_2 \equiv \sum_{i=1}^{n} \sum_{j=1}^{n} (M_{ii}M_{jj} - M_{ij}M_{ji}) = \sum_{\substack{i=1 \\ i \neq j}}^{n} \sum_{j=1}^{n} \lambda_i \lambda_j, \qquad (3.118)$$

$$\cdots\cdots\cdots\cdots\cdots\cdots\cdots\cdots\cdots\cdots\cdots\cdots$$

$$S_n \equiv \det(M) = \lambda_1 \lambda_2 \cdots \lambda_n,$$

where $\det(M)$ represents the determinant of M. The typical invariant S_j on the left-hand side is the sum of all principal minors of M of order j, and the right-hand side is the sum of all products of eigenvalues taken j at a time, $j = 1, 2, \ldots, n$. Because of the relationship between M_{ij} and L_{ij}, it can be easily shown that equations (3.118) remain valid if M_{ij} is replaced by L_{ij}. The relationship (3.99) between M and L is called a **similarity transformation**. Matrices that are related by a similarity transformation have the same eigenvalues.

Equations (3.117) and (3.118) represent the main results of this analysis.† The state of affairs with respect to compartment models may be summarized as follows. Suppose one compartment, say compartment 1, of a system is observed. This determines the number of compartments n, the n eigenvalues λ_j, and the n amplitudes A_{1i}. In terms of these quantities, we can write down $n - 1$ algebraic relations, which follow from (3.117), and the n algebraic relations (3.118). Altogether, these constitute $2n - 1$ equations for the n^2 unknown matrix elements M_{ij}. If a second compartment is also observed, the new data serve to confirm the values of the eigenvalues (or provide a larger data base for their determination), and provide $n - 1$ additional algebraic relations that follow from the equations (3.117) that the matrix elements must satisfy. In general, the total number of equations which will be found in this way for the M_{ij} will be less than n^2, unless all compartments are observed. Therefore, the equations constitute an **underdetermined** set, that is, there are more unknowns than equations to be satisfied. This implies that the compartmental model is not uniquely determined from the data and that there is a manifold of acceptable matrices or compartment models that is compatible with the observations. The possibilities are reduced by

---

† In order to recover the results of previous sections from these equations it is necessary to replace all diagonal elements L_{ii} by $-L_{ii}$.

the constraint equations (3.94), which the matrix elements of L must satisfy. These constraint equations can serve to provide upper and lower bounds for the turnover rates, as illustrated in problem 12.

Fortunately, there still remain compartment systems for which not only M but both L and V are uniquely determined even though not all of the compartments have been observed. Such systems occur when additional inferences can be made about the structure of the compartment system, as illustrated in the previous section.

SELECTED REFERENCE BOOKS

C. W. Sheppard (1962), *Basic Principles of the Tracer Method*, Wiley, New York.

D. S. Riggs (1963), *The Mathematical Approach to Physiological Problems*, Williams and Wilkins, Baltimore, Md.

A. Rescigno and G. Segre (1966), *Drug and Tracer Kinetics*, Blaisdell, Waltham, Mass.

G. L. Atkins (1969), *Multicompartment Models for Biological Systems*, Methuen, London.

J. A. Jacquez (1972), *Compartmental Analysis in Biology and Medicine*, Elsevier, New York.

K. L. Zierler (1962), Circulation times and the theory of indicator–dilution methods for determining blood flow and volume, in *Handbook of Physiological Society, Vol. I, Circulation*, W. F. Hamilton and P. Dow, eds., pp. 585–616, Amer. Physiol. Soc., Washington, D.C.

E. H. Wood, ed. (1962), *Symposium on Use of Indicator–Dilution Technics in the Study of the Circulation*, Monograph #4, Amer. Heart Assoc., New York.

PROBLEMS

1. In the indicator-dilution method by continuous infusion, a dye is injected continuously at a rate J_0 into a fluid reservoir of volume V. The fluid in the reservoir has a constant volumetric rate of influx or efflux equal to Q. Assume that the indicator–fluid mixture is homogeneous, and that homogeneity is attained instantly. Write down the differential equation governing the concentration of indicator c in the reservoir at time t. Solve it, with the concentration initially zero. Assuming that $c(t)$ was experimentally determined for a known value of J_0, and you had only a **planimeter** (a device for measuring the area under a given curve) available, how could you easily determine Q and V.

2. Consider a closed three-compartment catenary system with the associated graph illustrated in Figure 3.7. Let the compartment size $V_1 \to \infty$, and $L_{11} \to 0$, but assume that the actual flux of material leaving the compartment remains fixed, so that $K_{11} = L_{11}V_1$ approaches a finite constant. Also, assume that the concentration of labeled material initially in compartment 1 remains finite. Show that the compartment equations (3.58) reduce to the form of an open two-compartment system with a steady infusion of labeled material at a

rate J_1, namely,

$$\frac{d\rho_2}{dt} = -K_{22}x_2 + K_{23}x_3 + J_1, \qquad (3.119a)$$

$$\frac{d\rho_3}{dt} = K_{32}x_2 - K_{33}x_3. \qquad (3.119b)$$

3.* A constant infusion of radioactive ^{42}KCl was maintained in the vena cava leading into the heart of the dog (Conn and Robertson, 1955). During the time of the infusion, coronary sinus samplings were taken at regular 5 min intervals and the ^{42}K radioactivity was determined. It was found that the coronary sinus samplings could be fitted as a sum of exponentials, namely,

$$A_{10} + A_{11}e^{\lambda_1 t} + A_{12}e^{\lambda_2 t}, \qquad (3.120)$$

where

$$A_{10} = 57.5 \ \mu\text{Ci/liter},$$
$$A_{11} = -37.5 \ \mu\text{Ci/liter}, \ \lambda_1 = -0.0154 \ \text{min}, \qquad (3.121)$$
$$A_{12} = -20.0 \ \mu\text{Ci/liter}, \ \lambda_2 = -5.21 \ \text{min}.$$

These determinations were based on the averages of measurements taken in six dogs.

Assume that the myocardium is represented by a two-compartment system consisting of an extracellular or interstitial compartment (1), and an intracellular compartment (2). Assume that the compartment equations for the concentrations of radioactive material x_1 and x_2 in compartments 1 and 2, respectively, are of the form given in problem 2, namely,

$$\frac{dx_1}{dt} = -L_{11}x_1 + M_{12}x_2 + j_0,$$

$$\frac{dx_2}{dt} = M_{21}x_1 - L_{22}x_2, \qquad (3.122)$$

where j_0 is the constant infusion rate of radioactive ^{42}K per unit volume. The equilibration between coronary sinus blood and the extracellular compartment is assumed to be so rapid, that observations of the coronary blood effectively represents the extracellular compartment. Hence, we identify the observations given by equations (3.120) and (3.121) as representing $x_1(t)$. The mean value of the observed amount of myocardial potassium was $V_2 = 83.8$ mEq/kg of muscle (wet weight).

Determine as much as you can about j_0, the flux and clearance rates, and the compartment sizes.

Hint. Introduce new variables ξ_1 and ξ_2 which are the displacements of x_1 and x_2, respectively, from their equilibrium values $\bar{x}_1$ and $\bar{x}_2$.

4. The "mixing-cell" model for indicator-dilution experiments represents the flux through a physiological region schematically as a catenary sequence of n identical compartments (Newman *et al.*, 1951), as shown in Figure 3.11. The injection of a mass of tracer material ρ_0 takes place

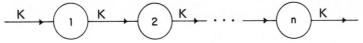

Figure 3.11. A "mixing-cell" model of an indicator-dilution experiment.

at time t_d before it reaches compartment 1. The observation of indicator is assumed to take place in compartment n or immediately following it. Each compartment is assumed to have the same volume V, and the volume flux through each is assumed to be K. Hence, the system of compartment equations for $t > t_d$ takes the form

$$V\frac{dc_1}{dt} = -Kc_1, \tag{3.123a}$$

$$V\frac{dc_j}{dt} = K(c_{j-1} - c_j), \qquad j = 2, 3, \dots, n. \tag{3.123b}$$

Let $n = 3$, and solve the equations for $c_j(t), j = 1, 2, 3$. Verify that the solution contains only one exponential decay constant given by $\lambda = -K/V$. Can the function $c_2(t)$ or $c_3(t)$ be expected to be a better fit to the typical observed indicator curve, Figure 3.1? Generalize the solution for the concentration in the last compartment when $j = n$.

Hint. Solve the equations sequentially.

5. Cholesterol, radioactively labeled with ^{14}C, was injected intravenously into human subjects, and the plasma was sampled subsequently for cholesterol activity over a period of ten weeks (Goodman and Noble, 1968). The time-dependent cholesterol activity $x_1(t)$ was found to be fitted quite well with a two-exponential curve, namely,

$$x_1(t) = A_{11}e^{\lambda_1 t} + A_{12}e^{\lambda_2 t}. \tag{3.124}$$

The mean values of the parameters appearing in the above expression, in a group of five persons, was determined to be as follows:

$$A_{11} = 1890 \text{ dis/min/mg}, \quad \lambda_1 = -0.125 \text{ per day}, \quad (3.125)$$
$$A_{12} = 760 \text{ dis/min/mg}, \quad \lambda_2 = -0.0148 \text{ per day}.$$

The amount of injected ^{14}C-cholesterol ρ_{10}, possessed an activity of 30 μCi. Assume a two-compartment model of cholesterol transport, with compartment 1 representing the plasma plus some rapidly exchanging tissues such as the liver, and compartment 2 representing an (unknown) leakproof tissue compartment. Assume that unlabeled cholesterol is supplied nutritionally to compartment 1 at a mean rate of 0.2 g/day, and is manufactured in both compartments. Infer quantitatively everything you can about the system.

6. Members of the genus **Euglena** are eucaryotic algae commonly found along the shores of marshes, ponds, and rivers. They are motile, unicellular, and in common with many other divisions of algae, flagellated. **Flagella** are, like cilia, motile hairlike projections from the surfaces of cells. They differ from cilia in their mode of motion. The species **Peranema trichophorum** is a euglenid flagellate which possesses two flagella, one pointing in the forward direction of motion, and the other in the trailing direction. The leading flagellum has a length $L = 63$ μm, the cell body being slightly longer. Following cell division, the new leading flagellum of a daughter cell is about 47 μm long, and it grows at a decreasing rate, approaching a final length L in an asymptotic manner. If the leading flagellum is amputated artificially by means of a fine tungsten needle controlled by a micromanipulator, it is observed that the flagellum regenerates at a more rapid rate than its normal growth rate, and later it grows essentially at its normal rate. Quantitatively, it is found from a semilog plot of $L - x$ versus time that $L - x$ can be represented by a sum of two exponential terms (see Figure 3.12), where x is the flagellar length at time t (Tamm, 1967). Show that this result follows from the solution of the equations:

$$\frac{d\xi}{dt} = k_1(L - \xi), \qquad\qquad t > 0, \qquad (3.126a)$$

$$\frac{dx}{dt} = k_1(L - x) + k_2(\xi - x), \qquad t > t_0, \qquad (3.126b)$$

where $\xi = \xi(t)$ represents the length of the flagellum at time t in its unamputated normal state, and $x = x(t)$ represents the length of the flagellum following amputation. In other words, normally the leading flagellum grows at a rate proportional to the difference between its asymptotic length L and its length at time t. When it is amputated, it

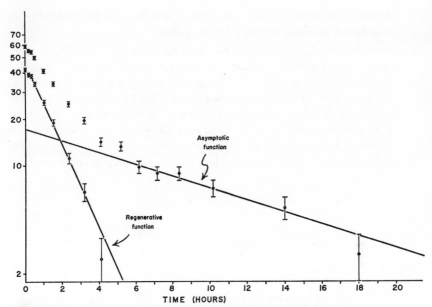

Figure 3.12. Semilog plot of the data (upper experimental points) representing the difference between the full grown length L of the leading flagellum and its length x at time t, or $(L - x)$ versus time, following amputation of the leading flagellum of a cell of the flagellate *Peranema trichophorum*. The ordinate unit of length is the micrometer. The age of the cell at the time of amputation, taken to be time zero in the figure, was $t_0 = 2$ hr. The straight line labeled "asymptotic function" represents the slowly decaying exponential term in the representation of the data points as a sum of two exponentials. The straight line labeled "regenerative function" represents the fast decaying exponential term, and the data points through which it runs represent the original data points after subtraction of the contribution of the slow exponential term. From Tamm (1967), with permission.

grows at this rate plus an additional rate proportional to the difference between the length it would have had at time t if it had not been amputated, and its actual length at time t. Because of amputation, $x(t_0) \equiv x_0 < \xi(t_0)$. Solve the equation system. Determine the values of x_0, $\xi(t_0)$, k_1, and k_2 from the data of Figure 3.12.

Hint. We infer from Figure 3.12 that for t large,

$$\log(L - x) \sim \log(L - \xi) \sim \log A + \lambda_1(t - t_0), \quad (3.127a)$$

$$\log[(L - x) - (L - \xi)] = \log(\xi - x) = \log B + \lambda_2(t - t_0), \quad (3.127b)$$

where $t_0 = 2$ hr, and, approximately,

$$A = 17 \ \mu\text{m}, \quad \lambda_1 = -0.087 \text{ per hour},$$
$$B = 46 \ \mu\text{m}, \quad \lambda_2 = -0.62 \text{ per hour}.$$

7. Derive expressions for B_{11}, B_{12}, B_{21}, and B_{22} for the soaking out experiment, in terms of measured quantities.

8. Consider a reaction for which the back reaction rate is negligible compared to the forward reaction rate. The graph representing a sequence of such reactions would have the appearance of "a string of pearls," as in Figure 3.11. Suppose that a radioactive label is introduced into the system which labels a precursor of the precursor, and the specific activity curves x_2 and x_1 of the product and precursor, respectively, are determined. Imagine they are plotted on the same scale as illustrated in Figure 3.13. How can we identify the **precursor-product relation**,

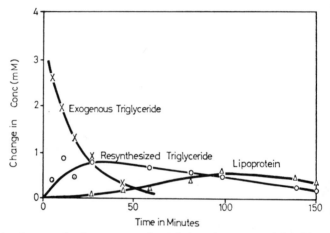

Figure 3.13. An example of precusor-product relations. Exogenous triglyceride was injected into the blood of a fasting individual, and the plasma subsequently sampled at suitable time intervals. The excess concentrations of exogenous triglyceride, resynthesized triglyceride, and lipoprotein were measured in each of the samples, as a function of the time. Reproduced from Atkins (1969), with permission of the author. After Hallberg (1965).

that is to say, which curve represents the product, and which curve represents the precursor?

Hint. Consider the sign of dx_2/dt with respect to the curves x_1 and x_2.

9. Show that for a closed n-compartment catenary system

$$K_{ij} = K_{ji} \tag{3.128}$$

for all values of i and j. Show that this relation is equivalent to the relation $L_{ij} = M_{ji}$.

10. During the replication of DNA in the growth of a population of *E. coli*

cells, thymine is incorporated into the DNA of the cells at a steady rate. In an effort to determine which of the thymine bearing molecules, thymidine monophosphate (TMP), thymidine diphosphate (TDP), or thymidine triphosphate (TTP), was a direct precursor of replicating DNA, a culture of cells was exposed to a radioactive tritiated thymine synthetic medium at $t = 0$. Subsequent measurements were made of the activities of samples of the above-mentioned candidate precursors and the DNA, taken from the cells, with the results shown in Figures 3.14 and 3.15 (Werner, 1971). Formulate a two-compartment

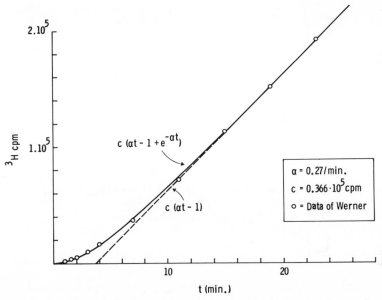

Figure 3.14.   The activity of ^{3}H-thymine in DNA is shown by the open circles as a function of the time. From Rubinow and Yen (1972), with the permission of Nature, redrawn from the work of Werner (1971).

system of equations to represent this system, consisting of a precursor compartment and a DNA compartment. Solve the system of equations. Does it appear that the solution can be fitted to the results shown in the figures? How would you determine whether one of the precursor candidates was a direct precursor?

Hint. Assume a steady-state influx of labeled thymine into the precursor compartment. Assume there is no exit of thymine from the DNA compartment.

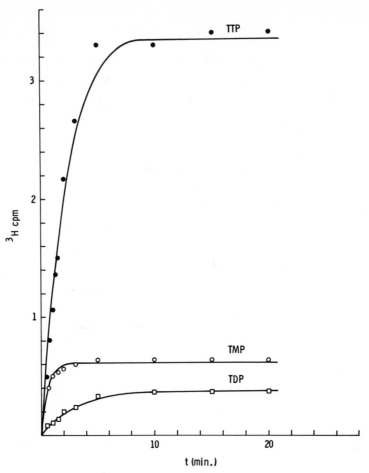

Figure 3.15. The activity of ^{3}H-thymine in the DNA precursor candidates TMP($\bigcirc$), TDP($\square$), and TTP($\bullet$). From Rubinow and Yen (1972), with the permission of Nature, redrawn from Werner (1971).

11.* In many tracer experiments, the labeled material is infused continuously into a compartment at constant infusion rates. Define the vector $J = \{J_i\}$ where J_i is the rate of infusion of labeled material into the ith compartment. Then, the compartment equations assume the form, in matrix notation,

$$\frac{d\rho}{dt} = L\rho + J. \tag{3.129}$$

What is the asymptotic value of the vector ρ as $t \to \infty$? How can the above equation be reduced to the standard compartment equation form, (3.92)? If L is a matrix of order n, how many terms are expected to appear in the expression for ρ_i?

12.* Consider a two-compartment system for which compartment 1 is initially injected with tracer material. Subsequent observation of this compartment yields the following expression for the fractional concentration as a function of the time,

$$\frac{x_1(t)}{x_1(0)} = f_1 e^{\lambda_1 t} + f_2 e^{\lambda_2 t}, \tag{3.130}$$

where f_1, f_2, λ_1, and λ_2 are known constants, and $f_1 + f_2 = 1$. Determine upper and lower bounds for L_{12}, L_{21}, and V_2/V_1, in terms of the known constants.

Hint. First determine $L_{12}L_{21}$. Utilize this result in conjunction with the upper bounds on L_{12} and L_{21} to obtain lower bounds for $(L_{12})^{-1}$ and $(L_{21})^{-1}$.

13. Show that the stationary state of the chemostat defined in Section 1.7, $(c,N) = (\bar{c},\bar{N})$, is stable.

Hint. Let $(c,N) = (\bar{c} + \xi, \bar{N} + \eta)$, and consider the linearized equations for ξ and η that result from equations (1.34) and (1.36).

14.* In the investigation of chemical reactions by means of chemical relaxation spectrometry, a reacting system which is in equilibrium initially is perturbed by a sudden change of an external parameter such as the pressure or temperature. The transient response of the system to its equilibrium state is then observed spectroscopically, and the dependence of the **spectrum of relaxation times** on the equilibrium concentrations is determined, as illustrated in Figure 3.16. For example, consider an enzyme E which interacts with a substrate S to form a complex C. Suppose that the complex C first undergoes a conformational change to a new state C' before it is capable of forming a product.

(*a*) Disregarding the catalytic formation of product, write down the rate equations governing the concentrations s, e, c, and c', respectively, in conformity with the graph of Figure 3.17. Eliminate the concentrations s and c from the rate equations, by making use of the conservation of mass of the substrate molecules and of the enzyme molecules. Find the equations for the equilibrium values $\bar{e}$ and $\bar{c}'$ of the concentrations of E and C'.

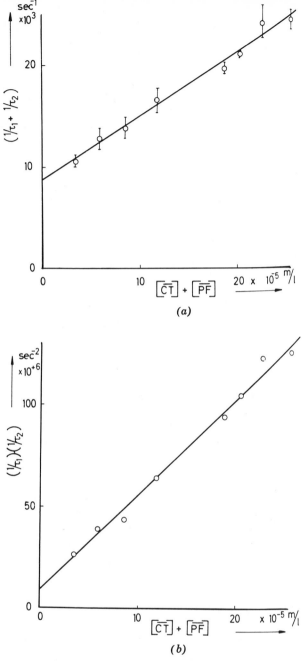

Figure 3.16. When the enzyme chymotrypsin reacts with proflavin in solution, two relaxation times τ_1 and τ_2 are observed when the solution is subjected, within a few seconds, to a temperature jump of 7° from 4.5°. In (a) is seen the dependence of the sum of the two reciprocal relaxation times, and in (b) is seen the product of the two reciprocal relaxation times, as a function of the sum of the equilibrium concentrations of chymotoypsin and proflavin, labeled $[\overline{CT}] + [\overline{PF}]$. From Havsteen (1967), with permission.

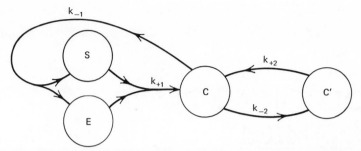

Figure 3.17. Graph representing an enzyme E interacting with a substrate S to form a complex C. The complex C can undergo a conformational change to a new state C'. The labels on the directed line segments of the graph indicate the reaction rate constants of the associated transitions.

(*b*)   Assume that, at $t = 0$, the concentrations of E and C' are subjected to small perturbations x and y, respectively, away from equilibrium, that is, let $(e, c') = (\bar{e} + x, \bar{c}' + y)$. Find the linearized differential equations satisfied by $x = x(t)$ and $y = y(t)$. Determine the **relaxation times** τ_1 and τ_2 in terms of the rate instants and the initial enzyme and substrate concentrations, where $\tau_{1,2} = 1/\lambda_{1,2}$, and λ_1 and λ_2 are the characteristic values associated with the differential equation system for x and y. Show how the rate constants can be found from the experimental information contained in Figure 3.16.

4

BIOLOGICAL FLUID
DYNAMICS

The cytoplasm of many cells is known to undergo a **streaming** phenomenon, that is, the cytoplasm circulates the cell interior in more or less well defined pathways. The flow of blood through the circulatory system of mammals, the swimming of fish, the flight of birds and insects, the motility of sperm, amoeboid motion, and the propulsion of bacteria and protozoa through liquids in a variety of ways are all examples of biological systems which involve the movement of a fluid (a liquid or a gas) in response to mechanical forces. To understand these phenomena in a theoretical way requires the understanding of the motion of the fluid medium.

The study of the motion of fluids in response to mechanical forces is called **fluid dynamics**. This is an advanced topic in physics or applied mathematics, and its detailed study would take us too far afield here, as well as involve us in problems of great mathematical complexity. We content ourselves with a brief excursion into it, including some results of particular interest in biology.

4.1. THE EQUATIONS OF MOTION OF A VISCOUS FLUID

The phenomena considered within the realm of fluid dynamics are **macroscopic**: Any small volume element of the fluid is supposed to be so large that it still contains a very large number of molecules. Accordingly, the fluid is regarded as a **continuous medium**. At any point $P = (x,y,z)$ of the medium, the thermodynamic properties of pressure p and mass density ρ are defined at any time t. In addition, if the fluid is in motion, the velocity of the fluid $\mathbf{v}$ is defined at a given point at time t. The velocity $\mathbf{v}$ is a vector quantity having three components, u, v, and w representing the components of the velocity in the x, y, and z directions, respectively. A common notation for $\mathbf{v}$ is $\mathbf{v} =$

156

(u,v,w). The fundamental problem of fluid mechanics is to determine the pressure p, the density ρ, and the velocity $\mathbf{v}$ of a fluid as functions of space and time. The variables p, ρ, and $\mathbf{v}$ are called the **dependent variables** of the problem, and the variables x, y, z, and t are called the **independent variables** of the problem. Thus, we write $p = p(x,y,z,t)$, $\rho = \rho(x,y,z,t)$, and similarly for u, v, and w. The fact that the density of a fluid can vary from point to point makes the fluid **compressible**. Fluids whose density either remains constant or changes in a negligible way are said to be **incompressible**. The transmission of sound through air or water is a property of these media that depends on their compressibility.

Fluids are "sticky" to greater or lesser extents, and this property is denoted by the term **viscosity**. As a result of viscosity, when a viscous fluid flows across a wall, the fluid in immediate contact with the wall is at rest. As we know from watching the flow of streams or rivers, the flow velocity is greater, the greater is the distance from the riverbank, and it reaches its maximum in the middle. One reason for this is that there exists a frictional force between neighboring elements of a viscous fluid. We say that there is a shearing stress between such elements. Fluids that are not viscous or **inviscid** cannot support shearing stresses. For **normal** or **Newtonian fluids**, the stress on a volume element of fluid is proportional to the rate of deformation or strain rate of the volume element, and this constant of proportionality is called the coefficient of viscosity μ.

The **equations of motion** for a fluid were first written down by Euler in 1755 for the case of an inviscid fluid. The equations of motion of a viscous fluid were developed in the next century and are called the **Navier–Stokes equations**. We shall consider the special case for which there is only one nonzero velocity component, u. Then the Navier–Stokes equations take the form

$$\rho\left[\frac{\partial u}{\partial t} + u\frac{\partial u}{\partial x}\right] = -\frac{\partial p}{\partial x} + \mu\left(\frac{\partial^2 u}{\partial x^2} + \frac{\partial^2 u}{\partial y^2} + \frac{\partial^2 u}{\partial z^2}\right) + f_x. \qquad (4.1)$$

If we consider a tiny volume element of fluid $\Delta x\,\Delta y\,\Delta z$ surrounding the point P, then the left-hand side represents the mass times the acceleration (per unit volume), and the terms on the right represent the forces acting on this element (per unit volume). The first term on the right is the force resulting from the pressure surrounding the fluid element, the second term represents the viscous force due to shearing stresses on the surface of the volume element, and the last term f_x represents any **body force**, such as the force of gravity, acting on the volume element in the x direction. In the latter case, f_x is expressed as ρg, where g is the acceleration due to gravity. More generally, with all components of $\mathbf{v}$ varying, equation (4.1) is replaced by three equations, because force or acceleration has three components.

In addition, the fluid element must satisfy an equation representing the law of conservation of mass flow. This equation expresses the fact that any change with time in the mass $\rho \, \Delta x \, \Delta y \, \Delta z$ of the infinitesimal volume element $\Delta x \, \Delta y \, \Delta z$ surrounding the point P arises as a result of a net efflux or influx of fluid across the bounding surface of the volume element (see Figure 4.1).

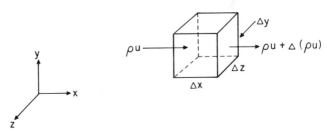

Figure 4.1. The mass flux into and out of a volume element of fluid surrounding the point $P = (x,y,z)$. Because only the x component of fluid velocity is assumed to be nonzero, there is only a flux across the area element $\Delta y \, \Delta z$. The increment in mass efflux over mass influx is $\Delta(\rho u) = (\partial(\rho u)/\partial x)\Delta x$.

Here, flux across a surface element is used in the technical sense meaning the rate of mass flow in the direction perpendicular to the surface element. Mathematically, this law is expressed as

$$\frac{\partial \rho}{\partial t} \Delta x \, \Delta y \, \Delta z = \rho u \, \Delta y \, \Delta z - \left[\rho u + \frac{\partial}{\partial x}(\rho u)\Delta x \right] \Delta y \, \Delta z,$$

where $\Delta y \, \Delta z$ is the cross-sectional area of the volume element through which the fluid flows. The left-hand side represents the increment per unit time of the mass of the infinitesimal volume element, the first term on the right-hand side represents the influx of fluid mass to the volume element, and the last term represents the efflux of fluid mass from the volume element. After canceling the common factor $\Delta x \, \Delta y \, \Delta z$, this equation becomes the **equation of continuity**,

$$\frac{\partial \rho}{\partial t} + \frac{\partial}{\partial x}(\rho u) = 0. \tag{4.2}$$

Finally, the pressure and density must satisfy a thermodynamic **equation of state**, which we write quite generally as

$$p = p(\rho). \tag{4.3}$$

The particular functional form of this relationship depends on whether the motional activity under consideration is isothermal, adiabatic, or otherwise.

The equations (4.1)–(4.3) constitute three equations for the three unknown functions p, ρ, and u. More generally, with the restrictive condition $v = w = 0$

removed, they constitute five equations for the five unknown functions p, ρ, and the three components of $\mathbf{v}$. These equations constitute a system of **partial differential equations**, because they invoke partial derivatives of the dependent variables. The equations must be supplemented by **boundary conditions** that express the conditions that the flow velocity and pressure must satisfy at the surface boundary of the fluid. For example, at a solid surface, the normal component v_n of the velocity $\mathbf{v}$ cannot penetrate the surface, so that

$$v_n = 0, \text{ on solid boundary.} \tag{4.4}$$

In addition, if the fluid is viscous, the tangential component v_t of the velocity $\mathbf{v}$ likewise vanishes at a solid surface, so that

$$v_t = 0, \text{ on solid boundary.} \tag{4.5}$$

The quantities p, ρ, and $\mathbf{v}$ must also satisfy an **initial condition** that specifies the initial state of the fluid as a function of position.

The system of partial differential equations, subject to prescribed boundary conditions and initial conditions, constitute the mathematical representation of a given fluid dynamic problem of interest. Even if the fluid is assumed to be incompressible so that ρ is constant, these equations are difficult to solve. One source of difficulty is the presence of nonlinear terms such as $u\partial u/\partial x$ in equation (4.1). The number of known solutions is relatively small, but approximation methods and numerical methods of solution are available. A principal approximation technique is the linearization of the equations, in a manner analogous to the discussion of the ordinary differential equations in Section 1.8.

4.2. POISEUILLE'S LAW

As an example of the application and solution of the Navier–Stokes equations, we shall consider the motion of a fluid through a long circular cylindrical tube of length L, radius a, one end of which is at pressure p_1 and the other end of which is at pressure p_2, where $p_1 > p_2$ (see Figure 4.2). Assume the flow is **steady**, so that all time derivatives appearing in the equations are zero. We seek a solution for which only the component tangent to the axis

Figure 4.2. Flow through a circular cylindrical tube of length L and radius a.

of the cylinder is nonzero. Denote this component of the velocity vector by u, and assume, for reasons of symmetry, that u does not depend on the polar angle in the plane perpendicular to the x axis. We shall also assume that the fluid is incompressible, so that, from equation (4.2),

$$\frac{\partial u}{\partial x} = 0. \tag{4.6}$$

Therefore, u depends on r only.

In the cylindrical coordinate system shown in Figure 4.2, it can be shown that the equation of motion (4.1), in the absence of body forces, takes the form

$$\mu \frac{1}{r} \frac{d}{dr} \left(r \frac{du}{dr} \right) = \frac{\partial p}{\partial x}, \tag{4.7}$$

where u and p are functions of position in the tube, that is to say, $u = u(r)$ and $p = p(x,r)$. Because there is no velocity or force in the radial direction, $\partial p / \partial r = 0$. Hence, p depends on x only, and the partial derivative $\partial p / \partial x$ in (4.7) can be written as a total derivative, $\partial p / \partial x = dp/dx$. Differentiating (4.7) with respect to x, and utilizing (4.6), we observe that

$$\frac{d^2 p}{dx^2} = 0. \tag{4.8}$$

Consequently, dp/dx is a constant, so that

$$p = p_1 + \frac{(p_2 - p_1)}{L} x, \tag{4.9}$$

which satisfies the boundary conditions $p = p_1$ at $x = 0$ and $p = p_2$ at $x = L$. Therefore

$$\frac{dp}{dx} = -\frac{p_1 - p_2}{L}. \tag{4.10}$$

Now we can return to (4.7) and, taking advantage of the fact that dp/dx is known and is a constant, solve for u. A first integration of (4.7) yields

$$r \frac{du}{dr} = \frac{1}{2\mu} \frac{dp}{dx} r^2 + A, \tag{4.11}$$

where A is a constant of integration, and dp/dx is given by equation (4.10). Dividing by r and integrating again, we obtain

$$u = \frac{1}{4\mu} \frac{dp}{dx} r^2 + A \log r + B, \tag{4.12}$$

where B is a second constant of integration. The physical condition that velocities cannot be infinite constrains us to set $A = 0$, because otherwise the velocity would become infinite at the center of the tube. The constant B is determined by the requirement that $u = 0$ at the tube wall $r = a$, according to equation (4.5). Thus, the solution u becomes

$$u = -\frac{1}{4\mu}\frac{dp}{dx}(a^2 - r^2),\tag{4.13}$$

which expresses the fact that the velocity distribution across the tube is parabolic. The maximum velocity occurs at the center of the tube (see Figure 4.3). Equations (4.9) and (4.13) constitute the solution to our problem.

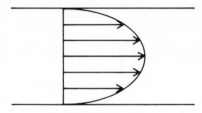

Figure 4.3. Velocity distribution of a Poiseuille flow across a tube. The arrows represent the magnitude and the direction of the velocity at a given position.

Knowing u, we can calculate the **discharge**, or volume **flux** of fluid Q, which passes any cross section of the tube per unit of time. Consider an annular element $2\pi r\,dr$. Passing through it in unit time is the volume $u2\pi r\,dr$. Hence, by summation, the total volume of fluid flowing through the cross section per unit time is

$$Q = \int_0^a u2\pi r\,dr.\tag{4.14}$$

Substituting (4.10) and (4.13) into this relation, we obtain the result

$$Q = \frac{\pi}{8}\frac{(p_1 - p_2)}{\mu L}a^4.\tag{4.15}$$

The discharge is seen to be proportional to the fourth power of the radius of the tube. Equation (4.15) is called **Poiseuille's formula** or **Poiseuille's law**, named in honor of the French physician who first investigated (circa 1840) in a quantitative manner the flow of water through glass pipes. Poiseuille's interest was the flow of blood through the vessels of the circulatory system, but he worked with water because of the difficulty at that time of preventing blood from clotting on exposure to air.

4.3. PROPERTIES OF BLOOD

Poiseuille's law is so well established experimentally, that it is often used in order to determine the viscosity coefficient of viscous fluids. When blood is examined in this manner, the viscosity coefficient of blood is found to be about five times the value for water, if the diameter of the tube is relatively large. Thus, at the normal physiological temperature of $37°C$, the viscosity of water μ_0 is 0.007 P (P = poise = dyne sec/cm^2, the c.g.s. unit of viscosity), and the viscosity of blood μ_B as determined by Poiseuille's law in large tubes is about 0.035 P.

In small arteries such as capillaries, the viscosity of blood as determined by Poiseuille's law approaches μ_p the viscosity of plasma, where $\mu_p = 0.012$ P at $37°C$. Here "small" refers to arteries whose diameter is less than ten times the largest diameter of a red cell, which is about 8 μm. The average diameter of mammalian capillaries is about the same as or slightly smaller than the red cell diameter. The red cells are able to pass through the capillaries because they are easily deformed, and are observed to go through such small capillaries with a parachute-like shape.

The fact that the effective viscosity coefficient of blood as determined by means of Poiseuille's law depends on the radius of the tube in which it is measured implies that blood is not a Newtonian fluid, for which μ is a constant. Rather, blood is said to be a **non-Newtonian fluid**. Fluids which have elaborate molecular structure, in particular those consisting of long chain molecules, are in general non-Newtonian. Thus, biological fluids such as cytoplasm can be expected to be non-Newtonian. Unfortunately, no adequate mathematical theory of such fluids exists at present.

When blood is subjected to centriguation in a centrifuge, it separates into a fluid called **plasma** and formed elements: **blood cells** and **platelets**. The blood cells are of two types, red cells or **erythrocytes**, and white cells or **leukocytes**. The latter are further identified as being either **granulocytes** (and further classified as being of **neutrophil**, **eosinophil**, or **basophil** variety), or **monocytes**, or **lymphocytes**. However, the overwhelming majority of the blood cells are erythrocytes. The volume concentration of red cells is called the **hematocrit**, which in normal physiological circumstances lies in the range 0.41–0.44. The total number of white cells is about 1/600 the total number of blood cells. Blood plasma is found to behave like a normal Newtonian fluid. Thus, the non-Newtonian nature of blood is a direct consequence of the fact that blood is a **suspension**, with plasma the suspending medium, and red blood cells for the most part being the suspended particles. There is as yet no theoretical basis for determining the viscosity of a suspension in terms of the viscosity of the suspending medium and the concentration of the suspended particles when the concentration is large, as is the case for blood. For most purposes, the blood can be treated theoretically as an ordinary Newtonian fluid with an appropriate "effective" viscosity coefficient that is constant.

The **specific gravity** of a red cell is defined as the ratio ρ_e/ρ_0, where ρ_e is the mass density of the erythrocyte, and ρ_0 is the mass density of water, under standard conditions. The specific gravity of red cells is about 1.06, while that of plasma is 1.03. Consequently, if blood is allowed to stand in a container, the red cells will settle out of suspension. They do so at a definite rate called the **erythrocyte sedimentation rate**, or ESR for short. The mathematical theory of this process is very simple, and is based on Stokes' law.

On the assumption that a rigid sphere is moving slowly with constant velocity through a viscous Newtonian fluid, Stokes (1851) calculated the motion of the fluid by assuming that the velocity of the sphere is slow. More precisely, it is necessary to state what "slow" means by introducing a reference velocity. Mathematically, this is accomplished by introducing the non-dimensional quantity $R = \rho v a/\mu$ called the **Reynolds number**. Here ρ is the fluid density, v is the velocity of the sphere, and a is the sphere radius. It can be shown that under the assumption $R \ll 1$, or, in dimensional terms, $v \ll \mu/\rho a$, the nonlinear terms in the equations of motion of the fluid can be sensibly neglected without causing a significant error in their solution. Then their solution is a relatively easy task. The resulting fluid motion under the general assumption $R \ll 1$ is called **laminar** flow. When $R \gg 1$, the laminar flow solution becomes unstable, and a different flow regime becomes established, called **turbulent** flow. The precise value of R at which instability occurs depends on the particular flow geometry. For example, if a dye is introduced into a Poiseuille flow at small Reynolds numbers, it will be seen to produce a thin thread parallel to the axis of the tube. The flow is laminar. At a critical Reynolds number of about 1000, the thread is seen to break up and the motion of the dye becomes irregular, the dye spreading throughout the tube. The flow has become turbulent, and Poiseuille's law no longer correctly describes it.

Knowing the fluid motion about a sphere, the force on the sphere can be deduced, and it is found that there exists a **drag force** $\mathbf{F}_D$ that opposes the motion of the sphere. The magnitude of the drag force is given by the expression

$$F_D = 6\pi\mu av. \tag{4.16}$$

This formula is known as **Stokes' law** and is in excellent agreement with observations and measurements of the slow motion of spheres through liquids. Equation (4.16) has been generalized to particles of arbitrary shape as $F_D = fv$, where f is called the **frictional coefficient**. In the case of the sphere, $f = 6\pi\mu a$. For a particle which is not a sphere, Stokes' law must be modified. This is accomplished by solving the fluid dynamic equations for the particular shape of the particle. Dimensional considerations show that the force will be proportional to μav where a is a characteristic length of the particle. Hence, only the factor 6π in Stokes' law is altered.

In the case of the red cell, which has a biconcave disk shape, it is not possible to solve the fluid equations for the flow past it. Therefore, an "equivalent" spherical shape is introduced for the red cell. Assume the red cell is settling or sedimenting under the influence of gravity, and let v be its velocity in the downward direction. By Newton's second law,

$$V\rho_e \frac{dv}{dt} = V\rho_e g - V\rho_p g - 6\pi\mu v a,$$

$$V = \tfrac{4}{3}\pi a^3.$$

(4.17)

Here, V is the volume of our hypothetical spherical red cell, ρ_e is its density, ρ_p is the density of plasma, and dv/dt is the acceleration of the sphere. The left-hand side of (4.17) is the mass times the acceleration of the sphere. This is equated to the sum of all the forces on the sphere which act in the vertical direction. The first term on the right is the force of gravity acting downward, and the second term represents the buoyant force on the sphere, acting upward, which equals the weight of the displaced plasma. The final term in (4.17) is the drag force on the sphere which acts upward because it opposes the motion. After a while, a steady rate of fall will be achieved when the acceleration is zero and v is constant. According to (4.17), this occurs when the right-hand side of (4.17) is zero, or when v equals v_s, where

$$v_s = \frac{2}{9\mu} a^2(\rho_e - \rho_p)g.$$

(4.18)

The steady velocity v_s with which the sphere moves is called the **sedimentation velocity** or **terminal velocity**. With the use of equation (4.18), some theoretical agreement can be attained between the observed and expected sedimentation rate of a dilute suspension of red cells by introducing an appropriate radius a of the "equivalent" spherical red cell.

In illness, ESR always increases dramatically. This is due to clumping, or formation of red cell aggregates called **rouleaux**. In effect, rouleaux formation increases the size of the sedimenting particle, and hence increases the value of a which appears in (4.18). Thus, the ESR is actually a measure of the tendency of red cells to clump or stick together, which depends on the surface properties of the red cell membrane. Such properties are profoundly affected by plasma proteins, especially fibrinogen.

Equation (4.18) can be utilized to determine the viscosity of cytoplasm. This is accomplished by injecting oildrops of a known density into a cell, and determining their terminal velocity under the influence of gravity by observation in a microscope. Because of surface tension, such drops assume a spherical shape, and the radius of the drop can be readily measured. By utilizing equation (4.18), with ρ_e and ρ_p denoting the density of the oil and the cytoplasm, respectively, the viscosity μ of cytoplasm is determined. Such determinations assume, of course, that cytoplasm can be considered to be a homogeneous Newtonian fluid.

Biochemists must often work with solutions of macromolecules, and the viscosities of such solutions vary greatly because of the presence of the solute molecules. The principal theoretical expression for the viscosity of such a solution stems from the work of Einstein (1906) who determined the effective viscosity of a dilute suspension of rigid hard spheres. To do this, he solved the Navier–Stokes equations at low Reynolds number for a sphere in a **linear shear flow**. Such a flow results from moving a plate parallel to another plate when there is a viscous fluid between the plates. If y is the distance from one plate measured in the perpendicular direction towards the other plate, then the velocity u in the fluid parallel to and in the direction of the moving plate is $u = u_0 y$, where u_0 is a constant.

If a solution is dilute enough, the disturbance due to one sphere does not affect the disturbance due to the other sphere. Hence, the disturbed flow of the dilute suspension is determined additively from the single sphere flow. Knowing this, the energy dissipated because of the fluid viscosity can be calculated, and compared to the dissipation that would result if no particles were present. By means of this comparison, the effective viscosity μ^* is determined to be

$$\mu^* = \mu(1 + \alpha c). \tag{4.19}$$

Here, μ is the viscosity coefficient of the suspending medium, c is the volume concentration of the suspended particles, and α is a numerical constant which in the case of spheres equals 2.5. For nonspherical particles, equation (4.19) remains applicable with a value of α that is always larger than 2.5. From equation (4.19) we see that the viscosity of a suspension is always increased relative to the viscosity of the suspending medium.

4.4. THE STEADY FLOW OF BLOOD THROUGH A VESSEL

The observed flow of blood through the superior vena cava of a dog shows a marked nonlinearity, as shown in Figure 4.4. Here the flux is plotted as a function of the difference between the pressure p_1 upstream from the vena cava and the pressure p_2 downstream in it. In this experiment, the upstream pressure was maintained constant while the downstream pressure p_2 was varied by artificial means. For small pressure differences, the flux Q is proportional to $p_1 - p_2$, just as in Poiseuille's law, but when the pressure difference becomes large, the flux somewhat paradoxically attains a maximum value and no longer increases. Actually, a similar result is obtained for the flux through a rubber tube. These results suggest that it is important to take into account the collapsible nature of elastic tubes, in order to try to understand them.

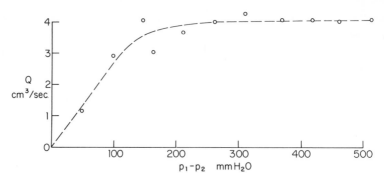

Figure 4.4. The steady flux of blood Q through the superior vena cava of a dog as a function of the pressure difference $p_1 - p_2$, where p_1 is the pressure in the jugular vein (upstream from the superior vena cava), and p_2 is the pressure applied to the peripheral end. Redrawn from Brecher (1952).

An artery or a vein is not a rigid tube but is rather an elastic tube. Consequently, the radius of such a tube is variable. Imagine such a vessel filled with fluid at rest and surrounded by fluid. Its radius will be determined by the **transmural pressure difference** between the interior and exterior pressures, as well as the tension in the wall of the vessel. Under normal circumstances, the thickness of a blood vessel wall is small compared to the resting radius of the vessel (when exterior pressure equals interior pressure). Consequently, to good approximation, we can treat the wall as a thin membrane for which the tension per unit length of tube and per unit thickness is denoted by t. Consider the mechanical equilibrium of half of such a long cylindrical vessel plus the fluid contained in it. Let the wall have a thickness h and radius r, while the interior fluid pressure is p, and the exterior pressure is p_0 (see Figure 4.5). The net downward force per unit length on this half cylinder is

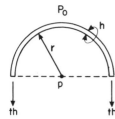

Figure 4.5. Cross section of one half of a long elastic cylinder containing a fluid with pressure $p > p_0$, the pressure exterior to the cylinder.

$2th$ and it is balanced by the net upward force per unit length, which is the pressure difference between the lower and upper surfaces times the diameter, $(p - p_0)2r$. Equality of forces yields the relation

$$th = (p - p_0)r, \qquad (4.20)$$

a result due to Young and to Laplace. The tension t that develops is a property of the elastic wall in its reaction to stretch. By determining experimentally the pressure needed to inflate such a tube to a given radius, by means of (4.21) a **tension-radius** diagram is developed, which relates the tension in the wall to the radius. The typical result of such measurements for mammalian arteries is shown for a section of the external iliac artery in man in Figure 4.6.

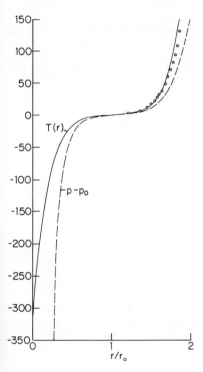

Figure 4.6. The tension-radius curve $T(r)$ for a section of the external iliac artery in man. The circles are experimental results of Roach and Burton (1957). Here $T(r) = th$, and $p - p_0$ is obtained from $T(r)$ by means of equation (4.20). The curve for $r/r_0 < 1$ indicates the presumed behavior of the artery under compression, on the basis of other observations of rubber tubing, and the *trachea*, and the *bronchus*, tubular elastic parts of the pulmonary airway system. The unit of $T(r)$ is grams per centimeter (here a gram is a weight unit, equal to 980 dynes), and the unit of length is $r_0 = 0.216\,\mathrm{cm}$. After Rubinow and Keller (1972).

Deformable materials that obey **Hooke's law** (such as steel) exhibit a linear relationship between the normal stress and the elongation per unit length. Robert Hooke, after whom this law is named, is also the seventeenth century scientist who first recognized and named the cell as a unit of living tissue. An infinitesimal volume element of an elastic body is subjected to interior (molecular) forces from neighboring volume elements, in addition to any external forces such as gravity that may be acting on it. These internal forces are proportional to the surface area of the volume element. The force per unit area acting on the surface of the volume element is called the **stress**. The direction of this force is not in general perpendicular to the surface, but is composed of a normal component (such as tension or pressure) called the **normal stress** or **principal stress**, and a tangential component or **shearing stress**. In response to stresses, the volume element undergoes a deformation called the **strain**. In response to a normal stress, the volume element undergoes an elongation in the same direction as the normal stress, which is called the **principal extension**. In addition, the surface element undergoes a change in its area, which is a contraction in response to tension, or an expansion in response to compression. On a macroscopic scale, such contraction is observed when a rubber band is stretched. In response to a shearing stress, the volume element undergoes a shape distortion, which, when expressed mathematically, is called the **shearing strain** or simply the **shear**.

For example, the tension across the end section of a bar or strip of elastic material is proportional to the principal extension or fractional amount of stretch in the direction of the tension, and the constant of proportionality is called **Young's modulus** or the **modulus of elasticity**, denoted by E. For the tension–radius curve exhibited in Figure 4.6, a linear relationship can be said to exist for only a limited range of variation of the radius. In such a case, a useful concept for representing the stress–strain law quantitatively is the **tangent modulus**, defined in terms of the tangent approximation to the curve in the neighborhood of a given point as $E = r(dt/dr)$, evaluated at the given point. For example, the tangent modulus in the neighborhood of the equilibrium value $r = r_0$ (when $t = 0$) is defined by the relation

$$t = E\left(\frac{r - r_0}{r_0}\right), \tag{4.21}$$

where $(r/r_0 - 1)$ is the fractional amount of stretch of the tube element in response to the applied tension t, and E denotes the tangent modulus. The larger arteries are known to consist primarily of two organic elastic materials, **elastin** and **collagen**. The linear portion of curve in the neighborhood of the equilibrium radius r_0 is known to be representative of the elastic behavior of elastin. The steeper portion of the curve is typical of the behavior of collagen. Thus, arteries in their elastic behavior are analogous to an article of clothing that has been "elasticized," such as a sock or girdle.

The steady fluid flow through an elastic tube of length L such as an artery or vein will now be considered (Rubinow and Keller, 1972). We assume the tube is long, and the pressure is a function of x alone, $p(x)$. The pressure has a value at the inlet position $x = 0$ given as $p(0) = p_1$, and a value at the outlet position $x = L$ given as $p(L) = p_2$. The jacket or external pressure of the fluid surrounding the tube is assumed to have the constant value p_0 (see Figure 4.7).

The cross sectional area at any point x is assumed to be circular and to adjust to the local transmural pressure difference at x in accordance with

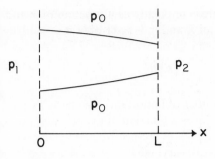

Figure 4.7. Flow through a thin elastic tube.

(4.20). Thus

$$p(x) - p_0 = \frac{th}{r},\tag{4.22}$$

where r is the radius of the cross-sectional area at x. Eliminating t between (4.21) and (4.22) yields the pressure–radius relationship

$$p - p_0 = \frac{Eh}{r_0}\left(1 - \frac{r_0}{r}\right),\tag{4.23}$$

where r_0 is the equilibrium radius of the tube for which $t = 0$. When the stress–strain law is nonlinear, as illustrated in Figure 4.6, t can still be eliminated between the nonlinear law and equation (4.22), and results in a pressure–radius relation different from equation (4.23). The dashed curve in Figure 4.6 illustrates the pressure–radius relation resulting in this case.

The flow through the tube is assumed to obey Poiseuille's law locally, that is, through any infinitesimal cross section of the tube, equation (4.15) holds true,

$$Q = -\frac{\pi}{8\mu}\frac{dp}{dx}r^4,\tag{4.24}$$

where r and p are related by (4.22), and t is given as a function of r, as in equation (4.21) or in a tension–radius diagram. From (4.23) or its suitable generalization, we see that r is a function of $p - p_0$, written as $r(p - p_0)$. We integrate (4.24) with respect to x, obtaining the relation

$$Qx = \frac{\pi}{8\mu}\int_p^{p_1} r^4(p - p_0)dp.$$

By a change in the integration variable to $p - p_0 = p'$, this becomes

$$Qx = \frac{\pi}{8\mu}\int_{p-p_0}^{p_1-p_0} r^4(p')dp'.\tag{4.25}$$

The above relation tells us the pressure p implicitly as a function of x and the parameter Q. To find Q, we set $x = L$ and $p = p_2$ to obtain

$$Q = \frac{\pi}{8\mu L} \int_{p_2 - p_0}^{p_1 - p_0} r^4(p')dp'. \tag{4.26}$$

From this result we can make a number of important inferences. We see that Q is proportional to L^{-1}, and that it is a function of the pre- and post-tubular pressure differences $p_1 - p_0$ and $p_2 - p_0$, respectively. When $p_1 - p_2$ is small and $r^4(p_1 - p_0)$ is not too rapidly varying.

$$Q \sim \frac{\pi}{8\mu L} r^4(p_1 - p_0)(p_1 - p_2), \tag{4.27}$$

so that the flux dependence is, like Poiseuille's law, proportional to the pressure difference $p_1 - p_2$. Also, as $p_1 - p_0$ remains fixed and $p_2 - p_0$ tends to $-\infty$, Q attains a constant value (assuming the integral exists, which means that it has a finite value),

$$Q \sim \frac{\pi}{8\mu L} \int_{-\infty}^{p_1 - p_0} r^4(p')dp'. \tag{4.28}$$

In such circumstances, Q is independent of any change in the downstream pressure p_2, and depends (for fixed μ and L) only on the pressure difference $p_1 - p_0$. This phenomenon, recognized empirically by physiologists, has been called the "vascular waterfall," although the name is misleading. In the other circumstance in which p_2 and p_0 are maintained fixed while p_1 is increased, the volume flux Q does increase without bound.

Thus, the theory yields qualitatively the principal features of the experimental results. By inserting in these formulas a functional representation of the particular tension–radius curve for an iliac artery shown in Figure 4.6 (obtained by least-square fitting of a polynomial), the integration of (4.26) can be performed explicitly and yields the theoretically predicted curves for the flux shown in Figure 4.8. The qualitative agreement of these curves with the experimental observations serves to corroborate the essential correctness of the theory.

The physical implication of these curves is that, if a collapsible tube is used as a drinking straw, then increasing the suction beyond a certain value (in the curves shown, this occurs for $p_1 - p_2 \approx 10$) does not appreciably increase the flux. Actually sucking too hard would lead to buckling which is not taken into account in the theory presented. Veins are known to collapse under normal physiological conditions, which is to say they buckle. However, the qualitative features of these curves would not be affected by such considerations.

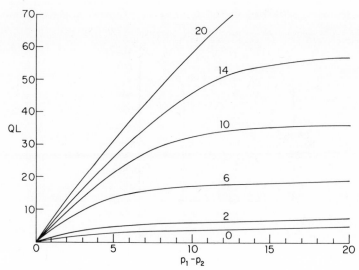

Figure 4.8. Curves based on theoretical calculations utilizing equations (4.26), (4.21), and $T(r)$ as given in Figure 4.6 (the extrapolated part of the curve representing $T(r)$ must also be utilized), for various fixed values of $p_1 - p_0$. The unit of pressure is $p(0) = r_0^{-1}$ g/cm $= 4.63$ g/cm^2. The unit of QL is $\pi r_0^3 L p_0/8\mu$. After Rubinow and Keller (1972).

4.5. THE PULSE WAVE

It has been known for a long time that the flow of blood is pulsatile as a consequence of the beating of the heart. The beating heart produces a pressure wave that travels through the blood, and this pressure wave is the pulse felt in the wrist. The wave is not the same as the acoustic waves associated with the beating heart that the physician hears in his stethoscope. Such acoustic waves are a consequence of the compressibility of the blood and of the living tissue surrounding it. Rather, the pulse wave exists even though the fluid is treated as incompressible. It owes its existence to the elasticity of the arteries and the coupling of arterial vibrations to the blood flow, as was first pointed out by Young (1808). We shall present here a simplified derivation of the velocity of this pulse wave, first given by Weber (1866) and Resal (1876).

Consider again an elastic tube through which fluid is flowing. This time we shall consider the fundamental fluid dynamic equations, assuming for the sake of simplicity that the fluid is both inviscid and incompressible. Let $A(x)$ be the cross-sectional area of the tube at the distance x, and designate the constant density by ρ, pressure by $p(x,t)$, and velocity parallel to the tube axis by $u(x,t)$. This velocity is assumed to be unvarying across a given cross-sectional area of the tube (see Figure 4.9). The net force in the positive x

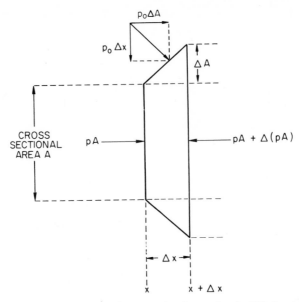

Figure 4.9. Volume element of fluid of cross sectional area A and width Δx at position x of a tube. The fluid is flowing in the x direction. The arrows represent the pressure forces acting on the element. Here, $\Delta A = (\partial A/\partial x)\Delta x$, and $\Delta(pA) = (\partial(pA)/\partial x)\Delta x$.

direction on the volume element is seen from the figure to be

$$pA + p_0 \frac{\partial A}{\partial x}\Delta x - \left[pA + \frac{\partial}{\partial x}(pA)\Delta x\right] = -\frac{\partial}{\partial x}[(p - p_0)A]\Delta x.$$

The mass $\rho A\,\Delta x$ times the acceleration of the volume element is given as in equation (4.1) by the expression

$$\rho A\,\Delta x\left[\frac{\partial u}{\partial t} + u\frac{\partial u}{\partial x}\right].$$

Hence, the equation of motion of a fluid element of cross-sectional area A and width Δx is

$$\rho A\left[\frac{\partial u}{\partial t} + u\frac{\partial u}{\partial x}\right] = -\frac{\partial}{\partial x}[(p - p_0)A]. \tag{4.29}$$

The equation of continuity for an incompressible fluid assumes the form

$$\frac{\partial A}{\partial t} + \frac{\partial}{\partial x}(Au) = 0. \tag{4.30}$$

In addition, a further relation must be supplied relating the cross-sectional area A and the pressure p. The proper relationship is determined by the equations of motion of the wall, which accounts for inertial and motional effects. Such an analysis was first made by Korteweg (1878). A simpler treatment is to neglect such effects and simply express the relationship between p and A as if the adjustment of the cross-sectional area to the pressure occurred instantaneously.

For example, let us assume that equation (4.23) holds, which we can rewrite in terms of the area A as

$$p - p_0 = \frac{Eh}{r_0}\left[1 - \left(\frac{A_0}{A}\right)^{1/2}\right], \tag{4.31}$$

where $A = \pi r^2$, $A_0 = \pi r_0^2$.

We shall now **linearize** equations (4.29)–(4.31) by assuming that u, $p - p_0$, and $(A - A_0)$, together with their derivatives, are small. In (4.31) we expand the right-hand side, regarded as a function of A, in a Taylor series about $A = A_0$. Neglecting all terms of second order in the small quantities, these equations become

$$\rho\frac{\partial u}{\partial t} = -\frac{\partial p}{\partial x}, \tag{4.32}$$

$$\frac{\partial A}{\partial t} + A_0\frac{\partial u}{\partial x} = 0, \tag{4.33}$$

$$p - p_0 = \frac{Eh}{2r_0 A_0}(A - A_0). \tag{4.34}$$

By differentiating (4.32) with respect to x and (4.33) with respect to t, we can eliminate u from these two equations and obtain the equation

$$\frac{\partial^2 A}{\partial t^2} = \frac{A_0}{\rho}\frac{\partial^2 p}{\partial x^2}. \tag{4.35}$$

From (4.34), $\partial^2 A/\partial t^2$ is proportional to $\partial^2 p/\partial t^2$, so that (4.35) can be expressed as an equation for p alone,

$$\frac{1}{c^2}\frac{\partial^2 p}{\partial t^2} = \frac{\partial^2 p}{\partial x^2}, \tag{4.36}$$

$$c = \left(\frac{Eh}{2\rho r_0}\right)^{1/2}. \tag{4.37}$$

Equation (4.36) is called the **wave equation** and is well known in the study of mathematical physics. It is readily seen that both A and u also satisfy it. The solution of (4.36), apart from any boundary conditions and initial conditions that are to be satisfied, can be written quite generally as

$$p = f(x - ct) + g(x + ct), \qquad (4.38)$$

where f and g are arbitrary functions. It is easy to verify this statement by direct substitution of equation (4.38) into equation (4.36). The function f represents a wave form that travels in the positive x direction with velocity c. This can be seen by examining the curves for the function $f(x - ct)$ as a function of x, at two different values of t (see Figure 4.10). In a similar manner

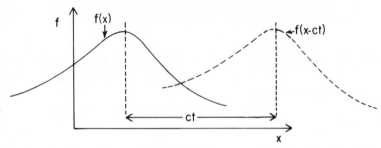

Figure 4.10. The representation of a wave form $f(x - ct)$ traveling in the positive x direction with velocity c.

it can be seen that the function $g(x + ct)$ represents a wave form that travels in the negative x direction with velocity c.

The meaning of this solution is that a perturbation in the pressure, say, at $t = 0$, of distributed amplitude $f(x)$, will propagate in time down the tube with velocity c and fixed shape $f(x)$. Similarly, if at $x = 0$, the position of the left ventricle, a periodic disturbance in the pressure above its mean value exists and is represented by $f(ct)$, then this disturbance will propagate down the aorta with velocity c. We emphasize that this wave is not a sound wave arising because of the compressibility of blood, but rather is a wave of the coupled blood-tube system, which depends primarily on the elastic properties of the tube. A similar wave phenomenon arises in the transmission of the **action potential** along the axon of a nerve cell: When a sufficiently large electrical potential difference is excited at some point across the membrane of the nerve axon, the potential (difference) propagates or travels with constant velocity along the length of the axon.

Equation (4.37) expresses the manner in which the velocity of propagation depends on the physical parameters of the tube and the fluid. This formula for the pulse wave velocity was first derived by Young (1808) and should therefore be called **Young's formula**, although it is usually referred to as the **Moens–Korteweg formula**. Let us apply the formula to the flow of blood through the **thoracic aorta**, the major blood vessel leaving the left ventricle of the heart. For it the vessel parameters, in dogs, have the approximate values, at a mean pressure of 100 mm Hg, $E \approx 4.3 \times 10^6$ dyne/cm^2, $h/a \approx 0.105$ (Bergel, 1961). Then, with $\rho \approx 1.06$ g/cm^3, we obtain the result $c \approx 4.6$ m/sec, which is in good agreement with observations (McDonald, 1968). The pulse wave velocity should not be confused with the actual velocity of a small element of blood in the aorta. For example, the mean velocity of blood in the aorta is approximately 30 cm/sec (see problem 1). The pulse wave is to be compared with ocean waves seen at beaches. For such waves, the wave form has a noticeable velocity in a horizontal direction toward the beach, but the actual motion of the ocean water (except where the waves actually "break" on the beach) is merely oscillatory in a vertical direction about a fixed position on the surface of the water. Because the direction of motion of the individual ocean water particles is perpendicular to the direction of motion of the ocean wave, the latter is said to be a **transverse** wave. By contrast, the direction of oscillation of the blood particles associated with the pulse wave is the same as the direction of propagation of the wave. Hence, the pulse wave is said to be a **longitudinal** wave. Acoustic waves are also longitudinal.

There remain many quantitative features of the flow of blood through the circulating system that have not been satisfactorily explained although their mechanisms are purely physical, and require for their understanding a knowledge of the motion of viscous fluids in interaction with elastic tubes. An example of such an unexplained phenomenon is the fact that the pulse wave changes its shape slowly as it travels through the major arteries.

What is the pressure and flow at any point in the circulatory system? An analog of Kirchhoff's laws for electrical circuits can be derived for the mammalian circulatory "circuit." These laws describe the current and voltage at any point in a given electrical circuit when the resistances and the voltage sources are known. This statement can be translated into hemodynamic terms by reading flow, pressure, circulatory network, and ventricles for current, voltage, circuit, and voltage sources, respectively. Such laws can be expected to be very useful with respect to the important problem of diagnosing the state of health of the vascular system, because pressure and flow depend on the parameters of the system, and the parameters reflect the state of health of the system.

4.6.   THE SWIMMING OF MICROORGANISMS

Spermatozoa, other microorganisms, and nematodes, worms that are frequently parasitic but also occur in a free-living marine state, propel themselves through water by generating waves of lateral displacement along their length. Superficially, the mechanism of swimming would appear to be the same as that for larger organisms such as snakes or fish, but actually, the dynamics of the associated fluid motion is quite different. This difference can be understood in terms of the Reynold's number $R = \rho v a / \mu$ associated with these motions, where v is a characteristic velocity and a is a characteristic length of the body. R is a measure of the ratio of the stress in the fluid due to inertia to the stress in the fluid due to viscosity. Alternatively, it is a measure of the relative importance in the equations of motion of the acceleration terms appearing on the left in equation (4.1), and the viscous stress terms appearing on the right. Thus, for the swimming of fish and snakes, R is of the order of 10^3 or more. This indicates that inertial forces are all important, and that viscous forces can be neglected by comparison.

When a propeller driven airplane flies through the air, two aerodynamic effects are usually distinguished, propulsion and resistance. The propeller develops a forward or propulsive force by imparting a backward momentum to the air. The body develops resistance because, in moving forward, it entrains some of the surrounding fluid, which imparts a forward momentum to the fluid. When the airplane is moving at a constant velocity, the net force on it is zero, and the forward momentum imparted to the fluid is exactly balanced by the backward momentum. Viscosity plays only a secondary role in these considerations. Airplanes are large bodies in motion with associated large Reynold's numbers. These same qualitative considerations apply to the swimming of fish, in that propulsive or oarlike units, the fins and the tail, can be distinguished from the resistance unit associated with the body. Furthermore, for snakes and fish, a large body of fluid in the neighborhood of the swimming body is noticeably in motion.

In the case of nematodes $R \sim 1$, and in the case of spermatozoa, $R \sim 10^{-3}$ or less, so that viscous forces are all important as compared to inertial forces. Here, the symbol $\sim$ is used in connection with numbers to denote "is the order of magnitude of." For example, for a bacterium swimming in water, $v \sim 0.01$ cm/sec, $\rho \sim 1$ g/cm^3, $a \sim 10^{-4}$ cm, and $\mu \sim 0.01$ P at 20°C, so that $R \sim 10^{-4}$. When R is small, it is no longer possible to consider the propulsion as being due to the separate effects of a propulsive unit and of resistance. It is still true that the net momentum imparted by the body to the fluid is zero, but the momentum of fluid in front of the body or just behind the body is a very small quantity. In fact, no displacement or disturbance of the fluid in the neighborhood of a swimming microorganism is discernible. Because viscous stresses in the fluid are large, and inertial stresses are negligible, the only force to be considered is that arising from the viscosity of the fluid.

In order for a microorganism to propel itself at a constant velocity, it must move in such a way that the viscous stress on it due to the surrounding fluid is zero. It accomplishes this by distorting its body in an appropriate manner. The appropriate manner in which it does so is the establishment of lateral waves of displacement along its length. It should be kept in mind that the viscous stress on a local part of its body is not in general zero. However, the net viscous force on the entire body in the direction of its uniform motion is zero. Also, the organism must expend energy in order to maintain the waves of displacement along its length.

The study of the motion of a fluid past a body in steady motion at small Reynold's numbers is simpler than that for large Reynold's numbers, because the inertial terms in the equations of motion, which are nonlinear, can be neglected. The study of such fluid motions resulting from the swimming of microorganisms was initiated by Taylor (1951, 1952). This study utilizes the fluid equations in three dimensions with the inertial terms neglected. The boundary conditions are that the tangential and normal velocity components relative to the microorganism are equal to zero on the boundary of the microorganism. The latter is assumed to possess a flagellum. The flagellum is idealized to be a long cylinder of length $2b$, with a circular cross section of radius a. The flagellum is assumed to be moving steadily in the positive x direction with velocity u. It is further assumed that the motion is **inextensional**, meaning that the axis of the flagellum does not change its length. Extensional motion occurs, for example, when earthworms translate themselves.

A lateral displacement wave given by the expression

$$y = \eta \sin k(x + ct) \qquad (4.39)$$

is assumed to be traveling with velocity c in the negative x direction down the length of the flagellum. Thus, the motion of the flagellum is assumed to be planar, that is to say, confined to the xy plane. Here η is the amplitude of the wave, and k is called the **wave number**. The **wave length** λ of the wave is defined as that distance for which $\sin kx = \sin k(x + \lambda)$. In other words, it is the distance (at a fixed time) at which the wave shape repeats itself. Because $\sin \xi = \sin(\xi + 2\pi)$, it follows that $k\lambda = 2\pi$, which is the fundamental relationship between the wave number and the wave length. Because of this relationship, the quantity $1/k = \lambda/2\pi$ is referred to as the **reduced wave length** We emphasize that equation (4.39) represents a planar wave in space and time, which can be viewed both as a spacially periodic wave at a fixed time, and a temporally periodic wave at a fixed position. Thus, we recall that the period T of the motion is defined by the condition $\sin k[x + c(t + T)] = \sin k(x + ct)$, so that

$$kcT = 2\pi.$$

The **natural frequency** v of the wave motion is defined by the relation

$$v = \frac{1}{T}. \tag{4.40}$$

Combining the last two relations, we infer the following fundamental relation between the frequency, wave length, and velocity of a periodic wave in space and time,

$$v = \frac{c}{\lambda}. \tag{4.41}$$

We shall not discuss the actual solution of the fluid dynamic equations for this problem, which is beyond our scope here. Rather we shall assume that the solution has been achieved, and indicate how it can be utilized to derive further information regarding the motion of the microorganism. We shall first show how the fluid dynamic force determines the propulsive force on the body. Taylor determined this force by integrating over the surface of the body a function representing the local stress in the fluid. We shall instead follow a simpler and physically more understandable derivation (Gray and Hancock, 1955).

Assume a spermatozoon consists of a head and a cylindrical tail, the latter being in motion as indicated above. We consider a small element ds of the tail, located at position (x,y) at time t. We assume that this element is inclined at an angle θ to the y axis (see Figure 4.11). The tangent of the angle of inclination θ to the x axis is the slope of the line $y = y(x,t)$ at time t, or

$$\tan \theta = \frac{\partial y}{\partial x} = k\eta \cos k(x + ct). \tag{4.42}$$

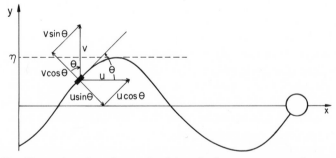

Figure 4.11. The motion of an element ds of the tail of a spermatozoon, shown as a heavy line segment, consists of a vertical velocity v in the positive y direction, and a horizontal velocity u in the positive x direction. These velocities can be decomposed into components which are tangential and normal to the element ds. After Gray and Hancock (1955).

The element ds has a vertical velocity v defined as

$$v \equiv \frac{\partial y}{\partial t} = ck\eta \cos k(x + ct). \qquad (4.43)$$

The total velocity of the element ds at time t consists of the vector sum of the velocity v in the vertical direction and the forward velocity u in the horizontal direction.

We consider the projections of these two velocities in the directions normal to the surface of the element ds, and tangential or lateral to it. It is in terms of the latter two motions that the fluid dynamic force on the body can be readily decomposed. Thus, these velocity components are, respectively (see Figure 4.11),

$$(v \cos \theta - u \sin \theta, v \sin \theta + u \cos \theta). \qquad (4.44)$$

As we have indicated, the fluid motion associated with the tail motion imparts to the tail a normal force N and a lateral or drag force L. These forces acting on the element ds are assumed to be proportional, respectively, to the local velocity components given above, and are in the directions that oppose the motion. This assumption is known from hydrodynamic studies to be valid for long slender bodies, and therefore is expected to be valid for a flagellum. Thus, the normal and tangential forces on the element ds are given, respectively, as

$$\begin{aligned} dN &= C_N(v \cos \theta - u \sin \theta)ds, \\ dL &= C_L(v \sin \theta + u \cos \theta)ds, \end{aligned} \qquad (4.45)$$

where C_N and C_L are called the **drag coefficients** associated with the normal and lateral motion of the element ds. They depend on the viscosity μ and other parameters of the problem. In the figure, dN is positive if the normal force is in the downward direction and dL is positive if the tangential force is directed to the left.

In order to determine the forward thrust or force dF that the fluid imparts to the element ds in the positive x direction, we project the forces dN and dL onto the x axis. Thus, with the aid of Figure 4.11, we see that

$$dF = dN \sin \theta - dL \cos \theta. \qquad (4.46)$$

By substituting (4.45) into (4.46), it follows that the forward thrust on the element ds is given by the expression

$$dF = [(C_n - C_L)v \sin \theta \cos \theta - u(C_N \sin^2 \theta + C_L \cos^2 \theta)]ds. \qquad (4.47)$$

We determine the total force or thrust on the tail of the microorganism by summation. For this purpose, we require the differential geometric relation,

$$ds = (dx^2 + dy^2)^{1/2} = \left[1 + \left(\frac{\partial y}{\partial x}\right)^2\right]^{1/2} dx. \tag{4.48}$$

Now, by substituting (4.42) and (4.48) into (4.47), utilizing the identity $\cos^2 \theta = 1/(1 + \tan^2 \theta)$, and summing over the length of the tail, we obtain the following expression for the total thrust F on the organism,

$$F = \int dF = \int_0^{2b} \frac{(C_N - C_L)v(\partial y/\partial x) - u[C_L + C_N(\partial y/\partial x)^2]}{[1 + (\partial y/\partial x)^2]^{1/2}} dx, \tag{4.49}$$

where v is given by (4.43), and $\partial y/\partial x$ is given by the last equality in (4.42).

If we assume that the tail wave always consists of an integral number of wavelengths, then because of the periodicity of the wave motion, we need only integrate over a single wave length of the motion. Let the number of waves over the length $2b$ of the tail be γ, so that

$$\gamma \equiv \frac{2b}{\lambda} = \frac{bk}{\pi}. \tag{4.50}$$

Furthermore, we note that from (4.42) and (4.43), that $v = c(\partial y/\partial x)$. Hence the total thrust on the tail becomes

$$F = \gamma\{[c(C_N - C_L) - uC_N]I - uC_LJ\}, \tag{4.51}$$

where I and J are defined as

$$I \equiv \int_0^{2\pi/k} \frac{(\partial y/\partial x)^2}{[1 + (\partial y/\partial x)^2]^{1/2}} dx, \tag{4.52a}$$

$$J \equiv \int_0^{2\pi/k} \frac{dx}{[1 + (\partial y/\partial x)^2]^{1/2}}. \tag{4.52b}$$

Note that the integrals I and J are positive quantities, because their integrands are always positive.

If the organism is to maintain a constant velocity u in the x direction, then the total force F must be zero. If we ignore the force on the head, then according to (4.51), F is zero only if the velocity u is given as

$$u = \frac{(C_N - C_L)I}{C_NI + C_LJ} c. \tag{4.53}$$

If we divide numerator and denominator of the right-hand side of this equation by C_LJ, we see that only the ratios C_N/C_L and I/J are required to determine the coefficient of c above.

It was found (Hancock, 1953) that for $ka \ll 1$,

$$C_N \gtrsim 2C_L. \tag{4.54}$$

By using this last result, we can write (4.53) as

$$u = \frac{I}{2I + J} c. \tag{4.55}$$

It can be readily shown that the integrals I and J depend only on the quantity $(k\eta)^2$, and are expressible in terms of certain special functions called **complete elliptic integrals** (see problem 12). It is sufficient for all practical purposes to approximate I and J as follows. Replace $\cos^2 k(x + ct)$ in the denominator of the integrals I and J by its mean value over one period. This value is determined as follows:

$$\frac{\int_0^{2\pi/k} \cos^2 k(x + ct)dx}{\int_0^{2\pi/k} dx} = \frac{1}{2\pi} \int_{kct}^{kct + 2\pi} \cos^2 \xi d\xi = \frac{1}{2\pi} \int_0^{2\pi} \cos^2 \xi \, d\xi$$

$$= \frac{1}{2\pi} \int_0^{2\pi} \frac{1}{2}(1 + \cos 2\xi)d\xi = \frac{1}{2}.$$

It follows with this replacement that

$$I \approx \frac{(k\eta)^2}{2[1 + \frac{1}{2}(k\eta)^2]^{1/2}} \frac{2\pi}{k},$$

$$J \approx \frac{1}{[1 + \frac{1}{2}(k\eta)^2]^{1/2}} \frac{2\pi}{k}. \tag{4.56}$$

Hence, $I/J \approx 1/2(k\eta)^2$, a result which is probably more accurate than either of equations (4.56). Combining the latter result, or equivalently equations (4.56), with (4.55) leads to the final result

$$u = \frac{1}{2} \frac{(k\eta)^2}{1 + (k\eta)^2} c. \tag{4.57}$$

This gives the velocity of propulsion u in terms of the velocity c, wave number k, and amplitude η of the wave motion produced by the organism. According to equation (4.57), the velocity of propulsion can never exceed $1/2$ the velocity of the wave motion. Equation (4.57) is an improvement on Taylor's principal result

$$u = \frac{1}{2}(k\eta)^2 c, \tag{4.58}$$

obtained under the assumptions that both $k\eta \ll 1$ and $ka \ll 1$ (where a is the flagellar cross-sectional radius). Of course, (4.57) reduces to (4.58) when $(k\eta) \ll 1$ and small quantities up to order $(k\eta)^2$ only are retained. It is important to emphasize that direction of motion of the spermatozoon through the fluid is opposite to the direction of motion of the wave along its tail. If the wave form $y = \eta \cos k(x - ct)$ were assumed instead of (4.39), the velocity of propulsion u would be in the negative x direction.

In Taylor's original derivation of (4.58), the value of ka was permitted to be arbitrary. The coefficient of c then included a complicated function of ka. When the assumption that $ka \ll 1$ was imposed, this function simplified to the expression

$$u = \frac{1}{2} k^2 \eta^2 \left[\frac{K_0(ka) - 1/2}{K_0(ka) + 1/2} \right] c, \qquad (4.59)$$

where $K_0(ka)$ is a **Bessel function of imaginary argument** of zero order. Its values for various values of the dimensionless argument ka can be found in tables. The function $K_0(ka)$ possesses a **logarithmic singularity** in the neighborhood of $ka = 0$, that is to say $K_0(ka)$ becomes infinite like $\log(ka)$ as ka approaches zero. Hence, for ka very small, the factor in square brackets in (4.59) is approximately unity, and (4.59) reduces to (4.58).

The more complicated case when the wave motion of the flagellum is not planar, but is described by a helical or spiral wave, was also considered (Taylor, 1952). In such a case, a torque is exerted on the head which therefore undergoes a rotational as well as a translational motion. It was found that the helical wave described by this equation produces a velocity of propulsion given by the following expression when $ka \ll 1$,

$$u = k^2 \eta^2 \left[\frac{K_0(ka) - 1/2}{K_0(ka) + 1/2} \right] c. \qquad (4.60)$$

By comparing this result with (4.57), it is inferred that for a given amplitude, the spiral wave propels the tail at twice the speed of the plane wave. This result is to be expected because a helical wave is composed of two plane waves at right angles to each other, and 90° out of phase, that is, in the xy plane and in the xz plane. If a helical wave is considered under the assumption that $ka \ll 1$ but with $k\eta$ permitted to be large (Hancock, 1953), then the velocity of propulsion is found to be

$$u = \frac{(k\eta)^2}{1 + 2(k\eta)^2} c, \qquad (4.61)$$

and this result is also in agreement with the aforementioned conclusion. For helical waves, too, it is seen that velocity of propulsion can never exceed $1/2$ the wave velocity.

The motion of the spermatozoa of the sea urchin *Psammechinus miliaris* was studied by Gray (1955). Such sperm are difficult to observe because of the very small radius of the tail. Nevertheless, some quantitative measurements could be made with the aid of high speed photography. He found that approximately planar oscillatory waves of the flagellum could be observed (see Figure 4.12). The average values for 89 sperm cells of some of the parameters of the theory were as follows:

$$a = 0.2 \ \mu\text{m}, \ \lambda = 24 \ \mu\text{m},$$
$$2b = 41 \ \mu\text{m}, \ v = 34.5/\text{sec},$$
$$\eta = 4 \ \mu\text{m}, \quad u = 191 \ \mu\text{m/sec}, \tag{4.62}$$
$$\gamma = 1.3.$$

We first note that the Reynolds number, assuming the medium is like water, is calculated to be $R \sim 2 \times 10^{-5}$. If the Reynolds number is defined in terms of the maximum vertical velocity $\eta c k = 2\pi\eta v$ instead of u, the Reynolds number is about four times larger. We also note that γ is not an integer, as

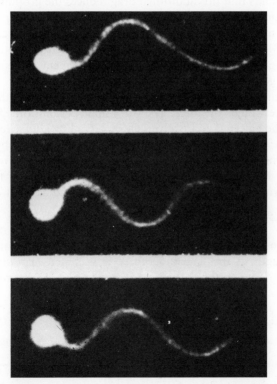

Figure 4.12. Photographs of spermatozoa of the sea urchin *Psammechinus miliaris* taken at an exposure 1/500 sec. The distance from the head to the end of the tail is approximately 40 μm. From Gray (1955), with permission.

required by the theory, but we shall assume that the error resulting from this discrepancy is negligible.

Using equation (4.41), c and k are calculated to be

$$c = 828 \ \mu\text{m/sec},$$
$$k\eta = 1.05.$$

We see that $k\eta$ is not very small compared to unity, so that Taylor's simple principal result for u is not applicable. We calculate u instead from equation (4.57), and find that the theoretically predicted value of u is

$$u = 217 \ \mu\text{m/sec}.$$

This value is in tolerable agreement with the observed value of the velocity of 191 μm/sec; a more remarkable agreement of the theory with the experimentally observed results is achieved if the additional drag due to the head of the sperm cell is taken into account (see problem 11). It is expected on physical grounds that this additional drag reduces the propulsive velocity u.

The velocities of microorganisms propelled by flagellar plane waves as observed by many investigators have been compared to the predictions of the theoretical formula (4.57), with the abovementioned correction of the additional drag of the head, as needed. The agreement in general has been found to be good (Holwill, 1966).

Suppose that a force F acts on a body that is moving with velocity v in the same direction as the force. The work done per unit time by the force is defined as the **power** P and is given as $P = Fv$. For a flagellum, the power associated with its motion is equal to the energy dissipated by the flagellar motion against the viscous drag forces. By utilizing the above definition in conjunction with equations (4.44) and (4.45), we infer that the power dP associated with the forces dN and dL acting on the element ds is

$$dP = C_N(v \cos \theta - u \sin \theta)^2 \ ds + C_L(v \sin \theta + u \cos \theta)^2 \ ds. \quad (4.63)$$

Substituting equations (4.42), (4.43), and (4.48) into (4.63) and integrating over the length of the flagellum, there is obtained in a straightforward manner analogous to the derivation of equation (4.51), the result (Carlson, 1959)

$$P = \gamma\{[C_N(u^2 + c^2) + 2(C_L - C_N)uc]I + C_L u^2 J + C_L c^2 M\}, \quad (4.64)$$

where I and J are defined by equations (4.52), and M is defined as

$$M \equiv \int_0^{2\pi/k} \frac{(\partial y/\partial x)^4 \ dx}{[1 + (\partial y/\partial x)^2]^{1/2}} = \frac{1}{k} \int_0^{2\pi} \frac{(k\eta)^4 \cos^4 \xi \ d\xi}{[1 + (k\eta)^2 \cos^2 \xi]^{1/2}}. \quad (4.65)$$

If we approximate I and J by equations (4.56), then to the same degree of approximation,

$$M \approx \frac{3(k\eta)^4}{8[1 + \frac{1}{2}(k\eta)^2]^{1/2}} \frac{2\pi}{k}. \tag{4.66}$$

The above relation utilizes the integral evaluation $(1/2\pi) \int_0^{2\pi} \cos^4 \xi \, d\xi = 3/8$. Substituting (4.50), (4.54), (4.56), and (4.66) yields the result

$$P = bC_N c^2 \left[\frac{u^2}{c^2} + \left(\frac{u^2}{c^2} + 1 - \frac{u}{c} \right)(k\eta)^2 + \frac{3}{8}(k\eta)^4 \right] \left[1 + \frac{1}{2}(k\eta)^2 \right]^{-1/2} \tag{4.67}$$

When we insert for u/c its value according to (4.57), we find that the power expended by the transverse wave propagating down the flagellum of length $2b$ is given as

$$P = \frac{bC_N c^2 (k\eta)^2 [1 + 9(k\eta)^2/8 + 3(k\eta)^4/8]}{[1 + (k\eta)^2][1 + (k\eta)^2/2]^{1/2}}. \tag{4.68}$$

If $k\eta \ll 1$, it is readily seen that (4.68) reduces to

$$P = bC_N c^2 (k\eta)^2. \tag{4.69}$$

Using these formulas, the energy dissipated per unit time by a single flagellum is estimated to be of the order of 10^{-7} erg/sec. This power must correlate with the biochemical power generated internally by the cell. The latter has been inferred to be about 10^{-6} erg/sec from metabolic data for the sperm of the sea urchin *Echinus esculentus* (Rothschild and Clelland, 1952), which have dimensions and motile properties similar to the sperm of *P. miliaris*. Thus, the right order of magnitude of the power is available from biochemical processes to account for the calculated dissipation.

We shall briefly discuss the drag coefficients C_N and C_L, which are utilized in the expressions for F and P. We noted already that for the determination of the propulsive velocity u, only the ratio C_N/C_L was needed. For a circular cylinder of length $2b$ and diameter $2a$, it has been shown (Cox, 1970) that for slow steady motion of the cylinder through a viscous fluid, under the assumption that the cylinder is long compared to its diameter so that $b \gg a$,

$$C_L = \frac{2\pi\mu}{\log(4b/a) - 3/2}, \tag{4.70a}$$

$$C_N = \frac{4\pi\mu}{\log(4b/a) - 1/2}. \tag{4.70b}$$

These equations also indicate that equation (4.54) is asymptotically correct when $b \gg a$. In fact, equation (4.54) is a well known result in the fluid dynamic theory of the steady motion of slender bodies. However, these results are not directly applicable to a cylinder undergoing wave motion.

The maintenance of the wave motion of a flagellum requires either a continual production of tensions and compressions in the interior of the flagellum along its length, or an appropriate oscillatory force produced in the cell at or near the point of attachment of the flagellum, which waves it. It is generally believed that an active contractile mechanism exists along the entire length of the flagellum of an eucaryotic cell, which is responsible for its motion. The basis of this belief is as follows.

If the flagellum is assumed to be a passive elastic filament which is being waved sinusoidally at the proximal end where it is attached to the cell, then an analysis of the equations of motion of the filament indicates that the amplitude of oscillation would decrease exponentially as the wave travels down the filament towards the distal end, due to the viscous drag of the surrounding fluid (Machin, 1958). In fact, the normal wave motion of sperm flagella exhibits no decrement but rather an increment in amplitude along its length. This is illustrated most strikingly in Figure 4.13 which shows a multiple exposure of a flagellum of a sperm cell from which the head has been removed. The figure also illustrates the fact that flagella waving in planar fashion do not move in a straight line but rather move in a circle. In relation to such motion, the forward velocity u of equations (4.44) et seq. refers to the tangential velocity at a point on the circular trajectory.

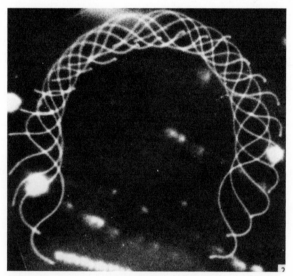

Figure 4.13.   Multiple exposures of a flagellum of a spermatozoon of the sea urchin *Lytechinus pictus* from which the head has been removed. The flagellum is swimming in a clockwise direction. The film was exposed at a frequency that is slightly slower than the frequency of oscillation of an element of the flagellum. From Brokaw (1970), with permission.

Further evidence for the attribution of the cause of flagella wave motion in eucaryotic cells to local contractile elements distributed along its length is the existence of a "**nine-plus-two**" structure in motile flagella and cilia. When a cross section of a flagellum is examined in the electron microscope, nine pairs of fibrils are found distributed in a regular array within an outer ring like the numbers on the face of a clock. In the center, two single fibrils are also seen. In addition, ATP is found in flagella. ATP is known to be a source of energy for both the contraction and lengthening of structures in both muscle and non-muscle cells. It is the major carrier of chemical energy in all living cells. The addition of ATP and Mg^{2+} to suspensions of isolated cilia causes them to undergo wavelike oscillations. Thus, the source of energy of flagellar movement is generally attributed to the **hydrolysis** of ATP. This is the degradative reaction of ATP with water to form adenosine diphosphate (ADP), an energy poor counterpart of ATP, plus inorganic phosphate. The actual mechanism of conversion of chemical energy to mechanical energy and the structural changes that take place internal to the flagellum that account for its motion are not known, and are at present matters of investigation and speculation.

In contrast to the flagella of eucaryotic cells, the flagella of procaryotic cells such as bacteria do not exhibit a nine-plus-two structure, but rather consist of a single helical fibril that protrudes through the bacterial membrane. The motion of bacterial flagella is passive. The bacterial flagella have no intrinsic mechanochemical activity. Rather, their motion is imparted to them by **basal granules** in the bacterial cell, to which flagella are attached. Basal granules do exhibit ATP activity. Thus, the motion of a flagellum in a procaryotic cell is most likely that of a passive elastic filament being waved at its proximal end. Flagellated bacteria are classified mainly into two types. Those having one or two flagella which originate from the pole or poles of the cell are said to possess **polar flagellation**. Those having numerous flagella along their sides are said to have **peritrichous flagellation**. Extreme examples of the latter type are *Proteus vulgaris* and *Proteus mirabilis*, which may possess over one thousand flagella (see Figure 4.14). Such bacteria grow and synthesize new flagella continuously.

SELECTED REFERENCE BOOKS

A. C. Burton (1965), *Physiology and Biophysics of the Circulation*, Year Book Medical Publishers, Chicago.

D. A. McDonald (1974), *Blood Flow in Arteries*, 2nd ed., Williams and Wilkins, Baltimore.

W. F. Hamilton and P. Dow, eds. (1962), *Handbook of Physiology*, Vol. I, Circulation, Amer. Physiol. Soc., Washington, D.C.

R. D. Allen and N. Kamiya (1964), *Primitive Motile Systems in Cell Biology*, Academic, New York.

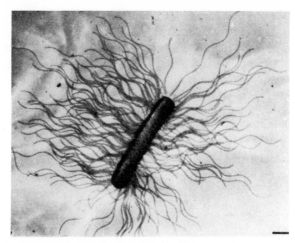

Figure 4.14.   An electron micrograph of a relatively short bacterium of *Proteus mirabilis*, possessing 176 flagella. The black line in the lower right hand corner represents a scale length of 1 μm. Reprinted, with permission from J. F. M. Hoeniger, "Development of flagella by *Proteus mirabilis*," 1965, *Journal of General Microbiology 40*, 42) Cambridge University Press.

M. A. Sleigh (1962), *The Biology of Cilia and Flagella*, Pergamon, Oxford.

J. Gray (1968), *Animal Locomotion*, Weidenfeld and Nicolson, London.

PROBLEMS

1. In a normal resting man, some circulatory parameters are given as follows (Folkow and Neil, 1971). The mean cardiac output is about 5.5 liter/min, and the mean radius of the thoracic aorta is about 1.1 cm. In the **systemic** circulatory system, the part of the entire circulatory system that is fed by the left ventricle, the mean radius of a capillary is about 3 μm. In the **pulmonary** circulatory system that is fed by the right ventricle, the mean capillary radius is about 4 μm. The mean pressure at the arterial end of the capillary bed is estimated to be about 30 mm Hg, and about 15 mm Hg at the venous end. The length of a capillary varies from 0.1 to 1 mm, with a "typical" length of about 0.75 mm. Calculate the mean blood flow velocity in the aorta and in a typical capillary of the systemic circulatory system. It appears that only 25 to 35% of the capillaries are open in resting conditions, while the remainder are collapsed or closed. Estimate the total number of capillaries in the body.

2. (Adapted from Murray, 1926) The suggestion has been made (see, for example, Thompson, 1942, pp. 948–957) that during embryogenesis, the angle of branching of a small artery is determined according to the

principle of minimization of energy dissipation. For steady flow in a tube, if the loss in kinetic energy of the blood in going from the inlet to the outlet is neglected, then the dissipated power P, according to the law of conservation of energy, is equal to the work done by the pressure forces at the inlet and outlet ends. Thus $P = Q \, \Delta p$, where Q is the flux through the artery and Δp is the pressure drop down its length. Assume a symmetric bifurcation occurs in an artery as shown in Figure 4.15, with the distance d fixed but the angle θ considered to

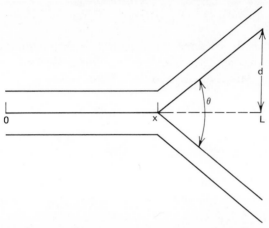

Figure 4.15.   Geometry of a hypothetical bifurcation in an artery. The direction of blood flow is from left to right.

be variable. Denote the flux Q and radius r in the upstream branch and downstream branches, respectively, by the subscripts 1 and 2. Write down the expressions for the conservation of flux and total power dissipated in the arterial segments shown in Figure 4.15. By minimizing the expression for the dissipated power, determine the origin x and angle θ of the bifurcation, for prescribed values of Q_1/Q_2 and r_1/r_2.

3. It has been advocated that velocity sedimentation in the earth's gravitational field be utilized as a means of fractionating mammalian cell populations which are heterogeneous in size. The size of a cell often correlates with its function. The widely used method of density gradient sedimentation equilibrium centrifugation (see Chapter 5) distinguishes cells by the criterion of cell density rather than size, and has the disadvantage that differentiated cells often have very similar densities. The suggested procedure has the advantage that cells of different sizes are readily distinguished.

The theory underlying the sedimentation of cells in the earth's gravitational field is provided by equation (4.18), suitably modified to take account of the asphericity of cells. This theory was applied to some observations of the sedimentation velocity of sheep erythrocytes (Miller and Phillips, 1969) in sucrose solution in a settling tube. The erythrocytes were considered to be oblate ellipsoids, with semi-major axis denoted by a, and semi-minor axis denoted by b. The Stokes' frictional coefficient for such a particle is given by the formula

$$f = f_0 \frac{(a^2/b^2 - 1)^{1/2}}{(a/b)^{2/3} \tan^{-1}(a^2/b^2 - 1)^{1/2}}, \qquad (4.71)$$

where f_0 is the frictional coefficient of a sphere with the same volume as the ellipsoid. The volume of an oblate ellipsoid is $(4\pi/3)a^2 b$. The following data characterized these observations.

Fluid	T (°C)	ρ (g^3/cm)	$\mu(P)$	v_S (mm/hr)
0.5–3% sucrose	4	1.009	1.62×10^{-2}	1.40
0.5–3% sucrose	22	1.007	0.99×10^{-2}	2.12

The mean major diameter $2a$ and volume of a sheep erythrocyte was found to be 4.8 μm and 31 μm^3, respectively. The mean erythrocyte density $\rho_c = 1.08$ g/cm^3.

(a) Show that the sedimentation velocity v_s of an aspherical cell can be expressed as

$$v_s = \frac{f_0}{f} \frac{2}{9\mu} (\rho_c - \rho) g r_0^2, \qquad (4.72)$$

where f is the frictional coefficient and ρ_c is the density of the cell, ρ is the fluid density, μ is the fluid viscosity, $f_0 = 6\pi\mu r_0$ is the Stokes' frictional coefficient of a reference sphere of radius r_0, and g is the gravitational constant.

(b) Find the theoretical values of v_s to compare with the observed values.

4. (Feldman, 1935) The cardiac apex, where the curvature is greatest and the radius of curvature is least, is the thinnest part of the heart. Similarly, the thickness of muscle in the uterus is greater at the fundus, where the radius of curvature is greater, than at the cervix, where it is less. Such observations have led to the suggestion that the thickness of a blood vessel has developed in proportion to the tension to which it is subjected. In the carotid artery of the sheep, the mean pressure is

40 mm Hg, and the diameter is 3 mm. In the carotid artery of the ox, the mean pressure is 60 mm Hg, and the diameter is 6 mm. The mean thickness of the carotid artery of the sheep is 0.616 cm.

(a) What is the expected value of the thickness of the carotid artery of the ox, according to the above suggestion?

(b) Assume that the transmural pressure difference in a human systemic capillary is 25 mm Hg. What is the tension in the capillary wall?

5. It was found that the tension–radius curve for a section of the iliac artery shown in Figure 4.6 could be fitted by the polynomial

$$T(r) = th = t_1 \left(\frac{r}{r_0} - 1 \right) + t_2 \left(\frac{r}{r_0} - 1 \right)^5, \qquad (4.73)$$

where $r_0 = 0.216$ cm, and according to the method of least squares, $t_1 = 13$ g/cm, $t_2 = 300$ g/cm. Transform the integral in equation (4.26) into a form in which the integration can be performed explicitly.

6. It is reasonable to suppose that the mechanism of ovulation by which the ovarian follicle ruptures and releases the ovum can be understood in mechanical terms (Rodbard, 1968). Assume the ovarian follicle is a sphere of liquid at pressure p surrounded by a thin elastic wall that obeys Hooke's law. Generally, for materials which obey Hooke's law, there is a maximum stress corresponding to a maximum rate of strain, beyond which the material will not restore itself following the relaxation of stress. This maximum is known as the **elastic limit**, and is often associated with rupture of the elastic material. Assuming the ovarian follicle ruptures when the elastic limit of its wall is attained at a radius $r = r_{max}$, find the pressure in its interior at which this occurs. Can this pressure be exceeded without rupture occurring?

Hint. Consider the mechanical equilibrium of a hemisphere of the follicle. Assume that the elastic wall is incompressible, so that the thickness varies with radius in such a manner that the material of the wall always occupies the same volume.

7.* In the derivation of Young's equation presented in Section 4.5, suppose that the compressibility of blood is taken into account so that the density of blood ρ is variable. Then the equations of motion for tube flow (4.32)–(4.34) are supplemented by an equation of state relating the pressure p and density ρ. The equation of state in its linearized version, for which $\rho - \rho_0$ and $p - p_0$ are assumed to be small, takes the form

$$p - p_0 = c_0^2(\rho - \rho_0), \qquad (4.74)$$

where ρ_0 is the value of the density when $p = p_0$, and c_0 is the velocity of sound in blood. Show that p still satisfies the wave equation and find the generalization of Young's formula for this case.

8. When the compressibility of the blood and the inertial motion of the arterial wall are taken into account in a theoretical manner, it is found that the velocity of propagation of a sinusoidal wave which is periodic in space and time depends on the frequency of the wave, as well as the properties of the arterial wall and the blood. For small values of the frequency, the velocity is independent of the frequency, and is given by the expression (Rubinow and Keller, 1971)

$$
c = c_y \left\{ \frac{1}{2} + \frac{m}{4}\left(1 + \frac{1}{K_0^2}\right) + \left(\left[\frac{1}{2} + \frac{m}{4}\left(1 + \frac{1}{K_0^2}\right)\right]^2 \right. \right.
$$
$$
\left. \left. - \frac{m}{4}\left[2(1 - \sigma^2) + \frac{m}{K_0^2}\right]\right)^{1/2} \right\}^{-1/2} .
$$
(4.75)

Here c_y is Young's velocity formula, equation (4.37), m and K_0 and are nondimensional parameters defined as $m = \rho_1 h / \rho r_0$, $K_0 = c_0(E/\rho_1)^{-1/2}$, ρ_1 is the mass density of the arterial wall, h is the wall thickness, and c_0 is the velocity of sound in blood. The parameter σ is called **Poisson's ratio** and is the ratio of the transverse contraction (change in length per unit length of a cross-sectional area element) to the longitudinal extension of a volume element of the elastic matter composing the wall. On theoretical grounds it can be shown that, as long as the elastic material contracts laterally when it is stretched, σ has a value between 0 and 1/2. The two parameters σ and E characterize homogeneous elastic materials. For arteries, $\sigma \approx 1/2$. Derive the correction to Young's formula when m is small by expanding the function c_y^2/c^2 in a Taylor series about $m = 0$, up to order m.

9.* Suppose we are given a pressure–radius curve of a thin tube obtained from measurements of the tube radius r following inflation of the tube at the pressure p. A tangent modulus E can be defined for such a curve by an extension of equation (4.23) to account for the transverse and longitudinal contraction of elastic elements. It is defined by

$$
E = \frac{(1 - \sigma^2)r^3}{h_0 r_0} \frac{dp}{dr},
$$
(4.76)

where h_0 and r_0 are the thickness and radius of the tube, respectively, when the pressure in the tube equals the external pressure, and the tube material is assumed to be incompressible.

(a) Figure 4.16 shows the tangent modulus defined essentially in this

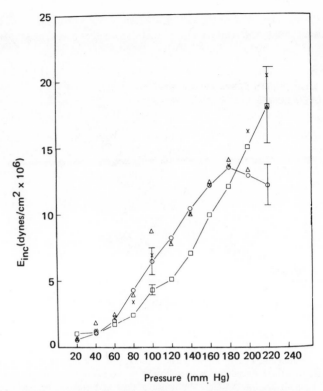

Figure 4.16. The tangent (or incremental) modulus E_{inc}, based essentially on the formula (4.76) and determined by static pressure–radius measurements of an inflated section of an artery, is shown as a function of the inflating pressure p in excess of the exterior pressure p_0. Four types of artery were investigated: □, thoracic aorta; △, abdominal aorta; × femoral artery; ○, carotid artery. From Bergel (1961), with permission.

manner as a function of the pressure p (in excess of the external pressure p_0) for the thoracic and abdominal aortas, and femoral and carotid arteries of the dog (Bergel, 1961). Assuming that an explicit functional representation of the curve for the thoracic aorta was available, say

$$E = E_0 + Ap^2, \qquad (4.77)$$

where E_0 and A are determined empirically, show how the pressure–radius curve could be reconstituted from it. Find h/r when $p = p_0$, knowing that the thickness–radius ratio of the tube $h/r = 0.105$ when $p = 100$ mm Hg.

(b)   Figure 4.17 shows the wave velocity c as a function of pressure

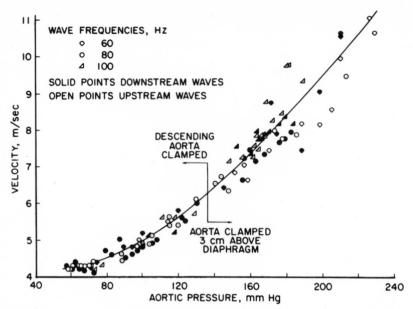

Figure 4.17.   The wave velocity of externally induced sinusoidal waves as a function of pressure in an occluded section of the thoracic aorta of the dog. Internal aortic pressures lower than normal were obtained by occlusion of the aorta upstream from the test section, and elevated pressures were obtained by occlusion of the aorta downstream from the test section. The temporal frequency of the induced waves is indicated in the figure (Hz = cycle/sec). Reprinted from M.B . Histand and M. Anliker, "Influence of flow and pressure on wave propagation in the canine aorta," 1973, *Circulation Research 32*, 524, by permission of the American Heart Association, Inc.

(in excess of the external pressure) of relatively high frequency waves excited artificially in the thoracic aorta of the dog (Histand and Anliker, 1973). Is this curve semi-quantitatively compatible with the curve shown in Figure 4.16 for the thoracic aorta, if Young's formula is assumed to be valid for these measurements? Assume, by a crude extrapolation of the curve in Figure 4.17 that $c = 3.8$ m/sec when the excess pressure is 20 mm Hg. Indicate how you could use the information provided in Figure 4.16 to predict a theoretical curve for the wave velocity as a function of pressure.

10.*   A sinusoidal pulse wave traveling through artery 1 of radius r_1 is incident on the junction of artery 1 with artery 2 of radius r_2. Reflected and transmitted waves are established in arteries 1 and 2, respectively. Assume that the blood is inviscid, and that the pressure associated with the incident, reflected, and transmitted waves are represented, respectively, as p_i, p_r, and p_t, where

$$p_i = A \sin \omega \left(t - \frac{x}{c_1} \right),$$

$$p_r = B \sin \omega \left(t + \frac{x}{c_1} \right), \qquad (4.78)$$

$$p_t = C \sin \omega \left(t - \frac{x}{c_2} \right),$$

A, B, and C are amplitude constants, and c_1 and c_2 are the pulse wave velocities in arteries 1 and 2, respectively. Let the junction be at $x = 0$, so that p_i and p_r are defined for $x < 0$ and p_t is defined for $x > 0$.
(a) Find the **reflection coefficient** R and **transmission coefficient** T for the wave at the junction (Karreman, 1952) in terms of the single parameter $\xi = r_2^2 c_1 / r_1^2 c_2$, where R and T are defined as

$$R = \frac{B}{A}, \quad T = \frac{C}{A}. \qquad (4.79)$$

(b) Find R and T for a junction as shown in Figure 4.15 for which the downstream side of the junction consists of two arterial branches.

Hint. Assume the pressure and flux are continuous at the junction.

11. (Gray and Hancock, 1955) Assume a spermatozoon has a spherical head which contributes a drag force to its steady motion, determined by Stokes' law. By including this force in the total force on a spermatozoon being propelled by planar wave motion of its tail, determine the theoretical expression for the velocity of propulsion, to replace equation (4.57). Apply the theory to the observed data concerning the motion of the spermatozoa of *Psammechinus miliaris* quoted in equations (4.62).

Hint. Utilize equation (4.70) to evaluate C_L.

12.* The complete elliptic integrals of the first and second kind, $K(m)$ and $E(m)$, respectively, are defined as follows:

$$K(m) \equiv \int_0^{\pi/2} \frac{d\xi}{[1 - m \sin^2 \xi]^{1/2}} = \frac{1}{4} \int_0^{2\pi} \frac{d\xi}{[1 - m \sin^2 \xi]^{1/2}},$$

$$E(m) \equiv \int_0^{\pi/2} [1 - m \sin^2 \xi]^{1/2} \, d\xi = \frac{1}{4} \int_0^{2\pi} [1 - m \sin^2 \xi]^{1/2} \, d\xi. \qquad (4.80)$$

Show that the integrals I and J of equation (4.52) can be expressed exactly in terms of them. Find the resulting modification of the expression for the velocity of a uniflagellated organism being propelled by a planar wave traversing its flagellum.

5

DIFFUSION IN BIOLOGY

A solution consists of a fluid, called the **solvent,** in which some matter has been dissolved, the **solute.** The composition of the solution is characterized by its mass concentration c, which is the mass of dissolved matter per unit volume of liquid. The water content of cells is approximately 80%, and water is the universal biological solvent.

One means by which the solute molecules are interspersed and transported through the solvent is **diffusion.** This mechanism is a consequence of the thermal motion of the individual solute molecules. The continuous motion of the solvent molecules produces a great many collisions with, say, a given large solute molecule. As a result, pressure fluctuations are produced which in turn impart to the solute molecule a jerky irregular path. This irregular path is called in the mathematical theory of probability a **random walk.** The result of this random walk is a net displacement of the molecule in some direction. The same phenomenon occurs in the case of suspended particles such as pollen in water. They are observed in the microscope to undergo dispersive motion which is called **Brownian motion** in honor of its discovery by the English botanist Robert Brown (1828).

We know that in the case of the classical chemical experiment in which a few copper crystals are placed in the bottom of a tall jar of water and allowed to stand undisturbed, eventually a uniform distribution of copper ions is produced, as is shown by the color imparted to the water by the copper. Because the uniform state may take many years to be produced, we commonly think of diffusion as being a slow process of transport. However, over the dimensions of a single cell or less, diffusion is a rapid and vital mechanism of transport.

5.1. FICK'S LAWS OF DIFFUSION

Consider a solution in which simple molecular diffusion is occurring, the fluid being otherwise at rest. The mechanism of transport of the solute is

196

governed only by concentration differences. We ask, what is the flux of solute particles going through a unit area in a unit amount of time? The material flux per unit area is an example of a **current density**, which is denoted quite generally by the letter j. For diffusion, this current density was given by Fick (1855) as

$$j = -D \frac{dc}{dx}. \tag{5.1}$$

Here we imagine that the concentration is varying from point to point and depends on position x only. This relation is called Fick's **first law of diffusion**. The coefficient D is called the **diffusion coefficient** and is a characteristic of the solute in the fluid. We see that it has the dimensions square centimeters per second. The significance of the minus sign in (5.1) is simply that the particle flow proceeds from a high concentration region to a low concentration region. Thus, if $c(x)$ is an increasing function of x, dc/dx is positive, and the material flux will be in the negative x direction.

More generally, the concentration c is a function of position of the point $P = (x,y,z)$ and the time t; hence we write $c = c(x,y,z,t)$. An infinitesimal section of area through P will have a mass current flowing through it with components in the three space directions. Thus, the current density is more properly denoted by the vector $\mathbf{j} = (j_x, j_y, j_z)$ where

$$j_x = -D \frac{\partial c}{\partial x}, j_y = -D \frac{\partial c}{\partial y}, j_z = -D \frac{\partial c}{\partial z}. \tag{5.2}$$

In expressing $\mathbf{j}$ in this manner, we have tacitly assumed that the solvent medium is **isotropic**: The diffusion properties in the neighborhood of any point are the same in all directions. In principle, D could be a function of position as well as of c. When D is neither, it is usually called the **diffusion constant**.

Consider a small cubical element of volume $dx\,dy\,dz$ surrounding the point P. The equation representing conservation of solute transport into and out of this volume element is the equation of continuity

$$\frac{\partial c}{\partial t} + \frac{\partial j_x}{\partial x} + \frac{\partial j_y}{\partial y} + \frac{\partial j_z}{\partial z} = 0. \tag{5.3}$$

We emphasize that the net diffusional flux of solute across the surface bounding the volume element, which is represented by the last three terms on the left-hand side of this equation, is separate from the flux due to bodily or bulk motion of the fluid as a whole. The latter motion is still represented by equation (4.2). Substituting equation (5.2) into (5.3) yields the result

$$\frac{\partial c}{\partial t} = \frac{\partial}{\partial x}\left(D\frac{\partial c}{\partial x}\right) + \frac{\partial}{\partial y}\left(D\frac{\partial c}{\partial y}\right) + \frac{\partial}{\partial z}\left(D\frac{\partial c}{\partial z}\right). \tag{5.4}$$

If D is constant, equation (5.4) simplifies to

$$\frac{\partial c}{\partial t} = D\,\Delta c, \tag{5.5}$$

$$\Delta \equiv \frac{\partial^2}{\partial x^2} + \frac{\partial^2}{\partial y^2} + \frac{\partial^2}{\partial z^2}. \tag{5.6}$$

If the concentration has a nonzero gradient only along the x axis, (5.5) reduces to

$$\frac{\partial c}{\partial t} = D\,\frac{\partial^2 c}{\partial x^2}, \tag{5.7}$$

called Fick's **second law of diffusion**. It is also called the **one-dimensional diffusion equation**. Its more general three-dimensional form (5.5) is usually referred to simply as the **diffusion equation**. The **differential operator** Δ that appears in it, and that is defined by equation (5.6), is called the **Laplace operator** or the **laplacian**. The three-dimensional generalization of the one-dimensional wave equation that we encountered in the previous chapter, equation (4.36), also contains the laplacian in place of the second space derivative.

The diffusion equation has exactly the same mathematical form as the **equation of heat conduction**. In the latter case, c is replaced by the **temperature** $T = T(x,y,z,t)$, and D is replaced by K, the **heat conductivity** of the medium. The difference between the diffusion equation and the wave equation that we encountered in Section 4.5 is that a first time derivative appears in the diffusion equation in contrast to the second time derivative appearing in the wave equation. This difference is profound and is a reflection of the fact that processes like diffusion and heat conduction are **irreversible** and dissipative in the thermodynamic sense: Entropy increases and energy is lost. Wave motion, on the other hand, is reversible and **conservative** (no energy is lost). Of course in real media, the energy associated with waves such as sound is ultimately dissipated into heat energy because of heat conduction and viscosity in the medium. Mathematically, such dissipative waves are represented by a more general wave equation in which both first and second time derivatives appear.

5.2. THE FICK PRINCIPLE

The equation of continuity, equation (5.3), quite generally is an expression of the **law of conservation of mass**. It states that, within a given infinitesimal volume element of a solution in which solute currents exist, whatever the cause of the currents may be, the rate at which matter accumulates or disappears within that region is equal to the net influx or efflux, respectively, across the surface bounding the infinitesimal region. This conservation law can be extended to a finite region. Thus, imagine an arbitrary region of a solution characterized at each point in the region by a concentration c and a current density $\mathbf{j}$, each of which varies from point to point in the region. Together they satisfy equation (5.3), assuming there are no sources (where matter can be created) or sinks (where matter can disappear) in the region. Let us integrate this equation over the volume of the region. It can be shown quite rigorously that the volume integral of the last three terms of equation (5.3) is expressible as a surface integral of the normal component of $\mathbf{j}$ at the bounding surface of the volume of the region under consideration. This mathematical result is called **Gauss' theorem**. It should be intuitively understandable that summing all the currents over the interior of the volume yields zero, because the interior currents only transport material from one region of the interior to another. The mathematical expression of this result is (see Figure 5.1)

$$\frac{\partial}{\partial t} \iiint_V c \, d\tau = - \iint_S j_n \, d\sigma. \tag{5.8}$$

Here $d\tau$ is a volume element ($= dx \, dy \, dz$ in cartesian coordinates), $d\sigma$ is a surface element in the surface S bounding the volume V, and j_n is the normal component of the vector $\mathbf{j}$, taken to be positive when the vector $\mathbf{j}$ points

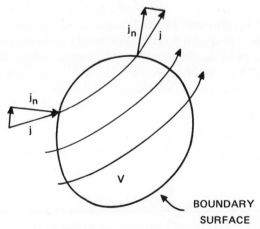

Figure 5.1. Illustration for the integration of the equation of continuity over a volume V with a boundary surface S. The normal current density component j_n is negative at the surface element on the left, and positive at the surface element on the right.

outward from the volume, and negative when it points inward. The triple integral sign on the left means that the integration takes place over the three space variables of the volume V, and the double integral sign on the right means that the integration takes place over two space variables describing the surface S. The quantity $\iiint_V c \, d\tau$ is just the mass m of solute contained within the volume V at time t. The integral on the right-hand side of (5.8) is the mass flux across the surface S at time t. The positive contribution of the right-hand side of (5.8) is just the rate at which mass flows into the volume V from the outside. Denoting the mass influx and efflux rates or currents by J_{in} and J_{out}, respectively, we may express equation (5.8) as

$$\frac{dm}{dt} = J_{in} - J_{out}. \qquad (5.9)$$

This equation is sometimes called the **Fick principle**. Usually, the volume V under consideration is a particular compartment of a biological system.

As an example of the application of the Fick principle, we shall consider the determination of cardiac output from oxygen consumption by the body. We recall that the cardiac output is defined as the steady rate at which blood is ejected by the heart. Of course, the blood flow and blood ejection from the heart is pulsatile and more or less periodic, over time spans which are large compared to the period or time interval between heart beats. Thus, in determining cardiac output, the time average of the flux from the heart over one or more periods is an implicit requisite. The cardiac output is an important medical parameter in determining the state of health of the heart.

Let the volume or compartment under consideration be the entire body, and let the solute mass in the given volume be the amount of oxygen in the body. Oxygen is consumed by the body at a steady rate and converted to carbon dioxide. The quantity dm/dt represents the steady oxygen consumption rate. Denote by K the steady cardiac output, or the mean volume of blood flowing from the heart into the aorta per unit time. Thus, K is the blood flow rate into the compartment as well as the blood flow rate leaving the compartment, or the mean rate of blood flow from the vena cava into the heart. Let c_a be the oxygen concentration per unit volume in the blood of the aorta or arterial side of the circulatory system, and let c_v be the oxygen concentration in the blood of the vena cava or venous side of the circulatory system. Then $J_{in} = Kc_a$, $J_{out} = Kc_v$ and (5.9) becomes

$$\frac{dm}{dt} = K(c_a - c_v). \qquad (5.10)$$

The oxygen consumption rate dm/dt, as well as c_a and c_v, can be determined experimentally, for example with the aid of tissue slices, so that equation (5.10) can be utilized to determine the cardiac output K.

As a second application, we consider the **nitrous oxide method** for determination of cerebral blood flow (Kety and Schmidt, 1948). In this method, an inert gas such as nitrous oxide at constant concentration is inhaled, starting at $t = 0$. The concentration of the gas in the arterial blood and in the venous blood leaving a given tissue is measured at frequent intervals. When the compartment or tissue under consideration is the brain, the concentrations of nitrous oxide in a convenient artery and in an internal (left or right) jugular vein are determined. The blood in one of the latter veins is assumed to represent venous outflow from the brain only. The typical observations of arterial nitrous oxide concentration c_a and internal jugular vein nitrous oxide concentrations c_v are illustrated in Figure 5.2.

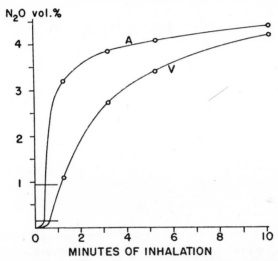

Figure 5.2. Typical curves showing the nitrous oxide (N_2O) volume concentration in arterial blood (labeled A) and in an internal jugular vein (labeled V) as a function of time. At $t = 0$, the inhalation of air containing 15% nitrous oxide was initiated. In measuring the nitrous oxide concentration, the mass of N_2O contained in a unit volume of blood was expressed as an equivalent volume of gas. From Kety and Schmidt (1948), with permission.

Let m now denote the amount of nitrous oxide contained in the cerebral blood at time t, and Q denote the steady rate of blood flow entering or leaving the brain. Then equation (5.10) applies with K replaced by Q, and we may integrate between the limits 0 and t to obtain the result

$$m(t) - m(0) = Q \int_0^t [c_a(t') - c_v(t')]dt'. \qquad (5.11)$$

The integral on the right represents the total area between the two curves labeled A and V in Figure 5.2, and represents the arteriovenous nitrous oxide concentration difference between time zero and time t.

When $t = \infty$, equilibrium is established and c_a and c_v both attain the constant asymptotic value c_0. This is the concentration of nitrous oxide in cerebral blood at equilibrium. If the nitrous oxide enters the cerebral tissue from the blood by diffusion, then c_0 is also the equilibrium concentration in the tissue. Thus,

$$m(\infty) = V_B c_0, \tag{5.12}$$

where V_B is the volume of the brain, the space occupied by the cells and the blood vessels. Recognizing that $m(0) = 0$, we find that in the limit as t approaches infinity, (5.11) becomes

$$\frac{Q}{V_B} = \frac{c_0}{\int_0^\infty [c_a(t') - c_v(t')]dt'}. \tag{5.13}$$

The expression on the right-hand side is determinable from the observations. Equation (5.13) shows that it is equal to the rate of blood flow to the brain, per unit volume of brain. The latter quantity is the principal inference to be made from the nitrous oxide method.

It is more usual to express the blood flow to the brain as blood flow per unit mass of the brain. Thus, if we multiply equation (5.13) by V_B/M_B, where M_B is the mass of the brain, it becomes

$$\frac{Q}{M_B} = \frac{m_0}{\int_0^\infty [c_a(t') - c_v(t')]dt'}, \tag{5.14}$$

where $m_0 = c_0 V_B/M_B$ is the total mass of nitrous oxide taken up by the brain at equilibrium, per unit mass of the brain. In the cited work, it was found on the basis of 34 observations in 14 healthy young men that the mean cerebral blood flow, expressed in units of volume flux per 100 grams of brain mass, was 54 ml/min/100 g.

5.3.   THE UNIT ONE–DIMENSIONAL SOURCE SOLUTION

We shall now discuss a solution of the one-dimensional diffusion equation, (5.7), which is of fundamental importance. The solution is

$$c = \frac{m}{(4\pi Dt)^{1/2}} e^{-x^2/4Dt}, \tag{5.15}$$

where m is a constant. It is easy to verify that (5.15) is a solution of the one-dimensional diffusion equation by substituting it into (5.7) and carrying out the indicated differentiations. What is the physio-chemical meaning of this solution? More mathematically, what are the boundary conditions and initial condition that it satisfies? To answer these questions, we observe that the total amount of solute diffusing in the medium is given by the expression

$$\int_{-\infty}^{\infty} c \, dx = \frac{m}{(4\pi Dt)^{1/2}} \int_{-\infty}^{\infty} e^{-x^2/4Dt} \, dx = m. \tag{5.16}$$

To obtain the last equality above, we have set $x^2/4Dt = \xi^2$ in the preceding integral and recognized that

$$\int_{-\infty}^{\infty} e^{-\alpha x^2} \, dx = \frac{1}{\alpha^{1/2}} \int_{-\infty}^{\infty} e^{-\xi^2} \, d\xi = \frac{2}{\alpha^{1/2}} \int_{0}^{\infty} e^{-\xi^2} \, d\xi = \left(\frac{\pi}{\alpha}\right)^{1/2}. \tag{5.17}$$

Thus, m represents the total amount of solute. The function c/m is plotted as a function of x for various values of Dt, in Figure 5.3. The curves can also

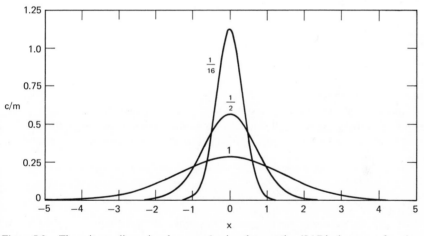

Figure 5.3. The unit one-dimensional source c/m given by equation (5.15) is shown as a function of x for various values of Dt. From J. Crank (1956), *The Mathematics of Diffusion*, The Clarendon Press, Oxford, with permission of the author and publisher.

be interpreted as representing c/m as a function of $x/\sqrt{D}$ for various values of t.

From the figure we note that as $t \to 0$, the function gets narrower and more peaked, although the area under it, equal to unity, remains the same. In the limit, it is a "function" which is infinite at $x = 0$ and zero for $x \neq 0$. This function, defined by a limiting process, is called the **Dirac delta function**, and represented symbolically as $\delta(x)$. In other words, the delta function is defined by the properties

$$\delta(x) = \begin{cases} \infty & x = 0, \\ 0 & x \neq 0, \end{cases} \tag{5.18a}$$

$$\int_{-\infty}^{\infty} \delta(x)dx = 1. \tag{5.18b}$$

It follows from the last equation that the delta function has the property

$$\int_{-\infty}^{\infty} f(x)\delta(x)dx = f(0), \tag{5.19}$$

where $f(x)$ is arbitrary. The delta function has many explicit representations, one of which is

$$\delta(x) = \lim_{t \to 0} \frac{1}{(4\pi Dt)^{1/2}} e^{-x^2/4Dt}. \tag{5.20}$$

Therefore, (5.15) describes the diffusion of a substance which initially has the distribution

$$c(x,0) = m\delta(x). \tag{5.21}$$

In other words, it represents the diffusion of a thin slab of solute placed at $x = 0$ in a medium of large (infinite) expanse. Because the solute mass is finite and the medium is of infinite extent, the concentration must satisfy the boundary conditions $c(\pm\infty,t) = 0$, as otherwise the solute mass would be infinite. We see that the solution (5.15) does have this property. This solution with $m = 1$ is called the **unit one-dimensional source solution**.

When the solute mass is placed initially at $x = \xi$, the unit one-dimensional source solution becomes

$$c(x,t) = \frac{m}{(4\pi Dt)^{1/2}} e^{-(x-\xi)^2/4Dt}. \tag{5.22}$$

With this fundamental solution, we can solve the problem of the diffusion of an arbitrary distribution of solute in space at $t = 0$. Thus, if $A(x)$ is the initial solute concentration distribution at $t = 0$ so that

$$c(x,0) = A(x) = \int_{-\infty}^{\infty} A(\xi)\delta(x - \xi)d\xi, \tag{5.23}$$

then the concentration distribution at all subsequent times is given as

$$c(x,t) = \frac{1}{(4\pi Dt)^{1/2}} \int_{-\infty}^{\infty} A(\xi)e^{-(x-\xi)^2/4Dt} \, d\xi. \tag{5.24}$$

It can be verified that $c(x,t)$ above vanishes as $|x| \to \infty$, and satisfies the one-dimensional diffusion equation (5.7) and the initial condition (5.23).

The right-hand side of (5.24) represents a summation of sources each of strength $A(\xi)d\xi$. The function $A(\xi)$ is called the **source strength density function**, or **source density**, or **source distribution function**.

5.4. THE DIFFUSION CONSTANT

Another purely symbolic representation of the Dirac delta function, which is nevertheless intuitively easy to understand, is

$$\delta(x) = \frac{d}{dx} h(x). \tag{5.25}$$

Here $h(x)$ is the **Heaviside unit step function**, defined as

$$h(x) = \begin{cases} 0, & x < 0, \\ 1, & x > 0. \end{cases} \tag{5.26}$$

$h(x)$ is discontinuous at $x = 0$. Clearly, $dh(x)/dx = 0$ for $x < 0$ and $x > 0$, but at $x = 0$ the derivative is not actually defined. However, we may consider the derivative there in a limiting sense as the derivative of a continuous function of x and a parameter, whose slope at $x = 0$ grows steeper as the parameter changes, until the slope becomes infinite (see Figure 5.4).

In fact, the diffusion process permits a very good physical realization of these concepts. As an example, consider a long column of solution of uniform concentration on top of which is another long column of pure solvent at $t = 0$. This is a common experimental arrangement for studying diffusion, and for determining the diffusion constant D. We ask, what is the concentration as a function of position at all future times? We represent the system mathematically by the one-dimensional diffusion equation, subject to the initial condition

$$c(x, 0) = \begin{cases} 0, & x > 0, \\ c_0, & x < 0, \end{cases} \tag{5.27}$$

where c_0 is a constant. The boundary conditions are that $c(\infty,t) \to 0$, $c(-\infty,t) \to c_0$. We say that c has a **step discontinuity** of amount $-c_0$ in going through the origin from the negative to the positive side of the x axis. Therefore, in view of our previous comments about the delta function representation,

$$\frac{\partial c}{\partial x}(x,0) = -c_0\delta(x). \tag{5.28}$$

We shall now consider $\partial c/\partial x$ as an unknown function of x and t whose value is prescribed at $t = 0$. Furthermore, we require on physical grounds

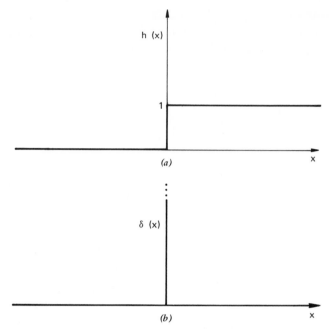

Figure 5.4. (*a*) The Heaviside unit step function is defined as $h(x) = 0$ for $x < 0$ and $h(x) = 1$ for $x > 0$. (*b*) The Dirac delta function $\delta(x)$, defined as $\delta(x) = 0$ for $x \neq 0$ and $\delta(x) = \infty$ for $x = 0$ can be considered in a limiting sense as the derivative of the Heaviside function.

that $\lim \partial c/\partial x$ as $x \to \pm \infty$ is zero. By differentiating the one-dimensional diffusion equation with respect to x, we see that $\partial c/\partial x$ also satisfies the diffusion equation. The latter equation for $\partial c/\partial x$, together with the initial condition (5.28), and the assumed boundary conditions, is a problem whose solution we already know from equation (5.15), namely,

$$\frac{\partial c}{\partial x}(x,t) = \frac{-c_0}{(4\pi Dt)^{1/2}} \, e^{-x^2/4Dt}. \tag{5.29}$$

In practice, $\partial c/\partial x$ or more conveniently $-\partial c/\partial x$, can be determined as a function of position and time by optical methods. From such an experimentally determined curve, we can measure the area under the curve, which equals the quantity c_0, as we have seen in our discussion of equation (5.16). The maximum height of the curve of $-\partial c/\partial x$ occurs at $x = 0$, and is denoted by $H = H(t)$. According to equation (5.29),

$$H(t) \equiv -\frac{\partial c}{\partial x}(0,t) = c_0(4\pi Dt)^{-1/2}, \tag{5.30a}$$

or

$$\left(\frac{c_0}{H}\right)^2 = 4\pi Dt. \tag{5.30b}$$

By measuring H as a function of the time, it is found that a plot of $(c_0/H)^2$ versus time yields a straight line from which D can be determined.

More accurate determinations of D can be made from the knowledge of $c(x,t)$. The latter is readily deduced from (5.29) by integration with respect to x, thus

$$c(x,t) = \int_\infty^x \frac{\partial c}{\partial \xi}(\xi,t)d\xi = -\frac{c_0}{(4\pi Dt)^{1/2}} \int_\infty^x e^{-\xi^2/4Dt}\, d\xi$$

$$= \frac{c_0}{\pi^{1/2}} \int_{x/(4Dt)^{1/2}}^\infty e^{-\eta^2}\, d\eta. \qquad (5.31)$$

In the first integral, the constant of integration has been determined by setting the lower limit of integration equal to $+\infty$ where c must be zero. To obtain the last integral, we set $\xi = (4Dt)^{1/2}\eta$ in the previous one. This solution is more commonly written as

$$c(x,t) = \frac{c_0}{2} \operatorname{erfc}(z), \qquad (5.32a)$$

$$z \equiv \frac{x}{(4Dt)^{1/2}}. \qquad (5.32b)$$

Here $\operatorname{erfc}(z)$ is the **error function complement**, defined as

$$\operatorname{erfc}(z) = 1 - \operatorname{erf}(z) = \frac{2}{\pi^{1/2}} \int_z^\infty e^{-\eta^2}\, d\eta, \qquad (5.33)$$

and $\operatorname{erf}(z)$ is the **error function** defined as

$$\operatorname{erf}(z) = \frac{2}{\pi^{1/2}} \int_0^z e^{-\eta^2}\, d\eta. \qquad (5.34)$$

It has the properties $\operatorname{erf}(-z) = -\operatorname{erf}(z)$, $\operatorname{erf}(0) = 0$, $\operatorname{erf}(\infty) = 1$. Both $\operatorname{erf}(z)$ and $\operatorname{erfc}(z)$ are tabulated functions. In Figure 5.5 is found a plot of the quantity $c/c_0 = (1/2)\operatorname{erfc}(z)$ as a function of z. Note that $c = c_0/2$ at $x = 0$ for all $t > 0$.

According to equation (5.32), the position x and time t appear in the function c exclusively in the combination $z = x/(4Dt)^{1/2}$. One consequence of this result is that c is constant when z is constant, or when x is proportional to $t^{1/2}$. That is to say, the distance traveled by a solute layer of fixed concentration is proportional to $t^{1/2}$, a relationship that is readily verified by observation.

What is the meaning of the diffusion constant from a molecular point of view? To provide some insight into the answer to this question, let us suppose a thin slab of solute mass is placed at $x = 0$ at time t, so that its concentration at subsequent times is represented by the one-dimensional source solution (5.15). The **mean square displacement** $\overline{x^2}$ of a diffusing solute

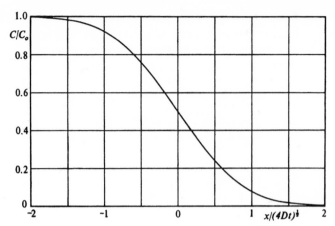

Figure 5.5. The function $c(x, t)/c_0$ given by equation (5.32a) is shown as a function of $z = x/(4Dt)^{1/2}$. The ordinate represents the concentration resulting from a thin slab of solute of unit amount placed at $x = 0$ at the time zero, called the unit one-dimensional source. The curve can be viewed as a representation of the concentration distribution as a function of distance, at a fixed time, with unit distance equal to $(4Dt)^{1/2}$. From J. Crank (1956), *The Mathematics of Diffusion*, The Clarendon Press, Oxford, with permission of the author and publisher.

molecule of the concentration distribution $c(x,t)$ is defined by the equation

$$\overline{x^2} = \frac{\int_{-\infty}^{\infty} x^2 c(x,t)dx}{\int_{-\infty}^{\infty} c(x,t)dx} = \frac{1}{(4\pi Dt)^{1/2}} \int_{-\infty}^{\infty} e^{-x^2/4Dt} x^2 \, dx. \qquad (5.35)$$

In obtaining the last expression on the right, we have substituted equation (5.15) into the preceding ratio, and recognized that the denominator there equals m, according to equation (5.16). The integrand on the right is an even function of x and therefore the integral can be written as twice the integral from 0 to ∞. With the aid of the abbreviation $\alpha = 1/4Dt$, the integral is evaluated to be twice the integral

$$\int_0^{\infty} x^2 e^{-\alpha x^2} \, dx = -\frac{d}{d\alpha} \int_0^{\infty} e^{-\alpha x^2} \, dx = -\frac{d}{d\alpha}\left(\frac{\pi}{4\alpha}\right)^{1/2} = \frac{\pi^{1/2}}{4\alpha^{3/2}}, \qquad (5.36)$$

so that

$$\overline{x^2} = 2Dt. \qquad (5.37)$$

Hence, D equals the mean square displacement of a molecule diffusing in one dimension during the time t, divided by $2t$. This equation tells us that the **root-mean-square (r.m.s.) displacement** $\tilde{x} \equiv (\overline{x^2})^{1/2}$ during the time t is given as

$$\tilde{x} = (2Dt)^{1/2}. \qquad (5.38)$$

Alternatively, we can interpret equation (5.37) as meaning that the time t it takes for an initially concentrated solute mass to become uniformly distributed over a region of dimension x is (see problem 2)

$$t = \frac{x^2}{2D}.$$ (5.39)

It is left as an exercise for the reader (see problem 3) to show that in three dimensions, the root-mean-square displacement of a diffusing molecule $\tilde{r} \equiv (\overline{r^2})^{1/2}$ during the time t is given by the formula

$$\tilde{r} = (6Dt)^{1/2}.$$ (5.40)

The values of the diffusion constant in water for some typically encountered biological molecules are displayed in Table 5.1. If we consider the largest molecule listed, tobacco mosaic virus (TMV) which has the smallest value

TABLE 5.1 MOLECULAR WEIGHTS AND
DIFFUSION CONSTANTS IN WATER AT 20°C[a]

Substance	M (g/mole)	D (cm^2/sec)
Glycine	75	9.335×10^{-6}
Sucrose	342	4.586×10^{-6}
Ribonuclease	13,683	1.068×10^{-6}
Serum albumin (bovine)	66,500	0.603×10^{-6}
Fibrinogen (human)	330,000	0.197×10^{-6}
Myosin	440,000	0.105×10^{-6}
TMV	$\sim 4 \times 10^7$	0.053×10^{-6}

[a] From K. E. Van Holde, *Physical Biochemistry*, © 1971, p. 89. By permission of Prentice-Hall, Inc., Englewood Cliffs, New Jersey.

of the diffusion constant, we calculate from equation (5.38) that the r.m.s. distance it diffuses in one second is about 0.3×10^{-3} cm $= 3$ μm. This shows that even the heaviest molecules normally found in cells travel across a cell in a second or so, and achieve a more or less uniform distribution in the cytoplasm. Hence, a particular biochemical reaction associated with cell growth which requires either two particular molecular species to interact, or the presence of one molecular species at the fixed position of another in the cell, is not likely to be limited by diffusion.

Deeper insight into the physical meaning of D was provided by Einstein (1905) in his theory of Brownian motion. He showed that if the solute molecule is large compared to the solvent molecule and is assumed to have a spherical shape, then

$$D = \frac{kT}{6\pi\mu a},$$ (5.41)

where k is **Boltzmann's constant**, T is the absolute temperature of the solution, μ is the coefficient of viscosity of the solution, and a is the radius of the solute molecule. Equation (5.41) is called **Einstein's relation**. Boltzmann's constant is given in terms of the universal gas constant R and **Avogadro's number** N_A, the number of molecules in a mole, as $k = R/N_A$. The combination of equations (5.37) and (5.41) that eliminates D is

$$\overline{x^2} = \frac{T}{3\pi\mu a} t, \tag{5.42}$$

a result that has been confirmed experimentally by direct observation (Perrin, 1910). Implicit in equations (5.37)–(5.42) is the assumption that t is large compared to the time between two successive collisions. For a large solute particle of arbitrary shape, Einstein's relation is generalized to read

$$D = \frac{kT}{f}, \text{ or } D = kTb, \tag{5.43}$$

where b is the **mobility** of the diffusing molecule, defined as $b = 1/f$, and f is the Stokes' frictional coefficient of the particle. For spheres, $b = 1/6\pi\mu a$.

Equation (5.43) or (5.41) shows that $D\mu$ is directly proportional to the absolute temperature of the solution. For a spherical solute molecule of radius a and density ρ, the molecular weight M can be expressed as

$$M = \frac{4}{3}\pi a^3 \rho N_A. \tag{5.44}$$

Thus the radius a is proportional to $M^{1/3}$, and we can eliminate it between equations (5.41) and (5.44) to obtain the relation

$$D = \frac{kT}{3\mu}\left(\frac{\rho N_A}{6\pi^2 M}\right)^{1/3}. \tag{5.45}$$

In words, the quantity $DM^{1/3}$ is constant at a fixed temperature.

Even if the solute molecule is not spherical, we know from dimensional considerations that the frictional coefficient f in equation (5.43) is proportional to some characteristic length of the molecule, which in turn is proportional to the molecular volume or $M^{1/3}$. Also, the density ρ of most large protein molecules is approximately constant (see entries for $\overline{v} = 1/\rho$ in Table 5.4). Hence, at a fixed temperature,

$$DM^{1/3} \approx \text{constant}, \tag{5.46}$$

for solute molecules that are large compared to solvent molecules. When the solute molecules are of the same order of magnitude or smaller than the solvent molecules, other theoretical considerations (Thovert, 1910) show that, at a fixed temperature,

$$DM^{1/2} \approx \text{constant.} \tag{5.47}$$

An alternative theoretical investigation of this problem by Danielli (Davson and Danielli, 1952; see also Stein, 1967) suggests that the constant on the right depends on the temperature in a more complicated fashion than that indicated by equation (5.45).

The experimental verification of equations (5.46) and (5.47) is illustrated in Figure 5.6, in which log D is shown as a function of log M for a variety of molecules diffusing in water at or near 20°C (Stein, 1962). The straight line labeled A represents equation (5.47), while the straight line labeled B represents equation (5.46).

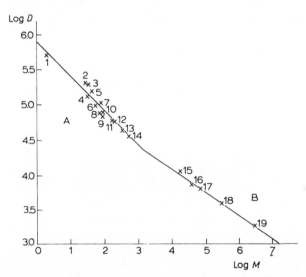

Figure 5.6. The log of the diffusion constant D as a function of the log of the molecular weight M for molecules diffusing in water at or near 20°C. The numbers represent the following molecular species: 1, hydrogen; 2, nitrogen; 3, oxygen; 4, methanol; 5, carbon dioxide; 6, acetamide; 7, urea; 8, *n*-butanol; 9, *n*-amyl alcohol; 10, glycerol; 11, chloral hydrate; 12, glucose; 13, lactose; 14, raffinose; 15, myoglobin; 16, lactoglobulin; 17, hemoglobin; 18, edestin; 19, erythrocruorin. From Stein (1962), with permission, based on data of Hitchcock (1947) and Davson and Danielli (1952).

5.5. OLFACTORY COMMUNICATION IN ANIMALS

Pheromones are "odor" chemicals released by animals and utilized for chemical communication with members of the same species. Transmission of the chemicals through the air is accomplished by diffusion, although the presence of winds may complicate the dispersal of the chemical. For example, the harvester ant **Pogonomyrmex badius** gives off a very volatile alarm substance under suitable provocation (Wilson, 1958). The substance originates in the mandibular glands of the ant. A sister ant, which detects the alarm substance, displays mild excitement and is attracted toward the source. At a greater concentration of the alarm substance (occurring when the distance of approach is closer, or at a later time at the original position of detection), the ant abruptly begins to run in circles. In this second state it sometimes appears to emit the alarm substance itself. This is the presumed manner in which an alarm communication system is established.

In order to make quantitative measurements of this system, experiments with harvester ants were performed (Bossert and Wilson, 1963) in which the puff of the alarm substance was simulated by crushing the head of a worker ant at one end inside a long closed tube, and removing it 3 sec later. The tube was 122 cm long and had a diameter of 2.6 cm. To detect the alarm, ants were placed at various positions in the tube. A screen placed 14 cm from the end of the tube containing the alarm prevented close approach of the ants to the source so that they were not stimulated to highest excitement. The location of, and time of detection by, the ants were observed. Some typical observations are shown in Table 5.2.

TABLE 5.2   DETECTION OF ATTRACTION OF ALARM SUBSTANCE FROM CRUSHED HEAD OF HARVESTER ANT[a]

Trial	Onset of Attraction	
	Distance (cm)	Time (sec)
1	18	60
	20.5	90
2	15	58
	18	91
	20.5	180
3	14	60
	18	100
4	16	81
	20.5	160

[a] From Bossert and Wilson (1963), with permission.

Because the experiments were carried out in a tube, we should solve the diffusion equation with boundary conditions appropriate to the tube geometry and the reflective nature of the tube walls. Rather than do this, we shall for simplicity adapt the one-dimensional solution with which we are already familiar, equation (5.15). In other words, we shall neglect the influence of the tube wall on the diffusion process, except insofar as it affects the source strength.

Thus, we assume that N molecules of the pheromone are placed at the origin at $t = 0$. The strength of the source is determined by the condition that the total number of pheromone molecules in the closed (semi-infinite) tube equals N, or,

$$A \int_0^\infty c(x,t)dx = N, \tag{5.48}$$

where A is the cross-sectional area of the tube and $c(x,t)$ is given by equation (5.15). Hence, by direct substitution,

$$c(x,t) = \frac{N}{A(\pi Dt)^{1/2}} e^{-x^2/4Dt}. \tag{5.49}$$

We assume that there is a threshold level of concentration K below which the animal species cannot detect the odor, and above which the odor is always detectable. Then, the condition,

$$c \geqq K, \tag{5.50}$$

determines a region of influence of the pheromone, inside of which the pheromone is detectable. Now c is a monotonic decreasing function of x at any time t. Therefore, there exists some finite time $\bar{t}$ beyond which equation (5.50) is no longer satisfied. Substituting (5.49) into (5.50) and utilizing the equality sign only determines a **distance of influence** $x(t)$. This distance can be visualized by imagining that the line $c = K$ is superimposed on Figure 5.3. The intersections of the curves with the line for $x > 0$ determine $x(t)$ for the times associated with the curves. We readily compute the distance of influence to be

$$x(t) = \begin{cases} [2Dt \log(\bar{t}/t)]^{1/2}, & 0 \leqq t \leqq \bar{t}, \\ 0, & t > \bar{t}, \end{cases} \tag{5.51}$$

$$\bar{t} = \frac{N^2}{\pi DA^2 K^2}. \tag{5.52}$$

Here $\bar{t}$ is called the **duration of influence** or **fadeout time**, because it is the time during which the pheromone is detectable. $x(t)$ is a function that begins at the origin, rises to a maximum value x_m, and then falls to zero again at

$t = \bar{t}$. The maximum distance of influence, determined by the condition $dx/dt = 0$, occurs when $t = \bar{t}/e$ and is $x_m = (2D\bar{t})^{1/2}$.

The volatile nature of the alarm chemical makes it probable that diffusion commences rapidly after ejection of the chemical. The measurements given in Table 5.2 represent experimental determinations of $x(t)$. These are assumed to be represented by equation (5.51). Therefore, it is possible to determine D and $\bar{t}$ from them. From (5.51), it follows that there is a linear relationship between $\log t$ and x^2/t, namely,

$$\log t = \log \bar{t} - \frac{1}{2D}\frac{x^2}{t}. \tag{5.53}$$

Hence, a plot of $\log t$ versus x^2/t permits the values of $\bar{t}$ and D to be readily inferred. N/K can then be calculated from equation (5.52). Such a calculation was performed for each trial reported in Table 5.2, representing a potentially different source strength in each case. The average values of D and N/K inferred from the data were

$$\begin{aligned} D &\sim 0.4 \text{ cm}^2/\text{sec}, \\ N/K &\sim 800 \text{ cm}^3. \end{aligned} \tag{5.54}$$

Of course there are various errors associated with the application of the theory, such as the inadequacy of the one-dimensional theory previously alluded to, the paucity of observations per trial, the difficulty of determining the precise recognition moment of the alarm chemical, the fact that the air was not perfectly still (if only because of ant motion), and so forth. However, the theory does provide a semiquantitative measure of the experimental data.

The fact that the ants run in the direction of the source indicates that in addition to detecting the presence of the alarm chemical, they are able to detect the direction from which it arises. This means that the gradient of concentration is also detectable, which is a rather different ability. Further applications of the theory have been made to the problem of chemical trails (the source is both continuously emitting and moving with a constant velocity through space), and to the problem of chemical trails in the presence of a steady wind. (Bossert and Wilson, 1963).

5.6. MEMBRANE TRANSPORT

As a simple application of Fick's laws of diffusion, we shall consider the diffusion of a solute into a cell. Such transport is mediated in a critical manner by a membrane surrounding the protoplasm of the cell called the **plasma membrane**. Structurally, the plasma membrane consists of a double layer of lipid molecules known as a **bilipid layer**. This layer contains a mosaic of globular proteins on both of its surfaces, some of which penetrate across the entire width of the lipid bilayer. The precise function of these proteins and their relation to the transport of molecules across the cell membrane is at the present time obscure. The cell membrane is about 50–100 Å thick (1 Å = 1 Angstrom = 10^{-8} cm).

The plasma membrane is **selectively permeable**, meaning that it permits the passage of some solutes but excludes others. It also permits the exit of waste products. Free living cells for the most part have **cell walls**, but these structures are porous and do not participate in the permeability property. Rather, the cell wall provides mechanical strength to the cell by protecting it from external injury. It also protects the cell from rupture due to swelling, called **lysis**. For example, a mammalian red cell, which does not possess a cell wall, undergoes such rupture, or **hemolysis**, when it is placed in a solution of pure water.

What is the mechanism of transport of solutes into cells? This question is exceedingly complex, and there is no simple answer to it. Broadly speaking, it has been found that three factors influence greatly the ability of some substances, called **permeants**, to penetrate into the interior of cells. These important characteristics of permeants are (a) lipid solubility, (b) size, and (c) charge or ionic strength. It has been known for a long time that lipid soluble molecules are more readily transported through cell membranes than insoluble ones. The mechanism of transport of such molecules is apparently diffusion. The justification for this remark will be indicated later.

Two notable exceptions to the lipid solubility characteristic are water and urea, which penetrate cells much more rapidly than can be understood on the basis of lipid solubility. These exceptions have led to the postulate that plasma membranes possess **pores**, through which water and other small water soluble molecules can permeate. The diameter of these pores are estimated to be from 4 to 8 Å. The diameters of the molecules of water and urea are 3 and 3.6 Å, respectively. Charged ions or **electrolytes** that are dissolved in a solution are strongly affected by the presence of electric fields existing in and near membranes. Understanding the transport of such particles by diffusion requires taking into account their electrical properties, and is called **electrodiffusion**. This is a more complex subject that we will not enter into. Rather, we shall confine our discussion to the permeation of **nonelectrolytes**.

The transport of molecules across cell membranes by diffusion, in accordance with Fick's law, is usually called **passive transport**. Many molecules, notably metabolites such as simple sugars and amino acids, are able to permeate the plasma membrane, but not in accordance with Fick's law. An example is galactose transport into *E. coli*, discussed in Section 1.4. In that example, if Fick's law were the controlling mechanism of transport, the intracellular galactose concentration would eventually attain the extracellular galactose concentration level, but would never become larger. As we recall, the internal concentration becomes 100-fold or so greater than the external concentration. This indicates that an "active" transport mechanism is at work, requiring energy, for concentrating galactose in the cell interior. Such mechanisms are referred to as **active transport**. A number of active transport mechanisms have been proposed, but from a detailed physico-chemical point of view, the mechanisms are not very well understood at present.

We shall mention here only the most well documented alternative to simple diffusion, which is called **facilitated transport**. In this mechanism, **carrier** molecules, which are mobile membrane components, possibly enzymes, are able to bind with an extracellular substrate and release the substrate molecules into the cells interior. The carrier molecule perhaps crosses the membrane's interior by diffusion. Facilitated transport is therefore not usually classified as an active transport mechanism, but rather a passive transport mechanism in which reaction kinetics is intimately involved. Facilitated transport is usually invoked to explain the empirically observed property of **saturation** of influx of some metabolites into cells, with increasing extracellular metabolite concentration. This phenomenon is usually seen for **hydrophilic** solutes, those that are readily soluble in water, but relatively insoluble in lipid. Presumably saturation occurs when the carrier molecules become saturated with substrate molecules, as in the Michaelis–Menten theory of enzymatic reactions. Facilitated transport may also be the basis for the observed parabolic dependence on a critical growth factor of the growth rate of a cell population in a chemostat, discussed in Section 1.7.

In what follows we confine ourselves to the mathematical theory of the diffusion process, governed by the diffusion equation. Even if it turns out in the future that most cases of solute transport are governed by active transport mechanisms, the knowledge of the diffusion process provides a standard for recognizing whether the controlling mechanism of transport is active or not, or whether diffusion is a contributing mechanism of transport over and above other mechanisms.

The simplest theory of transport into or out of a cell by diffusion results if we assume that the intracellular diffusion is so rapid that the solute concentration there is uniform in space. In most experiments of diffusion into cells, the volume of extracellular space is made so large that the solute concentration there is not significantly affected by any loss into cells, and is considered to be constant. Then the only rate-limiting step in solute transport across the membrane is the diffusion process through the cell membrane itself.

We consider as our idealized model of the cell membrane, a homogeneous lipid layer separating two aqueous phases, the cell interior and the cell exterior. At either of the two lipid–water interfaces, a discontinuity in solute concentration exists at equilibrium. This discontinuity is a consequence of the molecular barrier that exists for a solute molecule entering the lipid phase from the aqueous phase. Let c_0 represent the concentration on the outside of the cell, and $\bar{c}_0$ be the concentration inside the lipid layer and right next to the outer interface. Then the discontinuity between c_0 and $\bar{c}_0$ is represented by the lipid–water **partition coefficient** Γ, where

$$\bar{c}_0 = \Gamma c_0. \tag{5.55}$$

Normally $\Gamma < 1$. However, if a solute is more readily soluble in lipid than water, then $\Gamma > 1$. Similarly, if the subscript i denotes the inside of the cell, then

$$\bar{c}_i = \Gamma c_i, \tag{5.56}$$

where c_i is the concentration in the cell's interior, and $\bar{c}_i$ is the concentration inside the lipid layer and just next to the inner interface. Furthermore, we shall assume that the cell membrane thickness δ is small, so that the concentration gradient dc/dx through it may be approximated by $\Delta c/\Delta x = (\bar{c}_0 - \bar{c}_i)/\delta$. Here, x is measured positively outward from the cell. According to Fick's first law, the current density through the cell membrane is given by the expression

$$j = \frac{D}{\delta}(\bar{c}_i - \bar{c}_0), \tag{5.57}$$

where D is the diffusion constant of the solute in the lipid layer. The ratio D/δ is sometimes called the "permeability constant," but we shall use a more common definition, to be given later. Because measurements are actually made of c_i or c_0, while $\bar{c}_0$ and $\bar{c}_i$ are inaccessible to observation, we shall express j in terms of the former quantities as

$$j = \frac{\Gamma D}{\delta}(c_i - c_0). \tag{5.58}$$

Equation (5.58) is a good approximation for the current density as long as the concentrations c_i and c_0 are changing slowly or quasistatically.

Let m be the mass of solute contained in the cell. If we consider the cell volume and the solute mass it contains at time t, then the rate of increase of solute mass is given, according to equation (5.9), by the expression

$$\frac{dm}{dt} = -jA. \tag{5.59}$$

The minus sign indicates that when j is positive, which is the case when solute flow is outwards, m decreases. Here j is given by equation (5.58) and

A is the surface area of the cell. In deriving (5.57) from (5.9), we recall that, per unit area, j represents the net current or the net difference between outflux and influx. It should also be borne in mind that while there is a discontinuity in concentration at an interface, there is no discontinuity in current flow. Let c_i be uniform throughout the cell. Then we can express m as

$$m = c_i V, \tag{5.60}$$

where V is the volume of the cell. Substituting equations (5.60) and (5.58) into (5.59) yields the result

$$\frac{dc_i}{dt} = k(c_0 - c_i), \tag{5.61}$$

$$k = \frac{\Gamma DA}{\delta V}. \tag{5.62}$$

With $c_i(0) \equiv c_{i0}$ given, the solution to equation (5.61) is known from our previous studies to be

$$c_i(t) = c_0 + (c_{i0} - c_0)e^{-kt}. \tag{5.63}$$

The parameter k is seen to have the dimensions of sec^{-1}. If c_0 is zero and $c_{i0} \neq 0$, equation (5.61) expresses the fact that the solute leaves the cell at a fractional rate k, since $m^{-1}(dm/dt) = c^{-1}(dc/dt)$.

Equation (5.61) is recognized as having a familiar compartmental form. We see in its derivation a more fundamental basis for the application of a compartmental equation to transport across the membranes of cells. It is important to recognize that time-dependent concentration measurements that are averaged over space can only provide information about k and not about its component parts as expressed by equation (5.62). We emphasize that these components are almost all properties of the cell membrane exclusively: its area, the volume contained by it, its thickness, and the diffusion constant, which is a property of the solute molecule in the solvent contained within the membrane itself. The single exceptional property is the partition coefficient Γ for the lipid–water interface.

In practice, permeability constants are often determined by placing cells in a medium containing a radioactively labeled substrate with specific activity c_0, say. The intracellular specific activity c is then measured as a function of time, and k is determined as explained in Chapters 1 and 3. From k and independent determinations of the mean area A and mean volume V of the cells, the **permeability constant** P defined as

$$P = \frac{kV}{A} \tag{5.64}$$

is determined. By comparing (5.64) with (5.62), we see that

$$P = \frac{\Gamma D}{\delta},$$ (5.65)

so that P incorporates the effect of the lipid–water discontinuity at the inner and outer surface of the membrane. However, if diffusion is taking place through pores in the membrane so that the solute molecules are always in the aqueous phase, then it is unnecessary to make the distinction between c and $\bar{c}$, and $P = D/\delta$. In such a case, the quantity A is to be interpreted as representing the cross-sectional area of the pores in the membrane.

Care must be exercised in reading the biological literature in this regard, because values of the "permeability constant" are presented for molecules which pass through the membranes of cells by nondiffusional transport processes. Such "permeability constants" utilize special definitions unlike the one presented in the text. For example, suppose two experiments are performed, one for which $c_0 \neq 0$ and $c_i = 0$, and the other for which $c_0 = 0$ and $c_i \neq 0$. If it is found that c_i obeys equation (5.61) but with different values of k in the two cases, then **unidirectional permeability coefficients** P_i and P_0, respectively, can be ascribed to the membrane with respect to the solute in question. However, such a theoretical interpretation is not consistent with Fick's law.

It has been known for a long time (Overton, 1899; Osterhout, 1940) that many substances permeate cells to an extent which is proportional to their olive oil–water partition coefficients. In other words, we can say that Γ for a cell permeant is similar or proportional to its oil–water partition coefficient Γ_0. If we combine this result with equations (5.46), (5.47), and (5.65), we infer that at a fixed temperature,

$$PM^{1/2} = \text{constant} \cdot \Gamma_0, M \leq 100,$$ (5.66a)
$$PM^{1/3} = \text{constant} \cdot \Gamma_0, M > 100.$$ (5.66b)

The experimental verification of equation (5.66a) for a variety of cell types and solute molecules is the principal justification for the inference that passive diffusion is the mechanism of transport for these solute molecules. This is illustrated in Figure 5.7, which shows a plot of $PM^{1/2}$ versus Γ_0 for various small molecular weight nonelectrolytes permeating the diatom **Melosira** (Davson and Danielli, 1952).

Further evidence that some permeants enter cells by the mechanism of simple diffusion is given in Table 5.3. There the values of P and $PM^{1/2}$ are shown for some nonelectrolytes permeating the sulfur bacterium **Beggiatoa mirabilis**. The permeants listed all have a molecular weight of about 100 or less, and the values of $PM^{1/2}$ are seen to be constant. Consequently, it is

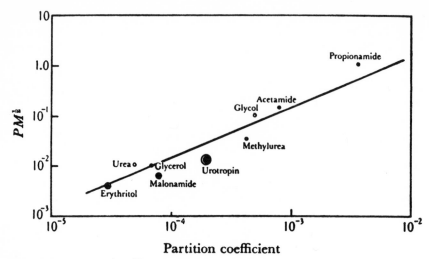

Partition coefficient

Figure 5.7. A plot of $PM^{1/2}$ versus the oil–water partition coefficient, for nonelectrolytes penetrating the diatom *Melosira*, displays a linear dependence in accordance with equation (5.66*a*). From Davson and Danielli (1952), with permission.

concluded that for these substances the lipid nature of the membrane is not significant, and that diffusion is taking place through aqueous pores in the membrane.

TABLE 5.3   VALUES OF P AND $PM^{1/2}$
FOR SUBSTANCES PERMEATING THE
BACTERIUM *BEGGIATOA MIRABILIS*[a]

Substance	P (cm/sec)	$PM^{1/2}$ (cm/sec)
Erythritol	0.838×10^{-5}	0.93×10^{-4}
Glycerol	1.060×10^{-5}	1.02×10^{-4}
Methylurea	1.170×10^{-5}	1.01×10^{-4}
Glycol	1.390×10^{-5}	1.10×10^{-4}
Urea	1.580×10^{-5}	1.22×10^{-4}

[a] From Davson and Danielli (1952).

It must be emphasized that many substances display transport characteristics into and out of membranes for which no theoretical explanation exists at present. For example, the action potential, which propagates along the membrane of the nerve axon, is known to be intimately associated with the transport of sodium and potassium ions across the membrane during the passage time. In the theory of Hodgkin and Huxley (1952), which so successfully describes the behavior of the electrical potential across the membrane as a function of position and time, this transport behavior is only accounted for in a phenomenological way. That is to say, the ion transport is described mathematically in accordance with experiment, but without any appeal to an underlying mechanism. The behavior of membranes is a subject of extensive biological investigation at the present time, and the reader is advised to consult the recent literature on the subject of membrane transport for further information.

5.7. DIFFUSION THROUGH A SLAB

We shall now consider in more precise fashion diffusion of a solute through a plane sheet or slab of thickness δ, the two sides of which are maintained at the concentrations c_1 and c_2, respectively (see Figure 5.8). We shall leave

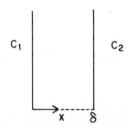

Figure 5.8. Diffusion through a slab of thickness δ, the left side of which is maintained at the concentration c_1 and the right side at the concentration c_2.

open for the moment the particular biological application of this problem.

In order to solve the diffusion equation in a unique way, it is necessary to prescribe boundary conditions for the concentration, as well as an initial condition describing the spatial distribution of the concentration at $t = 0$. For simplicity, we shall assume that initially there is no solute in the membrane. Thus, the mathematical statement of the problem is to find $c(x,t)$ given that it satisfies equation (5.7) in the region $0 < x < \delta$, and is subject to the

boundary conditions, $\qquad c(0,t) = c_1,$ $\qquad\qquad\qquad$ (5.67a)

$$c(\delta,t) = c_2, \qquad\qquad\qquad (5.67b)$$

and initial condition, $\qquad c(x,0) = 0.$ $\qquad\qquad\qquad$ (5.68)

We attempt to solve this problem by the **method of separation of variables**. In this method, we seek a solution of the partial differential equation for c, in the **separable form** of a function of x times a function of t, or,

$$c(x,t) = X(x)T(t). \qquad\qquad\qquad (5.69)$$

Substituting this expression into (5.7) and dividing by $c(x,t)$, we obtain the equation

$$\frac{1}{T}\frac{dT}{dt} = \frac{D}{X}\frac{d^2X}{dx^2}. \qquad (5.70)$$

The left-hand side is a function of t only, and the right-hand side is a function of x only. This equation can only be satisfied for all values of x and t provided both sides equal a constant, called the **separation constant**. We denote this constant without loss of generality by $-\lambda^2 D$. Then (5.70) becomes two equations,

$$\frac{1}{T}\frac{dT}{dt} = -\lambda^2 D, \qquad (5.71)$$

$$\frac{1}{X}\frac{d^2X}{dx^2} = -\lambda^2. \qquad (5.72)$$

Thus the functions X and T satisfy ordinary differential equations whose solutions are

$$X = Ax + B, \qquad\qquad \lambda = 0, \qquad (5.73a)$$
$$X = C\sin\lambda x + E\cos\lambda x, \lambda \neq 0, \qquad (5.73b)$$
$$T = Fe^{-\lambda^2 Dt}, \qquad (5.74)$$

where A, B, C, E, and F are constants. If we substitute these expressions into (5.69) we obtain the particular solutions

$$c = (Ax + B), \qquad\qquad \lambda = 0, \qquad (5.75a)$$
$$c = (C\sin\lambda x + E\cos\lambda x)e^{-\lambda^2 Dt}, \lambda \neq 0. \qquad (5.75b)$$

Note that the constant F is superflous in the expressions for c given above, and has been absorbed into the remaining constants A, B, C, and E. Since these constants are as yet undetermined, we retain the same symbols for them as in (5.73).

Let us apply the boundary conditions (5.67) to (5.75a). We can satisfy them by choosing A and B so that (5.75a) becomes

$$c = c_1 + (c_2 - c_1)\frac{x}{\delta}. \qquad (5.76)$$

However, (5.76) does not satisfy the initial condition (5.68). Here, the **linearity** of the diffusion equation comes to our rescue. All linear partial (and ordinary) differential equations satisfy the **principle of superposition**. According to it, if two particular solutions of the equation are found, then the sum of the two solutions is likewise a solution. Such a sum is called a **superposition of**

solutions. Therefore, let us seek a solution of the form given by the sum of (5.76) and (5.75b), that is,

$$c(x,t) = c_1 + (c_2 - c_1)\frac{x}{\delta} + (C \sin \lambda x + E \cos \lambda x)e^{-\lambda^2 Dt}, \quad (5.77)$$

where $\lambda \neq 0$. If now we apply the boundary condition (5.67a) to (5.77), we obtain the condition

$$c_1 = c_1 + Ee^{-\lambda^2 Dt}, \quad (5.78)$$

from whence it follows that $E = 0$. Applying the boundary condition (5.67b) to (5.77) leads to the condition

$$c_2 = c_2 + C \sin \lambda\delta \, e^{-\lambda^2 Dt}. \quad (5.79)$$

If we set $C = 0$ to satisfy this equation, $c(x,t)$ according to (5.77) is independent of t, and we are then unable to satisfy the initial condition (5.68) in general. However, equation (5.79) can also be satisfied for all values of t provided $\sin \lambda\delta = 0$. This condition in turn requires that λ have certain **characteristic values**, also called **eigenvalues**, given as

$$\lambda = \lambda_n \equiv \frac{n\pi}{\delta}, \qquad n = 1, 2, 3, \ldots. \quad (5.80)$$

The negative values of n that are permitted by the equation $\sin \lambda\delta = 0$ are redundant because $\sin \lambda_{-n}x = -\sin \lambda_n x$. Therefore, it is only necessary to consider positive values of n in (5.80). Thus, we see that there is a **denumerable** (meaning countable) infinite set of allowed values of λ given by (5.80). Corresponding to each allowed value λ_n, there is a separable solution of the form (5.77) having a particular constant C associated with it which we denote as C_n. Furthermore, by the principle of superposition, the sum of all these solutions is likewise a solution.

Hence, the most general solution of separable form that we can write down which is compatible with the boundary conditions is

$$c = c_1 + (c_2 - c_1)\frac{x}{\delta} + \sum_{n=1}^{\infty} C_n \sin \lambda_n x e^{-\lambda_n^2 Dt}, \quad (5.81)$$

where the summation extends over all positive integers. The as yet undetermined coefficients C_n must be chosen so as to satisfy the initial condition. With an infinite number of terms appearing in (5.81), it appears plausible that this condition can be satisfied, and indeed it is so. Imposing the initial condition (5.68) on (5.81) yields the requirement

$$-c_1 - (c_2 - c_1)\frac{x}{\delta} = \sum_{n=1}^{\infty} C_n \sin \lambda_n x. \quad (5.82)$$

This requirement is in fact a problem in **Fourier representation**: An arbitrary continuous function of x (or even a discontinuous function consisting of continuous pieces, with discontinuities at the junctions of the pieces), defined in a finite interval, can always be represented by an infinite trigonometric series called a **Fourier series**. In the above example, the left-hand side of equation (5.82) represents a particular continuous function defined in the interval $0 \leqq x \leqq \delta$. The Fourier representation problem associated with this function is to express it as an infinite sum of the sinusoidal functions $\sin \lambda_n x = \sin(n\pi x/\delta)$, which appear on the right-hand side. The sum is called a Fourier series.

The determination of the C_n is greatly simplified because of the **orthogonality** property of the trigonometric functions. Thus, if $n \neq m$, then

$$\int_0^\pi \sin nx \sin mx \, dx = \frac{1}{2} \int_0^\pi [\cos(n-m)x - \cos(n+m)x]dx$$

$$= \frac{1}{2} \left[\frac{\sin(n-m)x}{n-m} - \frac{\sin(n+m)x}{n+m} \right]_0^\pi = 0,$$

while if $n = m$, then

$$\int_0^\pi \sin^2 mx \, dx = \frac{1}{2} \int_0^\pi [1 - \cos 2mx]dx = \frac{\pi}{2} - \frac{\sin 2mx}{2m} \bigg|_0^\pi = \frac{\pi}{2}.$$

We can express these two results in a concise manner as

$$\frac{2}{\pi} \int_0^\pi \sin nx \sin mx \, dx = \delta_{nm} = \begin{cases} 0, & n \neq m, \\ 1, & n = m, \end{cases} \qquad (5.83)$$

where δ_{nm} is the **Kronecker delta**, and n and m are integers. From (5.83) and (5.80) it follows that

$$\frac{2}{\delta} \int_0^\delta \sin \lambda_n x \sin \lambda_m x \, dx = \frac{2}{\delta} \int_0^\delta \sin \frac{n\pi x}{\delta} \sin \frac{m\pi x}{\delta} \, dx = \delta_{nm}. \qquad (5.84)$$

Therefore, if we multiply both sides of equation (5.82) by $\sin \lambda_m x$ and integrate over x from 0 to δ, only one term of the series survives, the one for which $n = m$. Hence, the coefficients C_m are given by the expression

$$C_m = \frac{2}{\delta} \int_0^\delta \left[-c_1 - (c_2 - c_1)\frac{x}{\delta} \right] \sin \frac{m\pi x}{\delta} \, dx, \qquad m = 1, 2, \ldots. \qquad (5.85)$$

By explicit integration it is found that

$$C_m = \frac{2}{m\pi}(c_2 \cos m\pi - c_1), \qquad m = 1, 2, \ldots. \qquad (5.86)$$

The solution to the one-dimensional diffusion problem posed by equations (5.7), (5.67), and (5.68) can then be written as

$$c(x,t) = c_1 + (c_2 - c_1)\frac{x}{\delta} + \frac{2}{\pi}\sum_{n=1}^{\infty}\frac{c_2\cos n\pi - c_1}{n}\sin\frac{n\pi x}{\delta}e^{-n^2\pi^2 Dt/\delta^2}. \quad (5.87)$$

From this expression, we can readily calculate $D(\partial c/\partial x)_{x=0}$, which is the current density j_0 or rate at which the diffusing substance emerges at the interface $x = 0$ per unit area per unit time. We find that

$$j_0 = \frac{D}{\delta}(c_2 - c_1) + \frac{2}{\delta}\sum_{n=1}^{\infty}(c_2\cos n\pi - c_1)e^{-n^2\pi^2 Dt/\delta^2}. \quad (5.88)$$

The first term on the right-hand side represents the steady-state flux, while the remaining terms represent a transient flux which is significant only for short times.

Equation (5.87) is a useful representation of the solution except for very short times, when a great many terms contribute to it. However, because the exponential coefficient appearing in (5.87) and (5.88) is proportional to n^2, the terms in the series with large n decay very rapidly with time. Thus, to good approximation, we may retain only the first term in these series for which $n = 1$, and write

$$c(x,t) \approx c_1 + (c_2 - c_1)\frac{x}{\delta} - \frac{2}{\pi}(c_2 + c_1)\sin\frac{\pi x}{\delta}e^{-\pi^2 Dt/\delta^2},$$

$$j_0 \approx \frac{D}{\delta}(c_2 - c_1) - \frac{2D}{\delta}(c_1 + c_2)e^{-\pi^2 Dt/\delta^2}. \quad (5.89)$$

The parameter δ^2/D appearing in the exponent has the dimensions of time and is a characteristic property of the slab. We see from the last expression that for times that are large compared to δ^2/D, the current density is essentially equal to its steady-state value. This steady-state value was the one utilized in equation (5.57) in representing approximately the diffusion process into or out of a cell. The "quasistatic" change mentioned in the discussion following (5.57) can now be expressed more quantitatively as meaning one for which the fractional change in c_1 and/or c_2 is small in a time $t \sim \delta^2/\pi^2 D$, where δ represents the membrane thickness.

We return now to the question of the biological utility of the solution to the problem of diffusion through a slab. One application of it that we can make is to let the slab represent a cell in a large bathing solution of solute with fixed concentration c_0. In other words, the region $0 \leq x \leq \delta$ represents the cell, while the region $x < 0$ and $x > \delta$ represents the cell exterior. The thickness δ can be thought of as a characteristic length or size of the cell, equal to V/A, for example. Then, with $c_1 = c_2 = c_0$, the even-numbered terms in the solution (5.87) vanish, and (5.87) becomes

$$c(x,t) = c_0 \left\{ 1 - \frac{4}{\pi} \sum_{n=1,3,5...}^{\infty} \frac{1}{n} \sin \frac{n\pi x}{\delta} e^{-n^2\pi^2 Dt/\delta^2} \right\}. \tag{5.90}$$

With this solution we can deduce the **fractional equilibration**, which is the ratio of the solute mass in the cell $m = m(t)$ to the solute mass in the cell at equilibrium $m_0 \equiv m(\infty)$. Thus, with A representing the cross-sectional area of the slab or surface area of the cell,

$$\frac{m}{m_0} = \frac{A \int_0^\delta c(x,t)dx}{A \int_0^\delta c(x,\infty)dx} = 1 - \frac{4}{\pi\delta} \sum_{n=1,3,5,...}^{\infty} \frac{1}{n} e^{-n^2\pi^2 Dt/\delta^2} \int_0^\delta \sin \frac{n\pi x}{\delta} dx,$$

which becomes, because the last integral equals $2\delta/n\pi$,

$$\frac{m}{m_0} = 1 - \frac{8}{\pi^2} \sum_{n=1,3,5,...}^{\infty} \frac{1}{n^2} e^{-n^2\pi^2 Dt/\delta^2}. \tag{5.91}$$

Except for very short time intervals, this expression can be approximated by

$$\frac{m}{m_0} \approx 1 - \frac{8}{\pi^2} e^{-\pi^2 Dt/\delta^2}. \tag{5.92}$$

Since the fractional equilibration for an actual cell population can be readily measured, the above expression is useful for representing nonelectrolyte transport into cells for which the diffusion in the cytoplasm of the cell is rate-limiting rather than the transport through the membrane.

As a second application, consider again the problem worked out in the previous section of diffusion through a cell membrane, considered as a lipid layer. We could solve it according to the precise formulation of this section, instead of utilizing the steady-state approximation for j given by equation (5.57). However, the mathematical solution is somewhat more involved than the solution we have just presented, inasmuch as it requires the consideration of a coupled system of equations for the concentrations in the cell interior and in the membrane. Therefore, we shall not give the solution here but mention only the principal result: The fractional equilibration in the cell is given by an expression similar to equation (5.91), one plus a sum of a large (strictly infinite) number of exponential terms. If only the leading (most slowly decaying) exponential term is retained as in equation (5.92), then it is found that

$$\frac{m}{m_0} \approx 1 - \frac{3}{2} e^{-\Gamma ADt/\delta V}, \tag{5.93}$$

where m is the solute mass in the cell defined as $m = Vc_i(t)$, and $c_i(t)$, V and A are, respectively, the solute concentration, volume, and area of the cell. In deriving (5.93), it was assumed that $c_i(0) = 0$. The above expression is valid provided $\Gamma\delta A/V \ll 1$. Since V/A is a characteristic length of the cell which is large compared to the membrane thickness δ, and Γ is usually itself small compared to unity for permeants, this condition is readily satisfied. Therefore, the approximate treatment of Section 5.4 is justified when applied to lipid soluble cell permeants, except at very small times.

5.8. CONVECTIVE TRANSPORT: IONIC FLOW IN AN AXON

When a solute is in a moving liquid, it is entrained by the flow. The resulting motion of the solute is called **convective transport**. This transport is additional to the diffusive motion that the solute undergoes. Imagine a small cross-sectional area through which the fluid flows with a normal velocity u. If the fluid has a tangential velocity component, it does not contribute to the flux through the surface. The flux of solute matter flowing normally through the area per unit time is denoted j_{conv}, and is called the **convective current density**. Thus,

$$j_{conv} = cu, \tag{5.94}$$

where c represents the concentration of solute. The current density (5.94) also describes a solute that possesses a velocity u even when the fluid does not. Such velocities are established by external forces, for example, the gravitational force. Although the name "convective current density" is no longer apt for such a current density, we shall continue to use it because no other name is commonly used to describe it, and the mathematical theory is the same as that for true convective currents.

For an arbitrary three-dimensional geometry for which the flow velocity is a vector $\mathbf{u} = (u,v,w)$, the convective current density becomes a vector $\mathbf{j}_{conv}$ given by

$$\mathbf{j}_{conv} = c\mathbf{u}. \tag{5.95}$$

Combining this current density with the diffusional current density given by (5.2) yields the total current density in a solution in which convection and diffusion are both taking place. Substituting this total current density into the equation of continuity (5.3) leads to the **equation of convective diffusion**,

$$\frac{\partial c}{\partial t} + \frac{\partial}{\partial x}(cu) + \frac{\partial}{\partial y}(cv) + \frac{\partial}{\partial z}(cw) = D\,\Delta c, \tag{5.96}$$

if D is constant. Here $c = c(x,y,z,t)$. The velocity vector $\mathbf{u}$ must either be known and given, or this equation must be supplemented by the fluid dynamic equations determining **u**. When the velocity vector **u** is constant, (5.96) becomes

$$\frac{\partial c}{dt} + \mathbf{u} \cdot \nabla c = D \, \Delta c, \tag{5.97}$$

where the operator $\mathbf{u} \cdot \nabla$ is defined as

$$\mathbf{u} \cdot \nabla \equiv u \frac{\partial}{\partial x} + v \frac{\partial}{\partial y} + w \frac{\partial}{\partial z}. \tag{5.98}$$

We see that equation (5.97) is the diffusion equation plus an additional term representing the contribution of convection. In one dimension, equation (5.97) reduces to the **one-dimensional convective diffusion equation**,

$$\frac{\partial c}{\partial t} + u \frac{\partial c}{\partial x} = D \frac{\partial^2 c}{\partial x^2}. \tag{5.99}$$

In order to solve this equation, we shall introduce a **change of variables**, namely, instead of the independent variables x, t, we utilize ξ, τ defined by the equations

$$\begin{aligned} \xi &= x - ut, \\ \tau &= t. \end{aligned} \tag{5.100}$$

Let c be expressed as a function of ξ and τ, $c = c(\xi,\tau)$. Then, by the chain rule of differentiation,

$$\begin{aligned} \frac{\partial c(\xi,\tau)}{\partial t} &= \frac{\partial c}{\partial \xi}\frac{\partial \xi}{\partial t} + \frac{\partial c}{\partial \tau}\frac{\partial \tau}{\partial t} = -u\frac{\partial c}{\partial \xi} + \frac{\partial c}{\partial \tau}, \\ \frac{\partial c(\xi,\tau)}{\partial x} &= \frac{\partial c}{\partial \xi}\frac{\partial \xi}{\partial x} = \frac{\partial c}{\partial \xi}, \\ \frac{\partial^2 c}{\partial x^2} &= \frac{\partial^2 c}{\partial \xi^2}. \end{aligned} \tag{5.101}$$

By substituting (5.101) into the one-dimensional convective diffusion equation (5.99) we find that $c(\xi,\tau)$ satisfies the simpler one-dimensional diffusion equation

$$\frac{\partial c}{\partial \tau} = D \frac{\partial^2 c}{\partial \xi^2}. \tag{5.102}$$

Hence, solutions of the diffusion equation represented by $c(\xi,\tau)$ are readily transformed into solutions of the convective diffusion equation by means of the change of variables, equation (5.100). In other words, the solution of (5.99) can be written as $c(x - ut,t)$, where $c(x,t)$ satisfies the one-dimensional diffusion equation. Any solution $c(x,t)$ of the one-dimensional diffusion equation (without regard to boundary or initial conditions) becomes a solution of the one-dimensional convective diffusion equation $c(x - ut,t)$ merely by replacing x in $c(x,t)$ by $x - ut$.

As an application of the theory of convective diffusion, we shall consider the flow of ions inside an **axon**, a fiber of a nerve cell. The squid contains a giant nerve axon, which, when severed, isolated, and placed in a suitable salt solution, retains its bioelectric properties for many hours. When a small drop of radioactive ^{42}K is placed on a section of such an axon, a measurable quantity of the radioactive potassium enters the **axoplasm**, the cytoplasm of the axon. At $t = 0$, after a readily detectable quantity of radioactive ^{42}K has entered the axon section, the fiber is washed with sea water and placed in oil. Most of the ^{42}K then remains in the axoplasm, and its movement under the action of diffusion in an applied electric field can subsequently be studied. In response to the action of an electrical potential difference applied to the ends of the section, an electric field is established which tends to accelerate the ions. Because of viscosity, the potassium ions attain a mean terminal velocity, which we denote by u. Thus, in this example, the velocity u is the result of an applied force field, which acts on the solute molecules only. The velocity u is related to the applied potential gradient dV/dx and the **ionic mobility** β of potassium ions by the equation

$$u = -\beta \frac{dV}{dx}. \tag{5.103}$$

Note that the ionic mobility defined above is different from the mobility of molecules defined for equation (5.43), and has the dimensions of square centimeters per second per volt. The negative sign in (5.103) insures that a positively charged ion will move in the same direction as the applied electric field which is $-dV/dx$.

In experiments conducted as described above (Hodgkin and Keynes, 1953), the distribution of radioactivity as a function of distance at particular times was observed, and some typical observations are illustrated in Figure 5.9. In Figure 5.9, curve A represents the observed distribution of ^{42}K radioactivity with distance at $t = 0$, and curve B represents the observed distribution at a subsequent time $t = 37$ min. It was found that the initial distribution was represented very well by a Gaussian curve, namely,

$$c(x,0) = A_0 \exp\left[\frac{-(x - x_0)^2}{\alpha_0^2}\right]. \tag{5.104}$$

The fitting of the functional form (5.104) to the observed activity measurements was accomplished by the method of least squares, and was facilitated

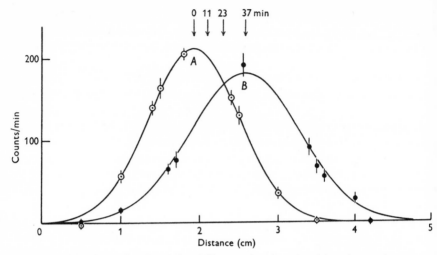

Figure 5.9.   The observed radioactivity of ^{42}K as a function of position along a giant nerve axon, at two different times. The curve labeled A represents the distribution of radioactivity at $t = 0$, and the curve labeled B represents the distribution at $t = 37$ min. A constant voltage gradient of magnitude -0.548 V/cm was maintained across the ends of the axon. The arrows indicate the distribution maxima at various times. The axon was taken from a specimen of the cuttlefish *Sepia officinalis*, a sea mollusk related to the squid. From Hodgkin and Keynes (1953), with permission.

by fitting log $c(x,0)$ to a quadratic function of x containing three parametric coefficients. From the values of the three coefficients in the quadratic function, the values of A_0, x_0, and α_0 were easily inferred.

By ignoring the cylindrical geometry of the axon, we can represent the subsequent transport and behavior of the potassium ion concentration by the one-dimensional convective diffusion equation (5.99). If only diffusion were taking place, then the subsequent concentration distribution, which we denote by $c_D(x,t)$, would be given by equations (5.23), (5.24), and (5.104) as

$$c_D(x,t) = \frac{1}{(4\pi Dt)^{1/2}} A_0 \int_{-\infty}^{\infty} \exp\left[-\frac{(\xi - x_0)^2}{\alpha_0^2} - \frac{(x - \xi)^2}{4Dt} \right] d\xi. \quad (5.105)$$

The integral appearing in the above equation can be readily evaluated by observing that it is of the form

$$\int_{-\infty}^{\infty} \exp[-(a\xi^2 - 2b\xi + c)]d\xi, \quad (5.106)$$

$$a = \frac{1}{\alpha_0^2} + \frac{1}{4Dt}, b = \frac{x_0}{\alpha_0^2} + \frac{x}{4Dt}, c = \frac{x_0^2}{\alpha_0^2} + \frac{x^2}{4Dt}.$$

We now "complete the square" in the exponent, writing

$$a\xi^2 - 2b\xi + c = a\left(\xi^2 - \frac{2b}{a}\xi\right) + c = a\left[\xi^2 - \frac{2b}{a}\xi + \left(\frac{b}{a}\right)^2 - \left(\frac{b}{a}\right)^2\right] + c$$

$$= a\left(\xi - \frac{b}{a}\right)^2 - \frac{b^2}{a} + c.$$

Next, let $\sqrt{a}(\xi - b/a) = \eta$, so that $\sqrt{a}\,d\xi = d\eta$, and the integral (5.106) becomes

$$e^{-b^2/a+c}\int_{-\infty}^{\infty} e^{-\eta^2}\frac{d\eta}{\sqrt{a}} = \sqrt{\frac{\pi}{a}}\,e^{-b^2/a+c}. \tag{5.107}$$

Hence, from (5.105)–(5.107), $c_D(x,t)$ becomes

$$c_D(x,t) = \frac{A_0\alpha_0}{(\alpha_0^2 + 4Dt)^{1/2}}\exp\left[-\frac{(x - x_0)^2}{\alpha_0^2 + 4Dt}\right]. \tag{5.108}$$

From the discussion following equation (5.102), we know that the solution to the convective diffusion equation $c(x,t)$ is obtained from $c_D(x,t)$ by replacing x by $x - ut$, that is,

$$c(x,t) = \frac{A_0\alpha_0}{(\alpha_0^2 + 4Dt)^{1/2}}\exp\left[-\frac{(x - ut - x_0)^2}{\alpha_0^2 + 4Dt}\right]. \tag{5.109}$$

Furthermore, we see that $c(x,t)$ does satisfy the initial condition (5.104) and the boundary conditions $c(\pm\infty,t) \to 0$. Hence $c(x,t)$ represents the required theoretical solution to the problem of the axoplasmic transport of the potassium ions.

According to equation (5.109), at any fixed time t subsequent to $t = 0$, the maximum in $c(x,t)$ occurs when the argument of the exponential function equals 0, or at

$$x = x_0 + ut. \tag{5.110}$$

In other words, the maximum in the distribution moves in the x direction with velocity u. Knowing u and dV/dx, we can infer the mobility β of potassium ions in axoplasm. By fitting the curve B to the experimental observations at time t in the same manner as described previously for time $t = 0$, both the maximum position $x_0 + ut$ and the quantity $\alpha_0^2 + 4Dt$ can be determined. Hence the diffusion constant D of potassium ions in axoplasm is also determined. This procedure provides an independent check on the value of D because D appears both in the exponent and in the amplitude of the Gaussian function which falls off like $(\alpha_0^2 + 4Dt)^{-1/2}$, according to equation (5.109).

Alternatively, the functional form (5.109) can be fitted to the data with only the two parameters D and u allowed to vary, so that unique "best" values of D and u are directly determined. The solid curve B in Figure 5.9 is based on equation (5.109) with values of D and β determined by the first procedure, for a given potential gradient.

Hodgkin and Keynes obtained in this manner the diffusion constant and mobility of potassium ions in axoplasm. At $18°C$, the mean values from 11 experiments were found to be

$$\beta = 4.9 \times 10^{-4} \, cm^2 \, sec^{-1} \, V^{-1},$$
$$D = 1.5 \times 10^{-5} \, cm^2 \, sec^{-1}. \tag{5.111}$$

These values are close to those obtained for potassium ions in sea water.

5.9. THE GAUSSIAN FUNCTION

In the one-dimensional unit source solution of Section 5.3 and again in Section 5.8 we have encountered in a physical context a function which is recognized as the **Gaussian function, normal distribution**, or "**bell-shaped curve.**" In probability theory it is introduced as the **normal density function**, which is defined as

$$f(x) = \frac{1}{\sqrt{2\pi} \, \sigma} \exp\left[-\frac{1}{2}\left(\frac{x - \mu}{\sigma}\right)^2 \right]. \tag{5.112}$$

This function is **normalized**, by which is meant that its integral over all x is unity,

$$\int_{\infty}^{\infty} f(x)dx = 1. \tag{5.113}$$

In measurements of various kinds in biology, such as the height or weight of humans, the hematocrit of normal human blood, or the volume of a cell, it is found that these quantities tend to have such a normal distribution. For example, if a thousand humans are chosen at random, their weights w are determined, and the number of people with a weight in the interval w to $w + dw$ is plotted as a function of w, then the resulting curve exhibits the general form of this normal density function. More generally, if x represents the measured quantity, then the fraction of observations having measured values falling between x and $x + dx$ is $f(x)dx$.

In the mathematical theory of probability, when events are described by the values of the continuous variable x, such that the probability of x taking on a value in the interval x to $x + dx$ is given by the probability $p(x)dx$, where $p(x) \geqq 0$, then x is called a **random variable** and $p(x)$ is called the

probability density function. An example of $p(x)$, which is frequently used as a model, is the normal density function. The statistical nature of the Brownian motion of large molecules underlies the intimate connection between the theories of diffusion and probability.

Returning now to the special case for which $p(x)$ is the normal density function $f(x)$, we see that it depends on the two parameters, σ and μ. The meaning of these two parameters is readily demonstrated. The **mean value** of x, denoted by $\bar{x}$, with respect to the function $f(x)$ is defined as

$$\bar{x} \equiv \int_{-\infty}^{\infty} xf(x)dx. \tag{5.114}$$

The expression on the right is also called the **first moment** about zero of the density function $f(x)$. The **nth moment** about zero of the density function $f(x)$ is the same expression with x replaced by x^n, where n is an integral. By substituting equation (5.112) into (5.114) above and performing the indicated integration explicitly, it is seen that

$$\bar{x} = \mu. \tag{5.115}$$

We say that μ is the **expected value** or mean value of x with respect to the density function $f(x)$, or equivalently μ is the first moment about zero of the density function $f(x)$. It is the value of x at the center of the curve.

The meaning of σ is seen by determining the **variance** of x, written var(x), which is defined with respect to the function $f(x)$ as follows:

$$\text{var}(x) \equiv \int_{-\infty}^{\infty} (x - \bar{x})^2 f(x)dx. \tag{5.116}$$

In words, the variance of x is the expected value or mean value of the square of the deviation of x from its mean. By integration by parts, it is found from equation (5.116) that

$$\text{var}(x) = \sigma^2,$$
or
$$\sigma = \sqrt{\text{var}(x)}. \tag{5.117}$$

σ is called the **standard deviation**. For $f(x)$, σ is the distance from the center of the curve to the inflection point at either side. The variance is a measure of the "spread" or "precision" of the density function about its mean. For example, by comparing equations (5.22) and (5.112), we see that for the simple diffusion process described by equation (5.22),

$$\sigma = \sqrt{2Dt}. \tag{5.118}$$

This means that "infinite precision" exists at $t = 0$, or complete concentration of the solute mass at the point $x = \xi$. The precision of localization of solute mass at the point ξ decreases with increasing time, as illustrated in Figure 5.3. At infinite time the precision is nil and the solute is uniformly distributed, or completely dispersed from its initial position ξ.

5.10. ULTRACENTRIFUGATION

A powerful instrument in the hands of the modern biologist is the **ultracentrifuge**. This is a device for spinning biochemical solutions at very large angular velocities, thereby subjecting any solute or suspension in the solution to a strong centrifugal force field. With it, the constituents of a cell are separated or **fractionated**. For example, the subunit composition of the enzyme aspartate transcarbamylase (Gerhart and Schachman, 1965) was first demonstrated with the essential aid of the ultracentrifuge. In addition, the molecular weights and diffusion constants or macromolecules are determined routinely.

Figure 5.10 displays the cylindrical symmetry of a sector-shaped container,

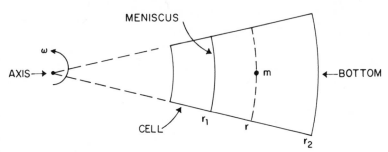

Figure 5.10.   Geometry of a sector-shaped centrifuge cell rotating with angular velocity ω in an ultracentrifuge. A single solute molecule of mass m is shown located at radius r.

called the **ultracentrifuge cell**, or **solution cell**, rotating with angular velocity ω. The depth of the cell (perpendicular to the plane of the paper) is large compared to its width. The cell is assumed to contain a solvent and a solute whose concentration is denoted by c. When the solution is inserted into the cell, the cell is normally held in a vertical position with the surface at $r = r_2$, called the **bottom**, on the downward side. Usually a small bubble remains in the cell, and the bubble-solution interface is called the **meniscus**. During centrifugation, the cell is put in place in the centrifuge, on its side, as indicated in Figure 5.10. Then the meniscus is at $r = r_1$, and the bottom of the container is at $r = r_2$. Typical dimensions of a cell are $r_1 = 6$ cm, and $r_2 = r_1 + 1$ cm. The angle subtended by the cell is usually about 2–4°.

Imagine for the moment that the solution contains a single solute molecule, and let us write down the balance of forces acting on it in the plane of the figure and in the radial direction. Thus by Newton's second law, after a terminal velocity u is attained so that the acceleration of the molecule is zero,

$$0 = m\omega^2 r - m_0\omega^2 r - fu. \tag{5.119}$$

Here m is the mass of the molecule, r is its radial position, u is the velocity in the radial direction (positive outwards), m_0 is the mass of an element of solvent having a volume equal to that of the molecule, and f is the frictional coefficient. The first term on the right is the centrifugal force on the particle due to rotation, acting outwards from the rotational center, the second term is the "buoyant" force on the particle, acting inwards, and the last term is the drag force on the molecule, also acting inwards. The buoyant force is so-called because the centrifugal force acts on the solvent as well as the solute, and is analogous to the buoyant force due to gravity, which acts on a body submerged in a liquid. In other words, the centrifugation process produces a force field, which can be thought of as an "effective gravitational force field." The "effective gravitational constant" of the centrifugal field, $\omega^2 r$, is normally very much greater than the true gravitational constant g. The reference to a drag force in equation (5.119) tacitly assumes that the solute molecule is large compared to the solvent molecule, so that the solute molecule is effectively a macroscopic body, and the solvent is considered to be a continuum. In the case of a sphere of radius a, we have seen that $f = 6\pi\mu a$, where μ is the coefficient of viscosity of the solvent.

Of course gravity is also acting on the particle, but the normal configuration of the ultracentrifuge makes the plane of rotation horizontal. Hence, the force of gravity is in a direction perpendicular to the plane of rotation, and so has no effect on the radial force equilibrium equation. We shall neglect the transient behavior of the particle during which it (and the ultracentrifuge cell) starts from rest and accelerates until it attains its steady angular rotational velocity ω. From (5.119), we see that the steady-state behavior is governed by the differential equation

$$u \equiv \frac{dr}{dt} = \frac{(m - m_0)}{f}\omega^2 r. \tag{5.120}$$

Since the length of the solution column $r_2 - r_1$ is small compared to the distance r_1 from the meniscus to the axis, we set $r \approx r_1$. Then the velocity dr/dt is approximately constant and given as

$$\frac{dr}{dt} \approx \frac{(m - m_0)}{f}\omega^2 r_1. \tag{5.121}$$

The solute molecule moves with this velocity until it rests on the bottom of

the cell, $r = r_2$. When applied to a solution, the above theoretical considerations neglect the interaction of one solute molecule with another, and can
therefore be expected to be useful only when the concentration of solute
particles is very small. For fractionation purposes, the solute is usually
allowed to accumulate at the bottom of the cell, where it forms a **pellet**, and
is easily removed from the remainder of the solution called the **supernatant**.

We wish to generalize the above considerations of the sedimentation of a
single particle in response to a centrifugal force field to the biologically
interesting case of a solution containing a single solute species. Such a
solution, containing one solute species and one solvent species, is called a
binary solution. In addition to being placed in the centrifugal field of an
ultracentrifuge, the solute is capable of diffusion. Then we can speak of a
solute flux or current due to diffusion, and a solute flux or current due to
sedimentation. The total solute current density j is given by the expression

$$j = -D \frac{\partial c}{\partial r} + cu, \tag{5.122}$$

where the first on the right represents the diffusional current density, and
the second term represents the convective current density or **sedimentation
current density**. Here the concentration c is considered to be a function of
the radial position r and the time t.

What is the velocity u associated with the solute flux? It is the velocity
of the solute particles resulting from the centrifugal field established by the
rotation of the ultracentrifuge cell. Therefore, we must utilize for u an
expression that is the generalization to a solute species of the velocity for
a single particle, equation (5.120). This expression is determined by thermodynamic considerations, according to which u is given by (5.120) with

$$m = \frac{M}{N_A},$$

$$m_0 = \frac{M}{N_A} \bar{v} \rho, \tag{5.123}$$

$$f = f(c) = f_0(1 + k_1 c + k_2 c^2 + \cdots).$$

Here M is the **molecular weight** or molar mass of the solute species, the
mass in grams of one mole of solute; N_A is Avogadro's number; $\bar{v}$ is the
partial specific volume of the solute, which is assumed to be constant ($m\bar{v}$
is the volume of solvent displaced by a molecule); ρ is the density of the
solution (solute plus solvent); and f is the frictional coefficient of the solute
molecules in the solution. f is concentration dependent as a consequence of
its dependence on the viscosity of the solution [see equation (4.19)]. In (5.123)
f_0 is the frictional coefficient in the limit as $c \to 0$, and is the same as the

frictional coefficient for a single solute molecule in the absence of other solute molecules. Thus, the velocity u appearing in the current density is written as

$$u = \frac{M}{N_A f}(1 - \bar{v}\rho)\omega^2 r. \tag{5.124}$$

The quantity s, called the **sedimentation coefficient**, is defined as

$$s \equiv \frac{u}{\omega^2 r}. \tag{5.125}$$

By substituting (5.124) into (5.125), we obtain the result

$$s = \frac{M(1 - \bar{v}\rho)}{N_A f}. \tag{5.126}$$

The quantity s is usually experimentally determined by measuring u and $\omega^2 r$. The customary unit of s is 1 Svedberg $\equiv 10^{-13}$ sec, denoted by the symbol S.

A useful form of equation (5.126) results if we eliminate f in it by means of the Einstein equation (5.43) relating the diffusion constant D to the frictional coefficient f, to obtain the result

$$M = \frac{sRT}{D(1 - \bar{v}\rho)}. \tag{5.127}$$

This equation is called the **Svedberg relation** (Svedberg and Pedersen, 1940), and permits the determination of the molecular weight of a macromolecule from measurements of s, D, and $\bar{v}$ in a solution of density ρ at temperature T. It is strictly correct for **ideal** solutions only, that is to say, in the limit of zero concentration, because s and D depend on c. In practice, measurements of s and D are often made for several different values of the concentration, at a fixed temperature. The values of s and D at zero concentration are then inferred from an extrapolation procedure applied to plots of s and D versus the concentration. The Svedberg equation is often written with the subscript 0 attached to s and to D, to emphasize the fact that the zero concentration limit of these quantities is to be inserted.

Proteins exhibit sedimentation coefficients which are usually in the range 1–20 S. Macromolecules with large molecular weights usually have correspondingly large sedimentation coefficients. For example, for ribosomes, $s_{20,w} = 83.0$ S, and for tobacco mosaic virus, $s_{20,w} = 198$ S. The subscripts on s indicate the standard conditions of water as a solvent at 20°C temperature. However, we stress that the sedimentation coefficient depends on the shape as well as the molecular weight of a molecule. For example, a highly elongated molecule possesses a larger frictional coefficient and hence a lower sedimentation coefficient than a compact molecule of the same molecular weight. These remarks are illustrated in Table 5.4, which lists the values s, D, and $\bar{v}$ for some representative proteins.

TABLE 5.4   VALUES OF THE SEDIMENTATION COEFFICIENT s, THE DIFFUSION CONSTANT D, THE PARTIAL SPECIFIC VOLUME $\bar{v}$, AND THE MOLECULAR WEIGHT M, OF SOME REPRESENTATIVE PROTEINS[a]

Protein	$s_{20, w}$ (S)	$D_{20, w}$ (10^{-7} cm^2/sec)	$\bar{v}_{20}$ (g/cm^3)	M (g/mole)
		Globular proteins		
Lipase, milk	1.14	14.48	0.7137	6,669
Myoglobin, horse heart	2.04	11.30	0.741	16,890
Albumin, horse serum	4.58	6.42	0.748	68,600
Old yellow enzyme, brewer's yeast	5.76	6.30	0.731	82,390
Lactate dehydrogenase, pig heart	6.93	5.95	0.741	109,000
Catalase, bovine liver	11.30	4.10	0.730	247,500
β-Galactosidase, E. coli ML 309	15.93	3.12	0.76	515,300
Glutamate dehydrogenase, bovine liver	26.60	2.54	0.750	1,015,000
		Fibrous proteins		
Histone I, calf thymus	2.03	5.13	0.74	36,890
β-Lipoprotein, human	5.9	1.7	0.97	2,663,000
Fibrinogen, human	7.2	1.2	0.710	501,800
Globulin, horse antipneumococcus	19.3	1.80	0.715	912,400

[a] The data are based largely on velocity sedimentation experiments. The characterization of a particular protein as globular or fibrous is based primarily on the value (not shown) of the ratio of the associated frictional coefficient f (see equation (5.126)) to that for an equivalent sphere. The subscripts 20 and w denote the temperature of 20°C and the solvent water, respectively. From *Handbook of Biochemistry*, 2nd ed., H. A. Sober, ed., © The Chemical Rubber Co., 1970. Used by permission of The Chemical Rubber Co.

We are now in a position to utilize the continuity equation so as to take into account the effect of the centrifugal field as well as diffusion on solute transport. We shall utilize the form of the equation appropriate to a cylindrical coordinate system so as to take advantage of the symmetry of the experimental configuration: There is no dependence of the concentration or the sedimentation coefficient on the depth z parallel to the axis of rotation, or on the angular position φ in the plane perpendicular to the rotation axis. In addition, the components of the current density $\mathbf{j}$ in any plane perpendicular to the radial direction are null. Thus, we apply the conservation of mass flow to the infinitesimal volume element shown in Figure 5.11. With j_r

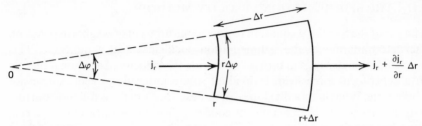

Figure 5.11. Illustrating the application of the equation of continuity to an infinitesimal volume element $r\,\Delta r\,\Delta\phi\,\Delta z$ in a cylindrical coordinate system. The direction of z is perpendicular to the plane of the paper. The current is positive in the outward direction and negative inwards, and is assumed to possess only a radial component j_r.

denoting the component of the current density in the radial direction, it follows that

$$r\,\Delta\varphi\,\Delta r\,\frac{\partial c}{\partial t} = r\,\Delta\varphi j_r - (r + \Delta r)\Delta\varphi\left(j_r + \frac{\partial j_r}{\partial r}\,\Delta r\right) = \Delta\varphi\,\Delta r\left\{-j_r - r\,\frac{\partial j_r}{\partial r}\right\},$$

to order Δr. After dividing by $r\,\Delta\varphi\,\Delta r$ and taking the limit as $\Delta r \to 0$, we obtain the continuity equation in cylindrical coordinates for a cylindrically symmetric flow:

$$\frac{\partial c}{\partial t} + \frac{1}{r}\frac{\partial}{\partial r}(rj_r) = 0. \tag{5.128}$$

Inserting in (5.128) the expression (5.122) for j_r and utilizing (5.125), we obtain the equation

$$\frac{\partial c}{\partial t} = \frac{1}{r}\frac{\partial}{\partial r}\left[Dr\,\frac{\partial c}{\partial r} - s\omega^2 r^2 c\right], \tag{5.129}$$

where $c = c(r,t)$. This equation was first derived by Lamm (1929) and is called the **Lamm equation**. In general it is necessary to be concerned about the possible dependence of D as well as s on c.

If a **multicomponent solution** is considered, one for which more than one solute component is present in the solvent, then it is necessary to represent the system by the **generalized Lamm equations**. These are coupled equations of the form of (5.129) in which c is replaced by the vector $\mathbf{c}$ and the diffusion coefficient D is replaced by a matrix. Each component of $\mathbf{c}$ represents one solute species. In biochemical practice, the solution is often a multicomponent system, although the simpler theoretical considerations based on a binary solution are usually utilized to describe it.

5.11. THE SEDIMENTATION VELOCITY METHOD

The use of the Svedberg equation to determine molecular weights depends on the determination of the sedimentation coefficient s of a molecule. This determination depends in turn on observing the velocity of a solute molecule in a centrifugal force field. Individual solute molecules are of course not observable. What is actually done can be described briefly as follows. Starting with a binary solution of uniform concentration in an ultracentrifuge cell, the ultracentrifuge rotor is operated at high speed, up to 70,000 r/min (revolutions per minute). At the same time, the concentration in the cell is observed by an optical method, based on either interferometry (**schlieren method, Rayleigh method**), or absorption (**ultraviolet absorption method**), as illustrated in Figure 5.12. The interferometric methods derive from the physical property that the index of refraction of a solution is proportional to the solute concentration, when the latter is small.

Under the influence of the centrifugal force field, the solute slowly sediments to the bottom of the cell leaving in its wake a clear region at the top of the cell. Between the clear region, or supernatant, and the solution region, or **plateau region**, there is a transition zone called the **boundary region**, or

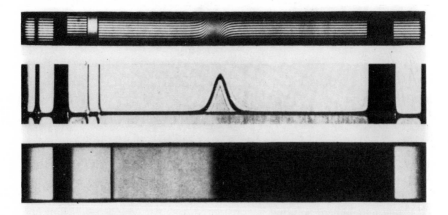

Figure 5.12. Illustrating the optical patterns obtained on viewing an ultracentrifuge cell during velocity sedimentation experiments with (reading from top to bottom) Rayleigh, schlieren, and ultraviolet absorption optical systems, respectively. From Schachman (1959), with permission.

simply the **boundary**. In addition to the boundary region and the plateau region, there is a **bottom region** in which the solute is accumulating. The concentration there increases rapidly with r as indicated schematically in Figure 5.13. The **sedimentation velocity method** or **"moving boundary" method** of determining s is based on the determination of the position of the boundary as a function of time. The utilization of the method requires a precise definition of the "position of the boundary." The theoretical basis for the determination of this position and the inferences to be drawn from it are provided by the Lamm equation.

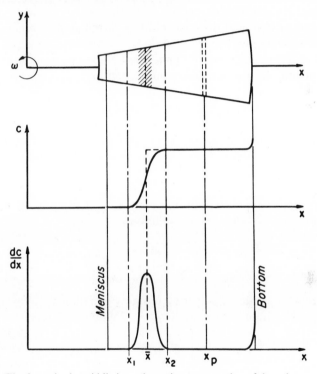

Figure 5.13. The figure in the middle is a schematic representation of the solute concentration as a function of radial distance (labeled x), with corresponding positions in the ultracentrifuge cell shown above, and the corresponding gradient curve shown below. The point labeled $\bar{x}$ in the figure denotes the boundary point $r_B(t)$. Reprinted, with permission, from R. Trautman and V. Schumaker (1954), *Journal of Chemical Physics, 22*, pp. 551–554.

Because D and s depend in general on c, the Lamm equation cannot be solved explicitly. However, it can be solved explicitly when one or another simplifying assumption is made. A comprehensive review of mathematical solutions of the Lamm equation under different conditions can be found in the work of Fujita (1962). Let us assume that s and D are constants, which are independent of the solute concentration, and denote them by s_0 and D_0, respectively. We shall present here some general consequences of the Lamm equation and its solution subject to one condition only (Goldberg, 1953), namely, that the ultracentrifuge cell contains a plateau region. That is to say, we assume that during some finite time interval following the initiation of centrifugation, there is a region between the meniscus and the bottom where $\partial c/\partial r$ vanishes. If we denote a generic point in the plateau region by $r = r_p$, then the existence of a plateau region can be expressed as

$$c(r_p,t) = c_p(t),$$ \hfill (5.130)

where $c_p(t)$ is the value of the concentration in the plateau region, and is a function of time alone.

If we substitute equation (5.130) into the Lamm equation (5.129), we find that $c_p(t)$ satisfies the ordinary differential equation

$$\frac{dc_p(t)}{dt} = -2s_0\omega^2 c_p(t).$$ \hfill (5.131)

Assume further that the ultracentrifuge cell has been prepared in the standard manner, so that the concentration satisfies the initial condition

$$c(r,0) = c_0.$$ \hfill (5.132)

It follows from equation (5.130) that $c_p(0) = c_0$. The solution of equation (5.131) subject to the latter initial condition is

$$c_p(t) = c_0 e^{-2s_0\omega^2 t}.$$ \hfill (5.133)

The reason c_p decreases with time is that the solute is accumulating in the bottom region.

Let us now integrate r times the continuity equation (5.128) with respect to r, between the limits $r = r_1$ and $r = r_p$. We obtain the relation

$$\int_{r_1}^{r_p} \frac{\partial c(r,t)}{\partial t} r \, dr = -\int_{r_1}^{r_p} \frac{\partial}{\partial r}(rj) dr = r_1 j(r_1,t) - r_p j(r_p,t).$$ \hfill (5.134)

The physical condition that the current density must vanish at the meniscus is expressed as

$$j(r_1,t) = \left(-D_0 \frac{\partial c}{\partial r} + s_0\omega^2 rc\right)_{r=r_1} = 0.$$ \hfill (5.135)

With the aid of this boundary condition and equation (5.130), equation (5.134) becomes

$$\frac{d}{dt}\int_{r_1}^{r_p} c(r,t)r \, dr = -s_0\omega^2 r_p^2 c_p(t).$$ \hfill (5.136)

Note that we have transferred the time derivative from inside the integral in (5.134) to outside the integral in (5.136).

In order to evaluate the integral appearing on the left-hand side of equation (5.136), we define the **boundary point** $r_B(t)$ of the sedimentation curve in the following manner. It is that value of r for which the area under the plateau curve $c_p(t)$, between the limits $r_B(t)$ and r_p, equals the area under the true sedimentation curve, between r_1 to r_p (see Figure 5.13). In other words it is the infinitely sharp position that the boundary would have if the solute did not diffuse. Mathematically, $r_B(t)$ is defined by the equation

$$\int_{r_1}^{r_p} c(r,t)r\,dr = \int_{r_B(t)}^{r_p} c_p(t)r\,dr. \tag{5.137}$$

By carrying out the indicated integration on the right-hand side above, we find that

$$\int_{r_1}^{r_p} c(r,t)r\,dr = \tfrac{1}{2}c_p(t)[r_p^2 - r_B^2(t)]. \tag{5.138}$$

If we insert this expression into (5.136), the latter becomes

$$\frac{1}{2}\frac{dc_p}{dt}[r_p^2 - r_B^2(t)] - c_p(t)r_B(t)\frac{dr_B}{dt} = -s_0\omega^2 r_p^2 c_p(t). \tag{5.139}$$

By combining this equation with (5.131), we obtain the following equation, which $r_B(t)$ must satisfy:

$$\frac{dr_B}{dt} = s_0\omega^2 r_B. \tag{5.140}$$

From (5.132) and (5.137) evaluated at $t = 0$, it follows that $r_B(0) = r_1$. Hence, the solution to equation (5.140) subject to the latter initial condition is

$$r_B(t) = r_1 e^{s_0\omega^2 t}. \tag{5.141}$$

A semilog plot of $r_B(t)$, which is observable as a function of time, permits the experimental determination of the slope $s_0\omega^2$ and hence s_0, since ω is known from the calibration of the ultracentrifuge rotor. If we eliminate the exponential term from the above expression by means of equation (5.133), it follows that

$$c_p(t) = \frac{c_0 r_1^2}{r_B^2(t)}. \tag{5.142}$$

In words, the concentration in the plateau region falls off in the time, inversely as the square of the distance of the boundary point $r_B(t)$ from the rotor. Alternatively, equation (5.142) states that the product $c_p(t)r_B^2(t)$ remains constant during the course of the ultracentrifugation process, as long as there is an observable plateau region. The result (5.142) is called the **square dilution rule**, or the **radial dilution law**.

The gradient of the concentration curve is shown schematically at the bottom of Figure 5.13. Such a curve is called a **sedimentation boundary curve**, or a **gradient curve**. The schlieren optical method observes this curve directly, as illustrated in Figure 5.12. For such observations, it is common practice to select the position of the boundary as that point at which the spatial gradient $\partial c/\partial r$ is a maximum. We will now show how the "true" boundary point $r_B(t)$, the one that is defined by equation (5.137) and satisfies equation (5.142) is expressible in terms of the gradient curve.

Thus, let us integrate the left-hand side of equation (5.138) by parts. We find readily that

$$\int_{r_1}^{r_p} c(r,t)r\,dr = \tfrac{1}{2}r^2 c(r,t)\Big|_{r_1}^{r_p} - \frac{1}{2}\int_{r_1}^{r_p} \frac{\partial c}{\partial r}r^2\,dr. \tag{5.143}$$

By substituting this result into equation (5.138), we see that

$$r_B^2(t) = \frac{r_1^2 c(r_1,t) + \displaystyle\int_{r_1}^{r_p} \frac{\partial c}{\partial r}r^2\,dr}{c_p(t)}. \tag{5.144}$$

Now, by definition,

$$c_p(t) = c(r_1,t) + \int_{r_1}^{r_p} \frac{\partial c(r,t)}{\partial r}\,dr. \tag{5.145}$$

Introducing this relation into the denominator in equation (5.144), we obtain the result

$$r_B^2(t) = \frac{r_1^2 c(r_1,t) + \displaystyle\int_{r_1}^{r_p} \frac{\partial c}{\partial r}r^2\,dr}{c(r_1,t) + \displaystyle\int_{r_1}^{r_p} \frac{\partial c}{\partial r}\,dr}. \tag{5.146}$$

This expression for $r_B^2(t)$ is especially useful if the boundary region does not extend back to the meniscus, so that $c(r_1,t) = 0$. Then (5.146) reduces to

$$r_B^2(t) = \frac{\displaystyle\int_{r_1}^{r_p} \frac{\partial c}{\partial r}r^2\,dr}{\displaystyle\int_{r_1}^{r_p} \frac{\partial c}{\partial r}\,dr}. \tag{5.147}$$

In words, $r_B(t)$ is expressible as the square root of the second moment of the gradient curve.

Velocity sedimentation is frequently utilized for the purpose of physically separating different macromolecular components of a solution. The procedure takes advantage of the different sedimentation coefficients of the different macromolecular species [see equation (5.126)]. Initially, the heterogenous components from a cell extract, say, are placed in a small layer on top of a solvent. The centrifugation is then commenced and continued for a period sufficiently long to permit the species with different sedimentation coefficients to reach distinct radial positions or **zones**, but not so long that all the components reach the bottom of the cell. Once the macromolecule species are in distinct zones, they are easily extracted separately.

It is usual to prepare the solvent initially so that it has a **preformed density gradient**, with the solvent density increasing in the direction of increasing radial distance in the ultracentrifuge cell. This solvent is commonly prepared from a sucrose solution, and is known as a **sucrose density gradient**. The purposes of the gradient is to help prevent convective disturbances from forming in the solution. Of course, the gradient is not stable itself under the action of diffusive forces. However, over the time scale of ultracentrifugal experiments of the order of several hours or less, little change in the gradient occurs, and it is effectively constant. This method of separation of macromolecules is called **zone centrifugation in a preformed density gradient**, or simply **zone centrifugation**. A variant procedure which does not utilize a preformed density gradient, but rather a salt solvent which forms a gradient as the centrifugation progresses, is known as **band centrifugation**.

5.12.* AN APPROXIMATE SOLUTION TO THE LAMM EQUATION

An explicit solution to the Lamm equation was first given by Faxen (1929) for the special case in which the cell is considered to be an infinite wedge, that is to say, $r_1 = 0$ and $r_2 = \infty$. The Faxen solution is applicable to a cell which is prepared initially so that there is a sharp boundary separating a solute free region and the solution at a point $r = r_0$ somewhere in the middle of the cell, or,

$$c(r,0) = \begin{cases} 0, & r_1 < r < r_0, \\ c_0, & r_0 < r < r_2. \end{cases} \tag{5.148}$$

In practice, the establishment of such an initial distribution of solute requires special effort. An ultracentrifuge cell prepared in this manner is called a **synthetic boundary cell**. Because the Faxen solution applies to a region of semi-infinite extent the solution does not represent a standard centrifuge cell near r_1 and r_2 where boundaries alter the concentration distribution.

We shall present here an approximate solution of the Faxen type that is considerably simpler than the original Faxen solution, but that nevertheless does not differ from it significantly. The approximate solution is applicable to velocity sedimentation experiments. The basic idea of the approximation is to represent the solution in the neighborhood of the sharp boundary accurately, and not to be concerned with what happens far from this region. We assume again that s and D are both constants, independent of c, and denoted by s_0 and D_0, respectively. The Lamm equation assumes the form

$$\frac{\partial c}{\partial t} = \frac{D_0}{r} \frac{\partial}{\partial r} \left(r \frac{\partial c}{\partial r} \right) - s_0 \omega^2 r \frac{\partial c}{\partial r} - 2 s_0 \omega^2 c. \tag{5.149}$$

The concentration $c = c(r,t)$ satisfies the initial condition (5.148).

We eliminate the last term in equation (5.149) by setting

$$c(r,t) = e^{-2 s_0 \omega^2 t} w(r,t). \tag{5.150}$$

By substituting the above expression for c into equation (5.149), we find that $w(r,t)$ satisfies the equation

$$\frac{\partial w}{\partial t} = \frac{D_0}{r} \frac{\partial}{\partial r} \left(r \frac{\partial w}{\partial r} \right) - s_0 \omega^2 r \frac{\partial w}{\partial r}. \tag{5.151}$$

The function w satisfies the same initial condition as does c. Now transform the variable r to the variable η, where

$$r = r_0 e^{\eta/2}. \tag{5.152}$$

The range $0 < r < \infty$ becomes, for the variable η, the range $-\infty < \eta < +\infty$. Furthermore, the boundary point $r = r_0$ becomes the origin of the variable η. With the aid of (5.152), we calculate

$$\frac{\partial w}{\partial r} = \frac{d\eta}{dr} \frac{\partial w}{\partial \eta} = \frac{2}{r} \frac{\partial w}{\partial \eta}. \tag{5.153}$$

Hence the equation for w considered as a function of η and t becomes

$$\frac{\partial w}{\partial t} = \frac{4 D_0}{r_0^2} e^{-\eta} \frac{\partial^2 w}{\partial \eta^2} - 2 s_0 \omega^2 \frac{\partial w}{\partial \eta}. \tag{5.154}$$

During the initial stages of the centrifugation process, most of the changes in concentration occur in the vicinity of the initial boundary where $r \approx r_0$ and, from (5.152),

$$e^{\eta/2} \approx 1, \eta \approx 0, e^{-\eta} \approx 1. \tag{5.155}$$

We shall restrict our consideration of equation (5.154) to the neighborhood of the boundary region and formally replace the factor $e^{-\eta}$ in (5.154) by unity.

Then equation (5.154) simplifies to

$$\frac{\partial w}{\partial t} = \frac{4D_0}{r_0^2}\frac{\partial^2 w}{\partial \eta^2} - 2s_0\omega^2\frac{\partial w}{\partial \eta}. \tag{5.156}$$

In terms of the variable η, the step discontinuity in c at $r = r_0$ becomes a step discontinuity in w at the origin, that is,

$$w(\eta,0) = \begin{cases} 0, & \eta < 0, \\ c_0, & \eta > 0. \end{cases} \tag{5.157}$$

Up to now, we have omitted reference to the boundary conditions to be satisfied by c and hence w. In the early stage of sedimentation in a synthetic cell, c must be zero at the meniscus r_1 and at the bottom of the cell r_2. We shall assume that the position of the boundary is initially, and at all subsequent times, far from either r_1 or r_2. Thus, our solution will not be expected to be a good approximation to the true solution if the initial position of the boundary is chosen to be at one of these two endpoints. We therefore assume that, in terms of the variable η, for η sufficiently far from $\eta = 0$ (the initial position of the boundary),

$$w(-\infty,t) = 0,$$
$$w(+\infty,t) = c_0. \tag{5.158}$$

Equation (5.156) is recognized as the one-dimensional convective diffusion equation with the convective velocity equal to $2s_0\omega^2$. If equation (5.156) were simply the diffusion equation, subject to the initial condition (5.157) and the boundary conditions (5.158), then the problem posed would be essentially the one we solved in Section 5.4, with the solution given by equation (5.32). There, however, the initial step discontinuity in c at the origin was negative (see equations (5.27) and (5.28)), while now the step discontinuity is positive. Let us denote the solution to the one-dimensional diffusion equation subject to the same conditions that w must satisfy, equations (5.157) and (5.158), by $c'(x,t)$. Then $c'(x,t)$ is obtainable from (5.32) by replacing x by $-x$, that is,

$$c'(x,t) = \frac{c_0}{2}\,\text{erfc}\left[\frac{-x}{(4Dt)^{1/2}}\right], \tag{5.159}$$

According to the discussion of Section (5.8), the solution c' given by (5.159) becomes a solution of the convective diffusion equation (5.99) if we replace x in c' by $x - ut$. Hence, the solution to equation (5.156) is obtained from (5.159) by means of the substitutions $x \to \eta - 2s_0\omega^2 t$, $D \to 4D_0/r_0^2$, namely,

$$w(\eta,t) = \frac{c_0}{2}\,\text{erfc}\left[r_0\frac{-\eta + 2s_0\omega^2 t}{(16D_0 t)^{1/2}}\right]. \tag{5.160}$$

From this result we can write down $c(r,t)$ with the aid of equations (5.152) and (5.150) as

$$c(r,t) = \frac{c_0}{2} e^{-2s_0\omega^2 t} \operatorname{erfc}\left[r_0 \frac{s_0\omega^2 t - \log(r/r_0)}{(4D_0 t)^{1/2}} \right]. \tag{5.161}$$

We know from equation (5.29) of Section 5.4 that the gradient of $c(x,t)$ as given by (5.32) is a Gaussian function. Similarly, we find by differentiating (5.161) with respect to r that the gradient curve is given as

$$\frac{\partial c}{\partial r} = \frac{c_0 e^{-2s_0\omega^2 t}}{(4\pi D_0 t)^{1/2}} \exp\left[-r_0^2 \frac{\{s_0\omega^2 t - \log(r/r_0)\}^2}{4D_0 t} \right]. \tag{5.162}$$

In deriving this equation, we have replaced a multiplicative factor r_0/r on the right-hand side by unity, because $r_0/r = e^{-n/2} \approx 1$, consistent with the approximation (5.155) by means of which equation (5.156) for w was obtained.

Equation (5.162) shows that $\partial c/\partial r$ is a Gaussian function with the mean position $\bar{r}$, which is determined by setting the argument in the square brackets equal to zero, namely,

$$\bar{r} = \bar{r}(t) = r_0 e^{s_0\omega^2 t}. \tag{5.163}$$

This expression for $\bar{r}$ has the same temporal dependence as does the boundary point $r_B(t)$, given in equation (5.141). The amplitude or height of the Gaussian function at its maximum is denoted by H and is given by (5.162) and (5.163) as

$$H \equiv \left(\frac{\partial c}{\partial r}\right)_{max} = \left(\frac{\partial c}{\partial r}\right)_{r=\bar{r}} = \frac{c_0 e^{-2s_0\omega^2 t}}{(4\pi D_0 t)^{1/2}}, \tag{5.164}$$

and is seen to be a decreasing function of time.

The value of the concentration in the plateau region is its value "far" from the boundary, or $c(\infty,t)$. From equation (5.161), this quantity is given as

$$c(\infty,t) = \frac{c_0}{2} e^{-2s_0\omega^2 t} \operatorname{erfc}(-\infty) = c_0 e^{-2s_0\omega^2 t}, \tag{5.165}$$

because $\operatorname{erfc}(-\infty) = 1 + \operatorname{erf}(\infty) = 2$. If we eliminate the exponential time factor in this equation by means of equation (5.163), we obtain the result

$$c(\infty,t) = c_0 \left[\frac{r_0}{\bar{r}(t)}\right]^2. \tag{5.166}$$

This equation is recognized to be the radial dilution law, except that the position of the boundary according to the approximate theory is $\bar{r}(t)$, the mean position of the Gaussian. For all practical purposes, there is a negligible difference between $\bar{r}(t)$ and $r_B(t)$ (see problem 18).

Let us examine to what extent this approximate solution represents the usual conditions encountered in a sedimentation velocity experiment. If the approximate solution (5.161) is compared to the exact Faxén solution for an infinite wedge (Fujita, 1962), it is found that the difference between the two is very small provided that each of the following conditions is satisfied:

$$\frac{r - r_0}{r_0} \ll 1, \; 2s_0\omega^2 t \ll 1, \text{ and } \frac{2D_0}{s_0\omega^2 r_0^2} \ll 1. \tag{5.167}$$

Representative values of the parameters in a typical sedimentation velocity experiment, are as follows:

$$r_1 = 6 \text{ cm}, r_2 = 7 \text{ cm}, r_0 = 6.5 \text{ cm}, s_0 = 5 \times 10^{-13} \text{ sec},$$

$$\omega = \frac{5\pi}{3} \times 10^3 \text{ rad/sec} \, (= 50{,}000 \text{ r/min}), \; D_0 = 5 \times 10^{-7} \text{ cm}^2/\text{sec}. \tag{5.168}$$

Inserting these values into the conditions (5.167), we see that for the maximum value of the radius $r = r_2$, $(r_2 - r_0)/r_0 = 0.077$. The duration of sedimentation velocity experiments is usually not greater than a couple of hours so that the maximum value of t is about 100 min. For this value, $2s_0\omega^2 t = 0.16$. The nondimensional parameter $2D_0/s_0\omega^2 r_0^2 = 0.0017$. Thus, the conditions (5.167) are easily satisfied for the duration of most sedimentation velocity experiments.

What if the initial concentration in the cell is prepared in the standard manner so that $r_0 = r_1$? Then the coincidence of the meniscus with the initial boundary point interferes with the concentration distribution described by equation (5.161), because of the boundary condition that j vanishes at the meniscus. The approximate solution given by (5.161) does not satisfy either of these conditions, and the solution is not valid in the neighborhood of these endpoints. However, after the moving boundary separates from the meniscus, the above solution does describe the subsequent motion adequately, until the moving boundary encounters the bottom region.

The sedimentation velocity method is also utilized to determine the diffusion constant. Thus, let us calculate the area B under a sedimentation boundary curve (again, assume that this Gaussian curve is distinct from and "far" from the cell boundaries r_1 and r_2):

$$B = \int_{-\infty}^{\infty} \frac{\partial c}{\partial r} \, dr = c(\infty, t) - c(-\infty, t) = c_0 e^{-2s_0\omega^2 t}, \tag{5.169}$$

where $c(\infty, t)$ is given by (5.165) and $c(-\infty, t)$ is null, according to (5.150) and (5.158). Now we combine this result with the height of the boundary curve as given by (5.164) to infer that

$$\left(\frac{B}{H}\right)^2 = 4\pi D_0 t. \tag{5.170}$$

Hence, a plot of $(B/H)^2$ as a function of the time yields a straight line whose slope determines the value of D_0, as illustrated in Figure 5.14. We note that equation (5.170) is independent of the sedimentation coefficient, although its derivation depends on the assumption that s is independent of concentration.

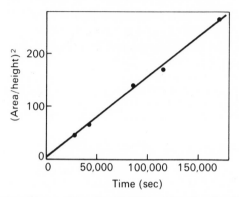

Figure 5.14. Illustrating the use of equation (5.170) to determine the diffusion constant. The data points (circles) apply to the solute *ovalbumin*, and are taken from Lamm and Polson (1936). Ovalbumin is the major protein found in the egg white of hens. The fact that the fitted straight line does not go through the origin is a consequence of the fact that at $t = 0$, the solute already extended over a small region surrounding the origin (as if diffusion had commenced at an earlier time). Reprinted from K. E. Van Holde, *Physical Biochemistry*, © 1971, p. 88. By permission of Prentice-Hall, Inc., Englewood Cliffs, New Jersey.

Finally, we remark that the Lamm equation, subject to the initial condition and boundary conditions of the standard cell and the assumption that s and D are constant, can be solved by the method of separation of variables (Archibald, 1938). From a practical point of view, the usefulness of the Archibald solution, which is expressed as an infinite series, depends on the rapidity of convergence of the infinite series, that is, the number of terms that need to be calculated to obtain c to a desired level of accuracy. Speaking qualitatively, it can be said that, in contrast to the Faxén type solutions, the Archibald solution is useful when Dt/r_1^2 is not small, because the series solution then converges rapidly. Hence, the Archibald solution is useful for small molecules and at later stages of the sedimentation process. Some progress has also been made in solving the Lamm equation approximately in certain cases for which s is an explicit function of the concentration.

5.13. SEDIMENTATION EQUILIBRIUM

In the **sedimentation equilibrium method**, the centrifugation of a solution is maintained for a very long time until an equilibrium distribution of the solute concentration is established inside the cell. In other words, the solute distribution consists entirely of the bottom region, where the forces due to the centrifugal field in the positive radial direction and the diffusional force directed backwards from the bottom of the cell have reached equilibrium. The angular velocity ω is usually maintained at a much slower rate than it is in the sedimentation velocity method, for example, at a speed $\omega = 10^4$ r/min. As a result, the effect of the backward diffusive force is enhanced, and the bottom region extends over a large radial part of the cell, so that the concentration and the concentration gradient are more easily measured. The concentration distribution over the radial distance in a cell, at several times during the course of a low speed sedimentation equilibrium experiment, is illustrated in Figure 5.15.

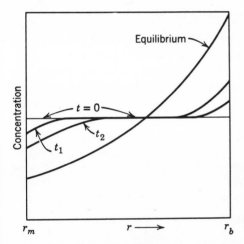

Figure 5.15. The concentration is shown schematically as a function of the radial distance r in an ultracentrifuge cell, extending from the meniscus at $r = r_m$ to the cell bottom at $r = r_b$, for successive times $t = 0, t_1, t_2,$ and ∞, during the course of a low-speed sedimentation equilibrium experiment. At $t = t_1$ and $t = t_2$, the bottom region extends over that portion of the curve lying above the initial $(t = 0)$ concentration level. From Tanford (1961), with permission.

The mathematical problem of solving the Lamm equation subject to the above mentioned conditions is considerably simplified by the equilibrium requirement, which makes $\partial c/\partial t = 0$. Then the concentration is a function of r alone, and the Lamm equation (5.129) becomes an ordinary differential equation for $c = c(r)$,

$$\frac{d}{dr}\left[Dr \frac{dc}{dr} - s\omega^2 r^2 c \right] = 0. \tag{5.171}$$

This equation can be integrated once immediately, with the result that the quantity in the square brackets, which is the radial current density, equals a constant. Because of the boundary conditions satisfied at the cell boundaries, that the current vanishes there (see equation (5.135)), the constant vanishes. Hence, a first integration of (5.171) leads to the relation

$$D \frac{dc}{dr} = s\omega^2 rc. \tag{5.172}$$

This equation, too, is easily integrated and yields the result, in the general case for which s and D are dependent on the concentration,

$$\int_{c_1}^{c(r)} \frac{D(c')dc'}{s(c')c'} = \omega^2 \int_{r_1}^{r} r'\,dr' = \tfrac{1}{2}\omega^2(r^2 - r_1^2), \tag{5.173}$$

where c_1 is the value of the concentration at the meniscus. The ratio D/s is given by equation (5.127) as

$$\frac{D}{s} = \frac{RT}{(1 - \bar{v}\rho)M}. \tag{5.174}$$

To integrate the left side of equation (5.173), we must know the dependence on the concentration of $\bar{v}$ and ρ. However, we shall not pursue the analysis of this case further, because when the solution is not dilute, equation (5.172) (as well as the Svedberg equation) must be modified. Rather, we shall assume here that the solution is dilute, so that D/s is independent of c, and equation (5.172) is applicable.

We rewrite equation (5.172), with (5.174) substituted into it, as

$$\frac{1}{c}\frac{dc}{dr} = \frac{(1 - \bar{v}\rho)\omega^2 M}{RT} r = 2AMr, \tag{5.175}$$

$$A = \frac{(1 - \bar{v}\rho)\omega^2}{2RT}. \tag{5.176}$$

Equation (5.175) is readily integrated with A considered to be constant to yield the mathematical solution

$$c = c_1 \exp[AM(r^2 - r_1^2)]. \tag{5.177}$$

While c_1 is not known, its value is not needed for the determination of M. Thus, equation (5.177) can also be written as

$$\log c = \log c_1 + AM(r^2 - r_1^2), \tag{5.178}$$

so that a plot of $\log c$ versus r^2 yields a straight line whose slope determines the value of M. Alternatively, a plot of $c^{-1}(dc/dr)$ versus r, or a plot of $r^{-1}(dc/dr)$ versus c, will also yield a straight line with slope $2AM$.

In utilizing the schlieren optical system to observe the concentration gradient distribution, concentration differences rather than the absolute values of the concentration at a given radius are inferred. In such a case, $r^{-1}(dc/dr)$

is plotted as a function of the concentration difference $c - c_m$, where c is the concentration at the position r, and c_m is the concentration at an arbitrarily chosen convenient reference position, near the meniscus, say. Here $c - c_m$ is obtained by integration of the observed concentration gradient, even though neither c nor c_m is known. The resulting curve is still a straight line, with the same slope as the line for which $c_m = 0$, because equation (5.175) is equivalent to the relation

$$\frac{1}{r}\frac{dc}{dr} = 2AM(c - c_m) + 2AMc_m. \tag{5.179}$$

Such a plot is illustrated in Figure 5.16.

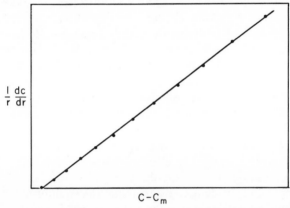

$$\frac{1}{r}\frac{dc}{dr}$$

$$C - C_m$$

Figure 5.16. A plot of the data from a sedimentation equilibrium experiment, in which the schlieren optical system was utilized, is presented in the form of equation (5.179). The slope of the straight line equals $2AM$. The ultracentrifuge cell contained a solution of the protein neurophysin, a carrier for some of the pituitary hormones. From Haschemeyer and Haschemeyer (1973), with permission.

In equation (5.178), c_1 can be determined in terms of c_0 and other parameters of the system by invoking the conservation of solute mass in the centrifuge cell (see problem 19). This determination permits the equilibrium distribution of a sedimentation experiment to be anticipated when the value of M is known approximately. Such knowledge is useful for adjusting the values of the parameters such as c_0 and ω, so as to optimize the experimental conditions of observation.

The sedimentation equilibrium method has the advantage over other methods that the range of molecular weights of the solutes that can be studied is very broad. One disadvantage it possesses is that, very often, considerable time must be expended in ultracentrifugation in order to attain equilibrium. With the aim of shortening this time, **transient state methods** of investigation have been introduced which study the approach to equilibrium of the concentration distribution (see Archibald (1947), Van Holde and Baldwin (1958), and problem 17).

We shall now consider how the equilibrium distribution of the concentration is modified by the presence of a density gradient in the solvent. Such a modification is the basis of the **density gradient sedimentation equilibrium method**, also called **isopycnic centrifugation** and **density gradient centrifugation**. The latter term can be confusing, however, because it does not distinguish the method clearly from the sedimentation velocity method called zone centrifugation, which also utilizes a density gradient. In this method, the ultracentrifuge cell contains a ternary solution consisting of the solvent, a dilute macromolecular solute, and a concentrated low molecular salt of high density. Under the influence of the centrifugation process, the salt redistributes itself and produces the density gradient. Because of the low molecular weight of the salt, the salt concentration is a slowly varying function of radial position in the cell. The salt is chosen so that the density variation of the solution between the meniscus and the bottom of the cell encompasses a value $\rho = 1/\bar{v}$, where $\bar{v}$ is the partial specific volume of the macromolecular solute component. In other words, the ternary solution is treated theoretically as if it were a binary solution consisting of the macromolecular solute and a solvent exhibiting a density gradient, which is the salt solution. We see that in density gradient centrifugation, unlike zone centrifugation, it is desirable to run the ultracentrifuge for a long time to ensure the establishment of the density gradient and the equilibrium condition.

We designate by r_0 the value of the radius at which $\rho = 1/\bar{v}$. Then, in the neighborhood of r_0, the density variation is, to good approximation, linear, so that we may write

$$\rho(r) = \frac{1}{\bar{v}} + (r - r_0)\left(\frac{d\rho}{dr}\right)_0, \tag{5.180}$$

where $(d\rho/dr)_0$ is the gradient of the density at $r = r_0$, and terms of order $(r - r_0)^2$ in the Taylor expansion of $\rho(r)$ about $r = r_0$ have been neglected. At equilibrium, the concentration c satisfies equation (5.175) with $\rho = \rho(r)$. For a small neighborhood of r_0, we replace the right-hand side of equation (5.175) by its linear expansion about $r = r_0$. Equivalently, substitute (5.180) into (5.175) and retain only terms to first order in $(r - r_0)$. Then (5.175) becomes

$$\frac{1}{c}\frac{dc}{dr} = -\frac{\omega^2 M \bar{v}}{RT}\left(\frac{d\rho}{dr}\right)_0 r_0(r - r_0). \tag{5.181}$$

This can be integrated readily to yield the solution

$$c = c(r_0)\exp\left[-\frac{\omega^2 M \bar{v} r_0}{2RT}\left(\frac{d\rho}{dr}\right)_0 (r - r_0)^2 \right].$$ (5.182)

Thus, at equilibrium, the macromolecule exhibits a Gaussian concentration distribution or "band" about its equilibrium value r_0. Note that the breadth of the band is proportional to $M^{-1/2}$.

Density gradient centrifugation was first introduce in order to investigate DNA, utilizing cesium chloride (CsCl) as the salt which produces the density gradient (Meselson, Stahl, and Vinograd, 1957). The resulting theoretical curve, based on equation (5.182), and the observed equilibrium distribution of DNA from the **bacteriophage** T4, a virus that attacks *E. coli*, is shown in in Figure 5.17. In this experiment, the DNA molecules suffered adventitious

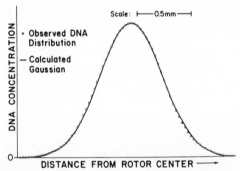

Figure 5.17. Equilibrium distribution of DNA from bacteriophage T4. The distribution is independent of the initial distribution of DNA, $\omega = 27{,}690$ r/min, duration of centrifugation is 80 hr, initial concentration of cesium chloride $= 7.7$ M, pH $= 8.4$. The theoretical curve is based on equation (5.182) with a molecular weight M for Cs–DNA salt of 18×10^6 g/mole. The DNA concentration maximum is 20 μg/ml, $1/\bar{v} = 1.70$ g/cm^3, and $(d\rho/dr)_0 = 0.046$ g/cm. From Meselson, Stahl, and Vinograd (1957).

breaking into approximately equal-sized fragments. Hence, the molecular weight quoted in the caption to the figure represents that of a fragment. The molecular weight of intact DNA from bacteriophage T4 is about 13×10^7 g/mole.

A mixture of radioactive ^{15}N-labeled DNA and nonradioactive ^{14}N-containing DNA forms two distinct bands in a cesium chloride density gradient, as illustrated in Figure 5.18. Meselson and Stahl (1958) subjected a culture of *E. coli* containing ^{15}N-labeled medium to an unlabeled ^{14}N medium at time zero. Successive generations of the bacteria were sampled,

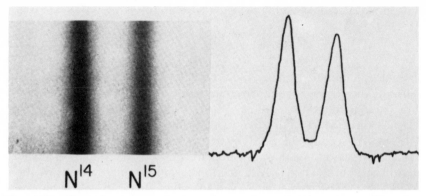

N^{14}　　N^{15}

Figure 5.18.　Illustrating the separation in an ultracentrifuge cell of ^{14}N–DNA and ^{15}N–DNA by means of density-gradient centrifugation. A lysate (the material taken from cells which have undergone lysis) from about 10^8 *E. coli* cells grown on ^{14}N medium, was mixed with a similar lysate from cells grown on ^{15}N medium, and placed in a CsCl solution. The solution was centrifuged for 24 hr at 44,770 r/min, at which time the ultraviolet absorption photograph shown on the left was taken. On the right is shown a microdensitometer tracing of the two bands in the photograph. The separation between the two peaks corresponds to a density difference of 0.014 g/cm^3. From Meselson and Stahl (1958).

their DNA isolated, and examined for activity after forming bands, as described above. They found, as expected, that initially, the DNA was in a band that was 100% labeled. After one generation, virtually all the DNA was found in a band that was only 50% labeled. In later generations, unlabeled bands were also observed. The importance of the 50% labeled band, in accordance with the Watson–Crick model of DNA duplication (Watson and Crick, 1953), is that it consists of DNA, which is a double stranded macromolecule, with one strand labeled and one strand unlabeled. Further, each daughter of a dividing cell receives one strand of the parental DNA, and one strand that is synthesized from the fresh medium, each strand acting as a **template** for the synthesis of the complementary strand. The cited experiment forms the basis of the principle that DNA replication is **semi-conservative**, that is to say, the strands are conserved through many duplications.

Further quantitative information regarding DNA derived from *E. coli*, and its molecular weight, was derived from the study of the DNA concentration distribution at equilibrium in a cesium chloride density gradient. The observed optical density distribution, which is proportional to the concentration of ^{14}N *E. coli* DNA, is illustrated in Figure 5.19*a*. Alongside it, in Figure 5.19*b*, is a semilog plot of the relative concentration c/c_{max} versus $(r - r_0)^2$, based on Figure 5.19*a*, where r_0 is the position of the maximum

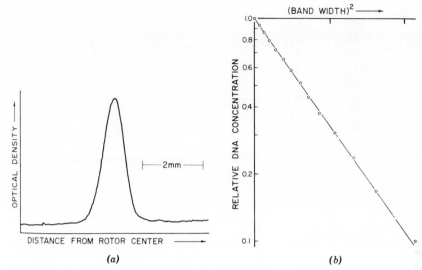

(a)   (b)

Figure 5.19. (*a*) The figure on the left is based on a tracing of an ultraviolet absorption photograph, and shows the optical density in the ultracentrifuge cell in the region of a band of ^{14}N *E. coli* DNA. The optical density is proportional to the DNA concentration. Centrifugation was carried out with ω = 31,410 r/min, initial CsCl concentration = 7.75 m, pH = 8.4, DNA concentration maximum $c_{max} \approx 50$ μg/ml, $1/\bar{v}$ = 1.71 g/cm^3, $(d\rho/dr)_0$ = 0.057 g/cm, and T = 25°C. (*b*) The figure on the right is a semilog plot of c/c_{max} versus the square of the band width $(r - r_0)^2$, based on the figure on the left, where c, r, and r_0 are defined as in equation (5.182), and $c_{max} = c(r_0)$. Unit distance on the abscissa represents 1 mm^2. From Meselson and Stahl (1958).

of the optical density recording. From the slope of the resulting straight line, the molecular weight of Cs–DNA formed from *E. coli* was inferred to be 9.4×10^6 g/mole (Meselson and Stahl, 1958).

As a final remark, we emphasize that in biological practice, the macromolecular solution in ultracentrifuge cells are most frequently not of binary type. Hence, considerations must be given to the heterogeneous or multicomponent character of the solutions, especially when they are not ideal. In such cases, the concentration of one solute influences the behavior of another solute. Even in the case when the multicomponent solution is ideal, the investigator is confronted mathematically with the problem of expressing the concentration distribution in, say, the sedimentation equilibrium method, as a sum of exponentials, analogous to the problem faced by the method of exponential peeling, as discussed in Section 3.7. In the present instance, the fitting procedure would be utilized for the purpose of establishing the **molecular weight distribution** in the solution.

5.14. TRANSCAPILLARY EXCHANGE

An important example of the process of convective transport is provided by the transport of diffusible, low molecular weight solutes across a capillary wall. Many substrates and products of tissue metabolism, such as sucrose and urea, are of such a nature. Assume the capillary is a long right circular cylinder of length L and radius a. The blood that flows through the model capillary will be considered here to be inviscid, and its velocity of constant magnitude will be denoted by u. The direction of flow is the positive x direction.

While the blood is flowing, a particular solute component of blood with concentration c is being transported across the capillary wall. This transport arises usually as the result of either of two driving forces. It may result from the difference in pressures on the inside and outside of the capillary wall. This pressure difference can be partly hydrostatic in origin, or it can be an osmotic pressure difference. The transport may also result from a concentration difference of the solute inside and outside the capillary wall. Of course in the latter case the capillary wall is recognized as a permeable membrane. The first mechanism is primarily responsible for maintenance and control of the plasma and interstitial fluid volumes. The second mechanism is primarily responsible for tissue metabolism by transcapillary transport of specific solutes. We shall consider here this latter passive transport mechanism.

We denote the concentration of the solute in the tissue surrounding the capillary by c_T, and the concentration of the solute in the capillary by c. We shall concern ourselves here with the steady-state process, so that $\partial c_T / \partial t = 0$, and $c = c(x)$ only. Also, we neglect diffusion inside the capillary, and assume that diffusion in the tissue is so rapid that it behaves like a homogeneous compartment. The solute diffuses across the capillary wall by virtue of the concentration difference $c(x) - c_T$. Consider a cross-sectional element of the capillary contained between x and $x + \Delta x$. Then conservation of mass flux into and out of an infinitesimal volume element of the capillary of cross-sectional area πa^2 and width Δx requires that (see Figure 5.20)

$$[uc(x + dx) - uc(x)]\pi a^2 = P[c_T - c(x)]2\pi a\,\Delta x,$$

where P is the permeability constant of the solute in the capillary membrane. The left-hand side represents the net mass flux across the two ends of the element, and the right-hand side represents the net mass flux across the perimetric area of the element, of length $2\pi a$ and width Δx. Here, outward flux is taken to be positive, and inward flux is taken to be negative. Dividing by Δx and taking the limit as $\Delta x \to 0$, there is obtained the equation

$$\pi a^2 u \frac{dc}{dx} = -2\pi a P[c - c_T] \tag{5.183}$$

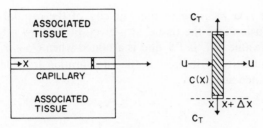

Figure 5.20. Steady-state flux through an element of a capillary, of width Δx and cross-sectional area πa^2.

It is customary to denote the steady volume rate of blood flow through the capillary by $Q = \pi a^2 u$, and the surface area of the capillary by $S = 2\pi aL$. Then (5.183) is written as

$$\frac{dc}{dx} = -\frac{PS}{QL}[c - c_T] \tag{5.184}$$

Let $c(0) = c_A$, the concentration in the artery feeding the capillary. Integration of (5.184) yields the concentration distribution in the capillary as

$$c(x) - c_T = (c_A - c_T)e^{-PSx/QL}. \tag{5.185}$$

The net mass current leaving the capillary to enter the tissue is denoted by J_T, and can be obtained by summing the negative of the right-hand side of (5.183) over the surface of the capillary. Equivalently and more simply, by conservation of mass, it is the difference between the mass influx Qc_A and the mass efflux Qc_V, where $c_V = c(L)$ is the concentration on the venous side of the capillary. Thus,

$$J_T = +2\pi aP \int_0^L [c(x) - c_T]dx = Qc_A - Qc_V. \tag{5.186}$$

From equation (5.185), setting $x = L$, we find that

$$c_V = c_T + (c_A - c_T)e^{-PS/Q}. \tag{5.187}$$

By substituting (5.187) into (5.186) we readily obtain the following expression for the mass current into the tissue,

$$J_T = (c_A - c_T)Q(1 - e^{-PS/Q}). \tag{5.188}$$

The coefficient of $(c_A - c_T)$ in the above equation is defined as the capillary **clearance** C, or

$$C = \frac{J_T}{c_A - c_T} = Q(1 - e^{-PS/Q}). \tag{5.189}$$

The clearance is a measure of the capacity of the capillary to transport matter by diffusion into the tissue, at a given blood flow rate Q. Note that the maximum value of C is PS, and is attained when $Q \to \infty$.

A related quantity of interest is the **equilibration fraction** or **extraction fraction** E, defined as

$$E = \frac{c_A - c_V}{c_A - c_T}. \tag{5.190}$$

The equilibration fraction is a convenient measure of the fraction of solute in excess of the concentration c_T, which passes through the capillary and enters the tissue. By substituting equation (5.187) into (5.190), we find that E is given theoretically as

$$E = 1 - e^{-PS/Q}. \tag{5.191}$$

Comparison of this expression with that for the clearance, equation (5.189), shows that E and C are simply related,

$$C = QE. \tag{5.192}$$

If $Q \gg PS$, $e^{-PS/Q} \sim 1 - PS/Q$, and the equilibration fraction is very small, so that $c_A \approx c_V$. From (5.188), the mass current into the tissue becomes

$$J_T = PS(c_A - c_T) + \cdots, \qquad Q \gg PS. \tag{5.193}$$

Since this expression involves only the diffusive characteristics of the capillary, it implies that the rate-limiting process governing the transport is diffusion. If $Q \ll PS$, the exponential term in (5.191) is negligible, $E \approx 1$, and therefore $c_V \approx c_T$. The mass current into the tissue in this case is

$$J_T = Q(c_A - c_T) + \cdots, \qquad Q \ll PS, \tag{5.194}$$

so that the rate-limiting process governing the transport is now the blood flow rate through the capillary.

The above theory is due to Renkin (1959). In comparing the theory of a single capillary to observations of mass flux into an actual tissue, we imagine that the microvascular bed of the tissue consists of a heterogeneous mixture of many capillaries with varying properties, and distributed more or less uniformly throughout the tissue. Hence, the model capillary can be thought of as representing an average capillary of the tissue. However, not all capillaries are open to flow at a given time. The number of open capillaries is generally presumed to be under the control of smooth muscle, namely the **precapillary sphincters**.

Figure 5.21 shows the results of experimental measurements of the clearance of the radioactive tracer material ^{86}Rb into dog muscle, for a given

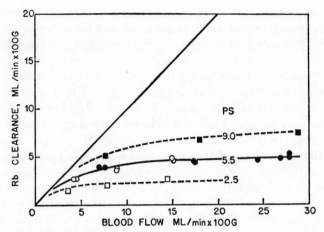

Figure 5.21. Capillary clearance C of ^{86}Rb in an isolated, blood perfused gracilis muscle of a dog, as a function of volume flux of blood through the tissue, in units of milliliters per minute per 100 grams of muscle tissue. The inferred values of PS are indicated, in the same units as those of Q. The solid straight line represents the limiting theoretical curve for PS approaching infinity, given by equation (5.189) as $C = Q$. The data points represent different experimental tissue preparations, as follows: ○, ●, controls; □, maximum sympathetic vasoconstriction; ■, maximum vasodilatation by motor nerve stimulation. From Renkin (1967), with permission, based on data from Renkin and Rosell (1962) and Renkin, Hudlicka, and Sheehan (1966).

flow rate Q. In these experiments, the diffusible tracer is infused continuously into the artery feeding the tissue, in order to achieve a steady state. The tracer ^{86}Rb enters the muscle cells and becomes diluted by the large quantity of nonradioactive potassium there. Because c_T cannot be measured directly, it is usually inferred from measurements of c_V made after the arterial infusion has been terminated. Actually, c_V does change very slowly with time even during the continuous infusion process, so that the system is in a quasi-steady state. The results shown in the figure conform to the theoretical prediction of equation (5.189), and permit the determination of PS. As is shown by the curves, different experiments yield different values of PS. In the experiments, the physiological state of the tissue was altered by vasoconstriction or vasodilation, achieved by nerve stimulation of arteriolar smooth muscle. The interpretation of the variation in PS that is offered is that the effective capillary surface S is altered by the opening or closing of precapillary sphincters.

For other diffusible substances such as simple sugars, it can be expected that the tissue concentration will not be negligible, and the problem of determining the tissue concentration c_T for such substances is serious. It has been suggested that the time-dependent uptake of diffusible substances by tissue be investigated by the initial injection of two tracers into the artery feeding the tissue, one of which is diffusible and the other is not. A comparison of the tracer dilution curves, of the kind shown in Figure 3.1, in the vein of the tissue for these two cases, permits the determination of a time-dependent extraction fraction $E(t)$. It can be shown that the initial value of the extraction fraction $E(0)$ also satisfies equation (5.191), so that the capillary permeability P can be inferred (assuming S is known). The above method has been called the **indicator diffusion method**, or the **early extraction method** (Crone, 1963; Martin and Yudilevich, 1964).

We shall not present here the theory of this method. There are numerous examples in the literature of other time-dependent mathematical models of transcapillary exchange. It is probably fair to state that measurements of time-dependent tissue uptake of diffusible substances are not sufficiently quantified at the present time, for the correctness of one or another of such theoretical models to be adequately tested.

SELECTED REFERENCE BOOKS

J. Crank (1956), *The Mathematics of Diffusion*, Oxford Univ. Press, London.

A. V. Luikov (1968), *Analytical Heat Diffusion Theory*, J. P. Hartnett, ed., Academic, New York.

M. H. Jacobs (1935), *Ergebnisse Biol.* **12**, 1 (reprinted as *Diffusion Processes*, Springer, New York, 1967).

E. O. Wilson (1971), *Insect Societies*, Harvard University Press, Cambridge, Mass.

A. C. Giese (1973), *Cell Physiology*, 4th ed., W. B. Saunders, Philadelphia.

H. Davson and J. F. Danielli (1952), *The Permeability of Natural Membranes*, 2nd ed., Cambridge Univ. Press, Cambridge (reprinted by Hafner, Darien, Conn., 1970).

W. D. Stein (1967), *The Movement of Molecules Across Cell Membranes*, Academic, New York.

H. Fujita (1962), *Mathematical Theory of Sedimentation Analysis*, Academic, New York.

R. H. Haschemeyer and A. E. V. Haschemeyer (1973), *Proteins*, Wiley, New York.

H. K. Schachman (1959), *Ultracentrifugation in Biochemistry*, Academic, New York.

K. E. Van Holde (1971), *Physical Biochemistry*, Prentice-Hall, Englewood Cliffs, N.J.

PROBLEMS

1. In a spherical coordinate system, a generic point P is specified by the radial distance r from the origin, and two angles, a polar (colatitudinal) angle ϑ, and an azimuthal (longitudinal) angle φ. A concentration distribution which is **spherically symmetric** has no angular dependence. For it, the diffusion equation assumes the form

$$\frac{\partial c}{\partial t} = D \frac{1}{r^2} \frac{\partial}{\partial r} \left(r^2 \frac{\partial c}{\partial r} \right), \tag{5.195}$$

where $c = c(r,t)$. A point mass of solute of magnitude m is placed in the solvent at the origin at $t = 0$. The mass subsequently diffuses according to the law

$$c(r,t) = \frac{m}{(4\pi Dt)^{3/2}} \, e^{-r^2/4Dt}. \qquad (5.196)$$

Verify that this expression is a solution of the diffusion equation. Also show, by integration by parts, that this solution has the integral property

$$\int_0^\infty c4\pi r^2 \, dr = m. \qquad (5.197)$$

Because of this integral property, the solution with m equal to unity is called the **unit three-dimensional source solution** of the diffusion equation.

2. A reasonable definition of the "extent of diffusion" in time t, following the placement of unit mass at the origin in one dimension, is that distance $x = \xi(t)$ at which the current density $j(x,t)$ is a maximum at time t. Find $\xi(t)$, and compare it with the expression for the r.m.s. displacement $\tilde{x}$ given by equation (5.38).

3. Find the r.m.s. displacement $\tilde{r}$ of a diffusing molecule in a concentration distribution represented by the unit three-dimensional source solution.

4. What is the theoretical value of the diffusion constant of hemoglobin in water at $20°C$? The globular hemoglobin molecule has a radius $a \approx 30$ Å (Perutz et al., 1960), and the Boltzmann constant $k = 1.38 \times 10^{-16}$ erg/deg (where erg = c.g.s. unit of energy = g cm^2/sec^2).

5. Find the solution to the one-dimensional diffusion equation in the region $x > 0$ subject to the condition that the concentration at $x = 0$ is being maintained at the value c_0 all the time.

   Hint. Find the initial condition and boundary conditions that the solution must satisfy. Make use of the properties of the solution given by equation (5.32).

6. Consider a solvent of half-infinite extent in the region $x > 0$, with a wall at $x = 0$. Assume that, in one dimension, a unit point mass of solute is placed at the position x_0 at time $t = 0$. Because of the presence of the wall, the current density must be zero there.
   (a) Find the concentration of solute $c(x,t)$.
   (b) Suppose the wall has the property that the solute molecules "adhere" to it on contact. The mathematical representation of this property is that the concentration vanishes at the wall, rather than the flux. Find the flux of solute onto the wall as a function of time, and

from it, calculate the total mass of solute that has adhered to the wall from $t = 0$ up to time t.

Hint. To solve the diffusion equation in the given region, introduce a point mass at the image point $x = -x_0$. The sign of the point mass is positive or negative, depending on the boundary condition at the wall.

7. In one dimension, let there be a slab of solute of uniform concentration c_0 and of thickness $2a$ at $t = 0$:

$$c(x,0) = \begin{cases} c_0, & -a < x < a, \\ 0, & |x| > a. \end{cases} \qquad (5.198)$$

Find the concentration $c(x,t)$ for all values of x and t.

8. Oxygen in muscle is utilized for the oxidative removal of lactic acid, and this fact affects its rate of diffusion into fatigued muscle. Hill (1928) suggested that a slab of muscle in contact with oxygen will possess a "recovered" oxygen zone in contact with the external oxygen, and an unrecovered "lactic acid" zone. Furthermore, the interfacial boundary between the two zones will advance with time into the lactic acid zone. Thus, imagine a half-infinite muscle tissue region $x > 0$ through which oxygen diffuses. At the external oxygen boundary $x = 0$, the oxygen concentration c is maintained at the constant value c_0. Let the oxygen zone be $0 < x < \xi$, where the boundary ξ between the oxygen zone and the lactic acid zone depends on the time t, so that $\xi = \xi(t)$, and $\xi(0) = 0$. The boundary conditions satisfied by $c = c(x,t)$ in the oxygen zone are that $c(0,t) = c_0$, and $c(\xi,t) = 0$.

    The velocity of the "advancing front," $d\xi/dt$, is assumed to be proportional to the oxygen flux at ξ, namely,

$$-D \left(\frac{\partial c}{\partial x} \right)_{x=\xi} = \lambda \frac{d\xi}{dt}, \qquad (5.199)$$

where λ is a constant of proportionality, called the "oxygen debt." (*a*)   Assume that the front advances slowly, so that a quasi-steady state of oxygen concentration is established behind it. In other words, in the oxygen zone, the concentration c satisfies the steady-state equation

$$D \frac{\partial^2 c}{\partial x^2} = 0, \qquad 0 < x < \xi. \qquad (5.200)$$

Find the velocity of the advancing front.
(*b*)*   Find the velocity if the oxygen concentration is not assumed to be in a quasi-steady state.

9. Assume that, when a worker harvester ant emits an alarm substance,

the ant attracts nearby workers that in turn release the alarm substance themselves. If the arrival of other workers is fairly uniform in time, we can assume that their emissions are represented by a source that is continuous in time. Let $c(r,t)$ represent the concentration of the alarm substance at time t at a distance r from the emitting ant. We suppose that $c(r,t)$ satisfies the diffusion equation with spherical symmetry, (5.195). Then the solution of this equation (Bossert and Wilson, 1963) corresponding to a source of strength $\eta(t)$ is

$$c(r,t) = \int_0^t \frac{2\eta(t_0)e^{-r^2/4D(t-t_0)}}{[4\pi D(t-t_0)]^{3/2}}\, dt_0. \tag{5.201}$$

(a) Show that the above expression formally satisfies the diffusion equation.

(b) Assume $\eta(t_0) = \eta_0$, a constant. This corresponds to all ants emitting the alarm substance with the same strength. By integration, show that

$$c(r,t) = \frac{\eta_0}{2\pi Dr}\, \text{erfc}\left[\frac{r}{(4Dt)^{1/2}}\right]. \tag{5.202}$$

(c) What is the limiting expression for $c(r,t)$ for large times? What is the maximum radius of influence for large times?

(d) As a hypothetical example, assume one worker reaches the alarm point every 5 sec. Let the threshold density for a receiver ant be denoted by K. The value of N_0/K for an individual ant releasing alarm substance in a puff was found to be about 1400 cm^3. Find the maximum radius of influence of the alarm, for large times, using the solution of part (c).

10. The mass current J due to steady-state diffusion of a solute through a slab of solution of width L, cross-sectional area A, diffusion constant D, and subjected to a concentration gradient $(c_2 - c_1)/L$ across its width, is given by the expression

$$J = \frac{AD}{L}(c_2 - c_1) = \frac{c_2 - c_1}{R}. \tag{5.203}$$

Here $R = L/AD$ is the **diffusive resistance** of the slab. This formula is analogous to Ohm's law relating the electric current I in a conducting wire of length L, cross-sectional area A, conductivity σ, and subjected to an electrical potential difference $V_1 - V_2$ across its ends:

$$I = \frac{A}{L}\sigma(V_2 - V_1) = \frac{V_2 - V_1}{R}. \tag{5.204}$$

Here R is the electrical resistance of the wire, equal to $L/A\sigma$. Comparison of the two expressions for the currents shows that the diffusion constant

D for the mass current is analogous to the electrical conductivity σ for the electrical current. For two electrical wires, of resistances R_1 and R_2, connected in series, the total resistance equals $R_1 + R_2$.

(a) Derive this formula for one-dimensional steady-state diffusion through two slabs in series, both of cross-sectional area A, and of widths L_1 and L_2, and diffusion constants D_1 and D_2.

(b) A solute is diffusing from an extracellular region through a cell. Assume the diffusion process is idealized as one-dimensional steady-state diffusion through three slabs, as shown in Figure 5.22. Show that

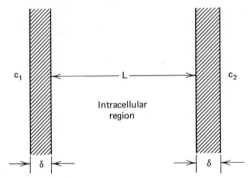

Figure 5.22.   The cell membrane is of width δ and permeability constant P. The solute diffusion constant in the intracellular region of width L is D.

the transport process through the cell can be represented as steady-state diffusion through a single slab of width L and associated diffusion constant D', where

$$\frac{1}{D'} = \frac{1}{D} + \frac{2}{PL}. \tag{5.205}$$

11. One method of determining the diffusion constant of a gas dissolved in solution, such as oxygen in water, is to measure the oxygen flux through a solution slab, one side of which is maintained at the partial oxygen pressure p_1, and the other side of which is maintained at the partial pressure p_2. Such a method is limited in its precision by the precision with which the solubility coefficient α is known, where α is the proportionality constant of Henry's law relating the solution concentration c and the partial pressure p at the air–solution interface, $c = \alpha p$. An alternative method which avoids this requirement is to investigate the transient state (Goldstick and Fatt, 1970), as follows.

Let one side of the aqueous slab be maintained at pressure p_1, and let the other side be buffered so that no oxygen flux is possible there. The initial value of the pressure in the solution is p_0. The oxygen partial pressure on the buffered side $p_b(t)$ is measured as a function of time at the buffered interface, as illustrated in Figure 5.23. Assume $c = c(x,t)$

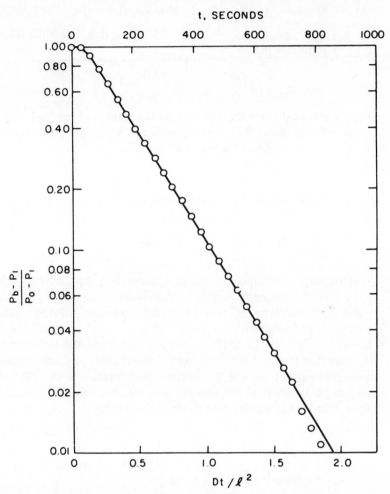

Figure 5.23. A representative decay curve is shown, in the transient state method for determining the diffusion constant of oxygen in water. The ordinate is plotted on a logarithmic scale. The circles represent experimentally determined values (abscissa scale at top of figure), and the solid line is a theoretical curve, based on a suitably chosen value of D (abscissa scale at bottom of figure). From Goldstick and Fatt (1970), with permission.

satisfies the one-dimensional diffusion equation subject to the conditions

$$\frac{\partial c}{\partial x} = 0, \qquad x = 0, \tag{5.206a}$$

$$c = c_1, \qquad x = L, \tag{5.206b}$$

$$c(x,0) = c_0, \qquad 0 < x < L. \tag{5.206c}$$

Here, $x = 0$ represents the buffered side of the slab. From the relation $c = \alpha p$, it follows that

$$\frac{p_b(t) - p_1}{p_0 - p_1} = \frac{c(0,t) - c_1}{c_0 - c_1}. \tag{5.207}$$

Hence, if we calculate the quantity on the right, and measure the one on the left, we can obtain an equation for D, which does not involve α. Solve for $c(x,t)$ and find the equation for D.

Hint.

$$\int_{\pi/2}^{\pi} \cos nx \cos mx \, dx = \int_{0}^{\pi/2} \cos nx \cos mx \, dx, n + m \text{ an even integer},$$

$$\frac{2}{\pi} \int_{0}^{\pi} \cos nx \cos mx \, dx = \delta_{nm}, n \text{ and } m \text{ integers}. \tag{5.208}$$

12. A stationary spherical cell of radius a is metabolizing a nutrient which is at uniform concentration c_0 in the surrounding medium initially. Solve the diffusion equation in spherical coordinates, given by equation (5.195), and find the nutrient concentration $c = c(r,t)$ in the surrounding medium as a function of the radial position r and the time t, $r > a$. Assume that the cell "instantaneously" metabolizes any nutrient molecules that enter it, so that the nutrient concentration at the cell wall is zero at all times. Find the nutrient flux into the cell as a function of time. What is the steady-state value of the flux?

    Hint. Let $c(r,t) = c_0 + u(r,t)/r$, and find the differential equation satisfied by u, together with the boundary conditions and initial condition that it satisfies. Use your knowledge of the solutions to the one-dimensional diffusion equation to find u.

13. **Morphogenesis** is the development of structure and form during embryological development in multicellular organisms. A classical concept (see, for example, Child, 1941) is that differentiation is initiated by the establishment of a gradient of a critical substance, in a given region of the embryo. Such a region is called an **embryonic field**, and

the substance has been called a **morphogen** (Turing, 1952). The magnitude of such fields appears to extend from about 50–100 cells (Wolpert, 1969). The question arises in connection with this concept as to whether diffusion is a sufficiently rapid mechanism for the establishment of such a gradient in the time normally allotted for embryogenesis. Crick (1970) addressed this question in the following simple manner.

Represent an embryonic field as a line of cells of length L, at one end of which there is a morphogen source, and at the other end, a morphogen sink. The transport of morphogen from the source, through the cells, to the sink is treated as a problem of diffusion in one dimension through a homogeneous medium. In principle, the establishment of a steady-state (time-independent) gradient across the line of cells takes an infinite time. Consequently, assume arbitrarily that the effect of the field is initiated when the morphogen concentration is within 1% of its final value at the midpoint of the line of cells.

(a) Assuming that the initial concentration $c = c(x,t)$ is null in the region $0 < x < L$, and that $c = c_0$ at $x = 0$, $c = 0$ at $x = L$, find an approximate expression for the time needed to establish the gradient.

(b) Assume that the diffusion of the morphogen through an individual cell is represented in the one-dimensional manner described in problem 10. Estimate the number of cells in an embryonic field, with the aid of the following estimates: the time necessary to establish a gradient is 10^4 sec (about 3 hr). The width L consists of a line of n cells, each of width d, and $d = 10–30 \, \mu m$. The diffusion constant D of the morphogen is about that for a medium-sized molecule, for example, a steroid, cyclic AMP, or ATP ($M = 507$). The viscosity of cytoplasm is about six times that of water. The permeability P of the morphogen with respect to the cell membrane is about that for glucose with respect to ascites cells at 37°C, or $P \approx 4.5 \times 10^{-4}$ cm/sec.

14. Artificial physical membranes have been studied because they perhaps represent simplified models of real biological membranes. For example, a thin mica sheet of thickness $L = 4.9 \, \mu m$ and containing 7.54×10^4 pores was investigated with respect to the flow of water through it (Bean, 1972). The pores were produced by exposing the mica sheet to a radioactive substance, which produces fission fragment tracks in it, and then etching the sheet in hydrofluoric acid. A pressure gradient $\Delta p/L$ was applied across the mica sheet, and the movement of radioactive tritium oxide, added to the water initially on one side of the sheet only, was followed. The tritium oxide traverses the pores by the mechanisms both of bulk flow and diffusion. The resultant volume flux of water Q is shown in Figure 5.24, as a function of the pressure difference Δp. Assume that the pores are all identical cylindrical tubes of

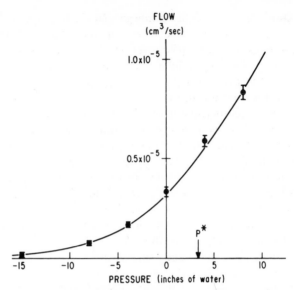

Figure 5.24. The ordinate represents observed flow of water through a mica sheet containing pores, as a function of the pressure difference Δp across the faces of the sheet. The pressure difference is represented on the abscissa, labeled "pressure". Reprinted from C.P. Bean (1972), in *Membranes*, Vol. I, G. Eisenman, ed., Marcel Dekker, Inc., New York, pp. 1–54, p. 22, by courtesy of Marcel Dekker, Inc.

length L and radius a. Assume also that the tritium oxide transport is in a steady state, and that the one-dimensional convective diffusion equation (5.99) is applicable to it. The mutual diffusion of different isotopes of the same substance is called **self-diffusion**. Let D be the **self-diffusion constant** for tritium oxide in water, and let u represent the mean velocity of flow through a tube, proportional to Δp.

Solve the equation for the concentration $c = c(x)$ of tritium oxide in the membrane, subject to the boundary conditions $c(0) = c_1$, $c(L) = 0$. Determine the theoretical expression for the tritium oxide mass current, and the associated volume flux of water. Find simplified expressions for the volume flux in the limiting cases, $|u| \ll D/L$, and $|u| \gg D/L$.

15. The convective diffusion equation has been used to model indicator-dilution curves (Norwich and Zelin, 1970), as illustrated in Figure 5.25. The model is based on the knowledge that the dispersion of indicator, which occurs because of the Poiseuille nature of the flow through a blood vessel, is equivalent to an effective longitudinal diffusion (Taylor, 1953). For example, assume that the aorta is a cylinder of fixed cross-sectional area A. Let a bolus of indicator of mass ρ_0 be injected into it at position $x = 0$, where x is longitudinal distance in the aorta. Let the point of observation of the indicator concentration be x_0. Find the

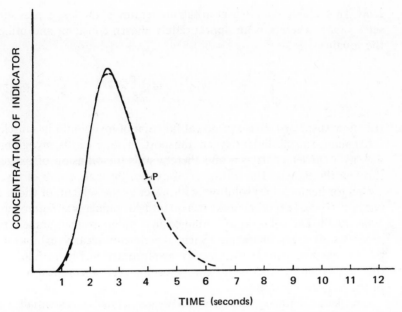

Figure 5.25. Illustrating the comparison between a typical experimentally determined indicator-dilution curve (solid line), and a theoretical curve based on the convective diffusion model of the problem (dashed line). The experimental curve was obtained from observation of cardiac outputs in anaesthetized dogs. The curve is terminated at the point P to avoid showing effects of recirculation of indicator, illustrated in Figure 3.1. From Norwich and Zelin (1970), with permission of Pergamon Press and the authors.

expression for the indicator concentration $c(t) = c(x_0,t)$ there, where $c(x,t)$, the indicator concentration at position x at time t, satisfies the one-dimensional convective diffusion equation (5.99). Assume that the bolus is concentrated at $x = 0$ initially, so that $c(x,0) = \rho_0\delta(x)/A$. Show that Hamilton's formula

$$K = \frac{\rho_0}{\int_0^\infty c(t)dt} \qquad (5.209)$$

is satisfied, where K is the cardiac output defined as $K = uA$, and u is the mean indicator particle velocity at x_0, defined as the average with respect to the distribution function $c(t)$ of the distance x_0 divided by the time t, or

$$u = \overline{\left(\frac{x_0}{t}\right)} = \frac{\int_0^\infty x_0 t^{-1} c(t)dt}{\int_0^\infty c(t)dt}. \qquad (5.210)$$

Hint. To evaluate the integral in the numerator of the last expression, set $t = \alpha \xi^2$, where α is an appropriately chosen constant, and utilize the equality

$$\int_0^\infty \frac{1}{x^2} e^{-x^2 - \beta^2/x^2} \, dx = \frac{\sqrt{\pi}}{2\beta} e^{-2\beta}. \tag{5.211}$$

16.* It is now suspected that a principal function of myoglobin in muscles is to enhance intracellular oxygen transport. In other words, myoglobin acts as a carrier of oxygen and thereby aids the diffusion of oxygen through the tissues. The transport of oxygen through a slab of myoglobin (or hemoglobin) solution, with water as the solvent, of width L has been studied experimentally in a quantified manner (see Wittenberg, 1966, 1970). The solution is contained in a millipore membrane, and separates two gas chambers. Such experiments clearly exhibit the phenomenon of facilitated transport, as illustrated in Figure 5.26, and the data from the experiments can be compared to theoretical models of the transport process.

One such mathematical model (Wyman, 1966) is essentially as follows. Let $c = c(x,t)$ denote oxygen concentration, $m = m(x,t)$ denote the free myoglobin concentration, and $u = u(x,t)$ denote the oxymyoglobin concentration, in the slab. Let k_+ and k_- denote the forward and backward reaction rates for conversion of oxygen plus myoglobin to oxymyoglobin. Then c, m, and u are assumed to satisfy the one-dimensional **reaction–diffusion equations** in the region $0 < x < L$,

$$\frac{\partial c}{\partial t} = D \frac{\partial^2 c}{\partial x^2} - k_+ cm + k_- u,$$

$$\frac{\partial m}{\partial t} = D_m \frac{\partial^2 m}{\partial x^2} - k_+ cm + k_- u, \tag{5.212}$$

$$\frac{\partial u}{\partial t} = D_m \frac{\partial^2 u}{\partial x^2} + k_+ cm - k_- u.$$

Here D is the oxygen diffusion constant, and D_m is the myoglobin diffusion constant, in water. Because the molecular weight of myoglobin is about 1000 times greater than that of oxygen, the diffusion constant of oxymyoglobin is essentially the same as that for myoglobin. Assume that initially, there is no oxymyoglobin present and the concentration of myoglobin is uniform, of amount m_0. The myoglobin and oxymyoglobin molecules are unable to cross the boundary of the slab, but the oxygen is able to do so. A fixed oxygen partial pressure difference

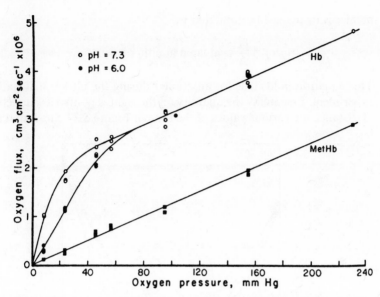

Figure 5.26. The upper curve labeled Hb represents the radioactive oxygen flux through a slab of hemoglobin solution at 25°C, and at two different values of pH, 7.3 and 6.0. The flux is shown as a function of the oxygen partial pressure, which was the same on both sides of the slab. The hemoglobin is maintained in the slab with the aid of a millipore membrane. One side of the slab contained radioactive ^{18}O as well as ^{16}O, while the other side contained only nonradioactive ^{16}O. The affinity of oxygen for hemoglobin is greater at pH 7.3 than at pH 6.0. The lower curve, labeled MetHb, represents the oxygen flux through a methemoglobin solution. Methemoglobin forms when hemoglobin has been exposed to potassium ferricyanide, which renders it incapable of combining with oxygen. Hence, the lower curve represents the pure oxygen diffusion flux through a solution of hemoglobin. From Hemmingsen (1965), with permission.

is maintained across the two sides of the slab. Find the nonlinear differential equation satisfied by the steady-state oxygen concentration $c = \bar{c}(x)$.

Hint. First, find the relationship between the steady-state concentrations $m = \bar{m}(x)$ and $u = \bar{u}(x)$ by considering the last two equations above. Second, find the relationship between $\bar{c}(x)$ and $\bar{m}(x)$ by considering the first two equations above.

17. The Archibald method (Archibald, 1947) for the determination of molecular weights depends on the observation that the Svedberg equation requires in it the ratio s/D only, rather than the values of s and D separately. Archibald observed that this ratio could be determined by measuring the concentration and its gradient at either the

meniscus or the cell bottom, that is,

$$\frac{s}{D} = (r\omega^2 c)^{-1} \frac{\partial c}{\partial r} \text{ evaluated at either } r = r_1 \text{ or } r = r_2. \quad (5.213)$$

This equation holds true for any time t during the ultracentrifugation experiment. Essentially the quantity on the right is plotted as a function of distance for various values of the time in Figure 5.27. Show that the

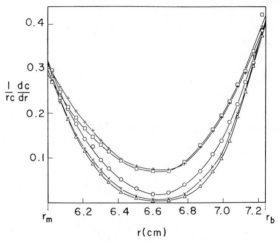

Figure 5.27. The function $(1/rc)(\partial c/\partial r)$ is shown as a function of r during the course of an ultracentrifugation experiment to determine the molecular weight of lysozyme. *Lysozyme* is an enzyme found in serum, which is capable of digesting the cell walls of bacteria. The experimental points on which the curves are based were computed from photographs taken at 427 ($\triangle$), 491 ($\times$), 619 ($\bigcirc$), 1061 ($+$), and 1084 ($\square$) min after the attainment of constant rotor speed. According to the theory of Archibald's method, the ordinate function has a common value, independent of the time, at either the position r_m of the meniscus, or the position r_b of the cell bottom. From Smith, Wood, and Charlwood (1956). Reproduced by permission of the National Research Council of Canada from the Canadian Journal of Chemistry, Volume 34, 1956, pp. 364–370.

value of $c(r_1,t)$ can be calculated, from the gradient curve observed in a standard cell, to be

$$c(r_1,t) = c_0 - \frac{1}{r_1^2} \int_{r_1}^{r_p} \frac{\partial c}{\partial r} r^2 \, dr, \quad (5.214)$$

where r_p is any point in the plateau region (Klainer and Kegeles, 1955).

Hint. Integrate equation (5.136), with the aid of either equation (5.131) or equation (5.133), with respect to time, between the limits 0 and t.

18. The gradient curve given by equation (5.162) is valid only in the neighborhood of $r = r_0$, and only when r_1 and r_p are far away from r_0. The expression can be simplified for some purposes by replacing the function $\log (r/r_0)$ appearing in the argument of the exponential function by its linear approximation in the neighborhood of $r = r_0$. Using the resulting simplified expression for $\partial c/\partial r$, calculate $r_B(t)$ from equation (5.147), and compare with $\bar{r}(t)$.

   Hint. For integration purposes, replace r_1 by $-\infty$ and r_p by $+\infty$.

19. For the expression (5.177) of the final concentration distribution in the equilibrium sedimentation method, find c_1 by utilizing the principle of conservation of mass.

20. The **motility** or "effective diffusion constant" M of *E. coli* cells has been measured in capillary tubes by determining their distribution as a function of time throughout the tube (Adler and Dahl, 1967). Initially, about 10^6 cells of a motile strain of *E. coli* K 12 were placed at one end of the tube. The cell concentration observed at subsequent times, as illustrated in Figure 5.28, resembles that for solute diffusion. The

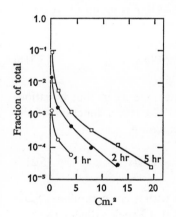

Figure 5.28. At time zero about 10^6 cells of a motile strain of *E. coli* K12 were placed at the origin in a capillary tube filled with medium. The fraction of the total number of cells is shown as a function of the squared distance from the origin, at three different times of observation. The ordinate is plotted on a semi-logarithmic scale. If a solute such as glucose were utilized instead of bacteria, the curves would be straight lines, in accordance with equation (5.15). The slopes of the rectilinear segments (at large distances) of the curves in the figure can be used to determine the motility M. Reprinted, with permission, from J. Adler and M. M. Dahl (1967), *a method for measuring the motility of bacteria and for comparing random and non-random motility*, J. Gen. Microbiol., 46, 161–173, Cambridge University Press.

mean value of twenty experimental determinations of the motility, inferred from the rectilinear segments of curves such as those shown in Figure 5.28, was $M = 0.25 \text{ cm}^2/\text{hr} = 0.69 \times 10^{-4} \text{ cm}^2/\text{sec}$.

Consider a large population N of motile *E. coli* cells sedimenting in a long tube of water of length L and cross-sectional area A, under the influence of the earth's gravitational field. The mass density ρ of an *E. coli* cell is greater than that of water, ρ_0, and is given as $\rho = 1.1 \text{ g/cm}^3$.

Assume that the cells have a concentration distribution c, which is independent of horizontal position, and depends only on the height x above the tube bottom and the time t. Consider the cells, with respect to their kinetic behavior, to be equivalent to large diffusing molecules with a diffusion constant M. Write down the differential equation and the boundary conditions satisfied by c. What is the equilibrium distribution $c = \bar{c}(x)$ of the cells? Let us arbitrarily adopt the convention that $E.$ $coli$ cells extend to that height x_0 for which precisely one cell occupies the height interval Δx_0, so that

$$\bar{c}(x_0)A\, \Delta x_0 = 1. \tag{5.215}$$

Find x_0, with $\Delta x_0 = 1$ mm, $A = 1$ mm^2, $L \sim \infty$, $a \sim 1$ μm, and $N = 10^6$.

$Hint.$ Use Stokes' law to determine the convective velocity of the $E.$ $coli$ cells.

21.* In the consideration of time-dependent capillary exchange, the concentration c in the cylinder representing a capillary is considered to be a function of the time t as well as longitudinal distance x in the capillary, so that $c = c(x,t)$. Instead of equation (5.183), c is assumed to satisfy the equation

$$\pi a^2 \frac{\partial c}{\partial t} + \pi a^2 u \frac{\partial c}{\partial x} = -2\pi a P(c - c_T), \tag{5.216}$$

with the parameters having the same meaning as in that equation. Furthermore, the tissue concentration is likewise a function of the time, $c_T = c_T(t)$, and is governed by the equation

$$V_T \frac{dc_T}{dt} = 2\pi a P \int_0^L (c - c_T)dx, \tag{5.217}$$

where V_T is the tissue volume associated with a given capillary. Such an assigned tissue volume has been called a **Krogh cylinder** (Krogh, 1919). The left-hand side above represents the time rate of change of solute mass in the cylinder, considered to be a spatially homogeneous compartment. The right-hand side above represents the rate of influx of solute matter from the capillary into the cylinder. Consumption of solute in the tissue is here neglected.

(a) Find the steady-state solutions $c = \bar{c}(x)$ and $c_T(t) = \bar{c}_T$ to equations (5.216) and (5.217), where $\bar{c}_T$ is a constant.

(b) Find an asymptotic solution to these equations that approaches the steady-state solution. Hence, seek a solution to the above two equations, to be valid for large times, of the form

$$c(x,t) = \bar{c}(x) + f(x)e^{-\lambda t},$$
$$c_T(t) = \bar{c}_T(1 - e^{-\lambda t}). \tag{5.218}$$

The boundary condition to be satisfied by $f(x)$ is that $f(0) = 0$. What is the transcendental equation satisfied by λ?

(c) Find an approximate solution to the transcendental equation when $\lambda \ll PSu/QL$. Discuss this approximate solution when $PS \gg Q$ and when $PS \ll Q$.

SOLUTIONS TO THE PROBLEMS

1.1. See discussion in Section 2.1.

1.2. Divide Hill's equation (1.80) by $(P_0 - P)V$ to obtain the relation

$$\frac{b}{V} = \frac{P_0 + a}{P_0 - P} - 1. \tag{S1}$$

From this we see that there is a linear relationship between $1/V$ and $1/(P_0 - P)$, which we can rewrite as

$$\frac{1}{V} = \frac{P_0 + a}{b} \frac{1}{P_0 - P} - \frac{1}{b}. \tag{S2}$$

Hence a plot of $1/V$ versus $1/(P_0 - P)$ will yield a straight line. From the V^{-1} intercept we obtain the value of $1/b$ and therefore b. The slope of the straight line has the value $(P_0 + a)/b$, from which we can determine a.

1.3. The net rate of energy gain of the primitive organism is assumed to be of the form

$$f(v) = E_0 - P_0 + E_1 v - P_1 v^2, \tag{S3}$$

where E_0, P_0, E_1, and P_1 are positive constants. This function has an extremum when $df/dv = 0$, which occurs at the positive value of the velocity $v = E_1/2P_1$. Because $f(v) \rightarrow -\infty$ as $|v| \rightarrow \infty$, we know that the extremum is a maximum. We see that $f(0) = E_0 - P_0$, so that the intercept at $v = 0$ is either positive, zero, or negative, accordingly as E_0 is greater than, equal to, or less than P_0 (see curves (a) and (b) in Figure S1.3). Viability requires that $f(v) > 0$ (shown as the shaded region under the curves).

Note that the maximum of $f(v)$ always occurs to the right of the origin, so that it is always of some advantage for the organism to swim, under the given assumptions. In case (a), the organism survives readily without motion. In case (b) when $E_0 = P_0$, suppose that the environment is variable so that E_0 fluctuates, then survival is marginal, and it is clearly advantageous to swim. When $E_0 > P_0$, swimming is a necessity in order to reach the shaded region where survival is possible. For survival in case (b), swimming must be maintained in the interval $v_{min} \leq v \leq v_{max}$, where v_{min} and v_{max} are the zeroes of the equation $f(v) = 0$.

278

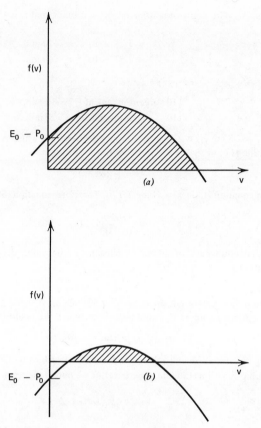

Figure S1.3. The function $f(v)$, as given by equation (S3), is shown as a function of v for the two cases (a) $E_0 > P_0$, and (b) $E_0 \leqq P_0$.

1.4. According to equation (1.6),

$$kt = \log\left(\frac{N}{N_0}\right). \tag{S4}$$

At time $t = T_{1/2}$, the population has decreased to one half of its initial value, or $N = N_0/2$. By substitution of these values into the above equation, we find that

$$kT_{1/2} = \log 1/2 = -\log 2, \tag{S5}$$

which is the desired result.

1.5. The disinfection process is represented by the equation

$$N = N_0 e^{-Kt}, \tag{S6}$$

where N is the number of survivors at time t, N_0 is the population of microorganisms exposed to the disinfectant at $t = 0$, and K is the fractional death rate per unit time. In the problem, $N_0 = 8 \times 10^5$ organisms/ml, $K = 0.9$/min, and $t = 10$ min, or 20 min. Therefore,

$$e^{-Kt} = \begin{cases} e^{-9} = 1.23 \times 10^{-4}, t = 10 \text{ min.} \\ e^{-18} = 1.52 \times 10^{-8}, t = 20 \text{ min.} \end{cases} \tag{S7}$$

Hence from (S6),

$$N \approx \begin{cases} 10^2 \text{ organisms/ml, at } t = 10 \text{ min,} \\ 10^{-2} \text{ organisms/ml, at } t = 20 \text{ min.} \end{cases}$$

1.6. For ^{14}C, $T_{1/2} = 5730$ yr. We wish to compute the time t for which $N(t) = N_0/3$. From either equation (1.6) or (S4), we require the time t such that

$$kt = -\log 3, \tag{S8}$$

where, as in equation (1.81), $k = -\log 2/T_{1/2}$. Therefore, equation (S8) yields the value

$$t = \frac{\log 3}{\log 2} T_{1/2} = \frac{1.0986}{0.6931} 5730 \text{ yr} = 9082 \text{ yr}.$$

1.7. The activity per unit volume of radioactive phosphate, c, has a time-dependence described by the equation

$$c = c_0 e^{-kt}, \tag{S9}$$

where the initial activity of c is given by $c_0 = 500 \ \mu\text{Ci/ml}$, $k = \log 2/T_{1/2}$, and $T_{1/2} = 14.3$ days. Suppose a volume V is administered at time t. We require that

$$Vc(t) = 250 \ \mu\text{Ci, at } t = 10 \text{ days.} \tag{S10}$$

Substituting (S9) into (S10) and solving for V, we find that $V = 0.812$ ml.

1.8. By substituting equation (1.8) into equation (1.6), we obtain the result

$$\frac{N}{N_0} = e^{(t \log 2)/T} = (e^{\log 2})^{t/T} = 2^{t/T}, \tag{S11}$$

since $2 = e^{\log 2}$.

1.9. For wild type cells, the population at time t, $N(t)$, is given by equation (1.82), with $N_0 = 10^3$, and $T = 106$ min. Here, let $t = 0$ represent the instant at which a mutant is born. Similarly, for the mutants, the population number at time t, $N_m(t)$, is given as

$$N_m(t) = 2^{2t/T}. \tag{S12}$$

The initial number of mutants is unity, and the mutant generation time is $T/2$. The time t at which the two cell population are equal is determined by the condition $N_m(t) = N(t)$, or from equations (1.82) and (S12),

$$2^{t/T} = 10^3. \tag{S13}$$

By taking the common logarithm of both sides of the last equation, we find

$$t = \frac{3}{\log_{10} 2} T = 1056 \text{ min.}$$

The latter time equals 9.967 T, or approximately 10 times the generation time of the wild type cell.

In other words, one proliferating cell becomes, in 10 generations, exactly 2^{10} cells =

1024 cells. For every generation interval of the wild type, the mutant population passes through two generations, for a net gain of one generation. After 10 generations of the wild type, there is a gain of 10 generations for the mutant, representing an approximately 1000-fold gain in cell number. If the two types are equal in number, the mutants will outnumber the wild type by a factor of 1000, in another 10 generations.

The total population when the two generations are equal is given by $2N(t)$, where $N(t)$ is given by equation (1.82), and t satisfies equation (S13). Hence, the total population is

$$2N_0 \times 10^3 = 2 \times 10^6 \text{ cells.}$$

1.10. (a) A linear relationship on log–log paper is represented mathematically as

$$\log_{10} y = m \log_{10} x + b = \log_{10} x^m + b, \tag{S14}$$

where the constants m and b represent the slope and intercept, respectively, of the straight line. By applying the inverse log operation to both sides of (S14), we find that

$$y = 10^b x^m. \tag{S15}$$

Exponents are always dimensionless, and therefore so are m and b.

(b) Let x and y^* denote the weights of the clawless body and the claw, respectively. Assume there is no error in x, and ascribe all the errors in measurement to y^*. We have emphasized this assumption by the use of the asterisk. For the six listed measurements, with i running from 1 to 6, we readily compute with the aid of a calculator and/or a table of common logarithms,

$$\sum \log_{10} x_i = 16.512, \sum \log_{10} y_i^* = 13.641,$$
$$\sum (\log_{10} x_i)^2 = 47.098, \sum \log_{10} x_i \log_{10} y_i^* = 40.098.$$

These quantities are needed in utilizing the method of least squares to find the best linear relationship between $X = \log_{10} x$, and $Y^* = \log_{10} y^*$. By substitution of the above values into equations (1.15), with x_i and y_i there replaced by X_i and Y_i^*, we obtain the following two equations for m and b:

$$47.098\, m + 16.512\, b - 40.098 = 0,$$
$$16.512\, m + 6\, b - 13.641 = 0. \tag{S16}$$

These equations have the solution

$$m = 1.54, b = -1.96. \tag{S17}$$

The straight line based on these values, together with the quoted data points are shown in Figure S1.10. With these values and equation (S15), we find the desired relation between y and x to be

$$y = 0.0106\, x^{1.54} \tag{S18}$$

Based on more extensive data, it was found that $y = 0.0073\, x^{1.62}$ (Thompson, 1942).

1.11. According to the logistic law, equation (1.30), the population number is represented by the equation

$$N = \frac{N_e}{1 + Ae^{-kt}}, \tag{S19a}$$

$$A = \frac{N_e}{N_0} - 1. \tag{S19b}$$

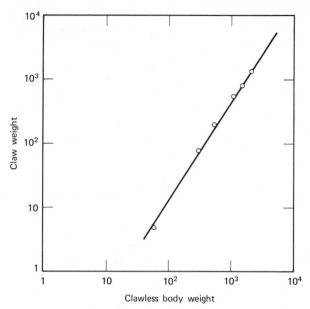

Figure S1.10. The relative weights of the claw and the clawless body in the fiddler crab are indicated by circles, in a log–log plot. The solid line is based on equation (S14) with the values of m and b given by equations (S17).

Equation (S19a) is equivalent to the following linear relationship:

$$\log_{10} \left(\frac{N_e}{N} - 1 \right) = \log_{10} A - \frac{k}{2.3026} t. \tag{S20}$$

From the last two data points of the table, we determine by inspection that $N_e = 381$. From the first entry in the table for $t = 0$, we see that $N_0 = 2$. Hence, A is determined by means of equation (S19b), and we determine the value of k by means of the least-square method, as follows. We exclude from our data the observations for $t = 0$, 7, and 8, as they have been utilized for the above-mentioned determination of N_e and N_0. Let $Y_i^* \equiv \log (N_e/N_i - 1)$, where $N_i = N(t_i)$ and i runs from 1 to 6, corresponding to the observations running from $t = t_1 = 1$ day to $t = t_6 = 6$ days. By applying the least-square method to equation (S20) with k variable, we obtain the first equation in (1.15) with $b = \log_{10} A$ fixed. With the values

$$\sum t_i = 21, \sum t_i^2 = 91, \sum Y_i^* t_i = -33.26, \text{ and } \log_{10} A = 2.278,$$

we solve for k and find that

$$k = 2.3026 \left\{ \frac{\log_{10} A \sum t_i - \sum Y_i^* t_i}{\sum t_i^2} \right\} = 2.05/\text{day}. \tag{S21}$$

1.12. According to assumption, the fetal weight $w = w(t)$ satisfies the equation

$$\frac{dw}{dt} = kw^{2/3}, \quad t > t_0. \tag{S22}$$

Here t_0 represents the time at which the lag phase of prenatal development ends, with $w = w_0$ at $t = t_0$. By multiplying equation (S22) by dt and integrating from t_0 to t, we obtain the solution

$$w^{1/3} = w_0^{1/3} + \frac{k}{3}(t - t_0). \tag{S23}$$

Let $k \equiv 3w_0^{1/3}/\tau$, which defines the characteristic time τ. After dividing (S23) by $w_0^{1/3}$, we see that it is equivalent to the expression

$$\left(\frac{w}{w_0}\right)^{1/3} = \frac{1}{\tau}[t - (t_0 - \tau)], \tag{S24}$$

which is the same as equation (1.83) if we set $a = w_0/\tau^3$ and $t' = t_0 - \tau$. Note that, according to the above theory, the observed lag time t' of Figure 1.18 represents the actual lag time t_0 minus the characteristic time $\tau = 3w_0^{1/3}/k$.

1.13.   The application of the method of least squares to the problem posed is simplified if the observations $c^*(t_i)$ are first converted to $y^*(\tau_i) \equiv \log c^*(\tau_i)$, where $\tau_i = t_i - t_0$. The error E is defined as

$$E = \sum_{i=1}^{n} [y(\tau_i) - y^*(\tau_i)]^2, \tag{S25}$$

where, from equation (1.84),

$$y(\tau) = \log c(\tau) = \log c_0 + v \log \tau - \lambda\tau, \qquad \tau > 0. \tag{S26}$$

Minimizing E with respect to the parameters $\log c_0$, v, and λ leads to the equations

$$0 = \tfrac{1}{2}\frac{\partial E}{\partial \log c_0} = n \log c_0 + v\sum \log \tau_i - \lambda\sum\tau_i - \sum y^*(\tau_i),$$

$$0 = \tfrac{1}{2}\frac{\partial E}{\partial v} = \log c_0 \sum \log \tau_i + v\sum(\log \tau_i)^2 - \lambda\sum\tau_i \log \tau_i - \sum y^*(\tau_i)\log \tau_i, \tag{S27}$$

$$0 = \tfrac{1}{2}\frac{\partial E}{\partial \lambda} = -\log c_0 \sum\tau_i - v\sum\tau_i \log \tau_i + \lambda\sum\tau_i^2 + \sum y^*(\tau_i)\tau_i,$$

where $\sum$ represents the summation from $i = 1$ to $i = n$. The consequence of the conversion of the data to $y^*(\tau_i)$ can now be observed: The above three equations are linear and are readily solved for $\log c_0$, v, and λ.

1.14.   The solution to the differential equation for $k(t)$, equation (1.86b), that satisfies the initial condition is readily found to be

$$k = k_0 e^{-\alpha t}. \tag{S28}$$

If we substitute this expression into equation (1.86a) for $V(t)$ and integrate, we find that

$$\int_{V_0}^{V} \frac{dV}{V} = \int_0^t k_0 e^{-\alpha t}\, dt,$$

or

$$\log \frac{V}{V_0} = \frac{k_0}{\alpha}[1 - e^{-\alpha t}]. \tag{S29}$$

This result is equivalent to equation (1.85), the Gompertz growth law. In equation (1.87), set $\log(\bar{V}/V) = w$. Then $dV/dt = -V\,(dw/dt)$, and the equation for $w = w(t)$ becomes

$$\frac{dw}{dt} = -\alpha w. \tag{S30}$$

The solution to this equation subject to the initial condition $w(0) = w_0 \equiv \log(\bar{V}/V_0)$ is

$$w = w_0 e^{-\alpha t}. \tag{S31}$$

In terms of V, equation (S31) becomes

$$\log\left(\frac{\bar{V}}{V}\right) = \log\left(\frac{\bar{V}}{V_0}\right) e^{-\alpha t},$$

or

$$V = \bar{V} \exp\left[-\log\left(\frac{\bar{V}}{V_0}\right) e^{-\alpha t}\right]. \tag{S32}$$

A comparison of the above expression with the Gompertz growth law shows that they are identical, provided that $\bar{V} = V_0 \exp(k_0/\alpha)$.

1.15. Because of the first assumption, $q = 0$. As a result of the second assumption, it follows from equation (1.35) that $k(c) = k_m c/K$, because c is always less than or equal to c_0. Hence, equations (1.34) and (1.36) take the form

$$\frac{dN}{dt} = \frac{k_m}{K} cN,$$
$$\frac{dc}{dt} = -\frac{k_m}{yK} cN. \tag{S33}$$

From the ratio of these two equations, it follows that

$$\frac{dN}{dc} = -y,$$

which, when integrated, yields the relation

$$N - N_0 = -y(c - c_0). \tag{S34}$$

Here we have satisfied the initial conditions that at $t = 0$, $N = N_0$, and $c = c_0$. If we utilize equation (S34) to eliminate c from the first of equations (S33), we obtain the result

$$\frac{dN}{dt} = \frac{k_m}{Ky} [c_0 y + N_0 - N]N. \tag{S35}$$

This equation is the logistic growth equation, as is verified by comparison with equation (1.22).

1.16. (a) (1)   When the birth rate equals the death rate in the host population, $k_b = k_d$, so that $\beta = 0$. The Lotka–Volterra equations assume the form

$$\frac{dx}{dt} = -\alpha xy,$$
$$\frac{dy}{dt} = y(-\delta + f\alpha x). \tag{S36}$$

Taking the ratio of these two equations, we obtain the equation

$$\frac{dy}{dx} = -f + \frac{\delta}{\alpha x}. \tag{S37}$$

From this equation we infer that dy/dx is $(+, 0, -)$, accordingly as x is $(<, =, >)$ $\delta/f\alpha$. The trajectories in the xy plane are shown schematically in Figure S1.16a. Because dx/dt is negative, x decreases with time, as indicated by the arrowheads in the figure. Note that even though the host population is dying out, the predator population is able to increase in size until x decreases to the critical value $\delta/f\alpha$, after which it too begins to die out.

(a) (2) If none of the parasites' eggs hatch, then $f = 0$. The Lotka–Volterra equations become

$$\frac{dx}{dt} = x(\beta - \alpha y),$$
$$\frac{dy}{dt} = -\delta y. \tag{S38}$$

From the last equation we see that the parasite population decreases exponentially. From the first equation, we see that asymptotically, the host population will increase exponentially with time because the term αxy becomes negligible compared to βx. The trajectories in the xy plane are determined by the equation

$$\frac{dy}{dx} = \frac{-\delta y}{x(\beta - \alpha y)}. \tag{S39}$$

We infer from this equation that dy/dx is $(+, \infty, -)$ accordingly as y $(>, =, <)$ β/α. The trajectories are illustrated schematically in Figure S1.16b. The host population is at a minimum when $y = \beta/\alpha$.

(b) To investigate the stability properties of the Lotka–Volterra equations in the neighborhood of the equilibrium point $(x,y) = (0,0)$, we linearize the functions of x and y appearing on the right hand of the equations. These functions, as they stand, are already in the form of Taylor series in the neighborhood of the origin. (Think of x and y as being replaced by $x - 0$ and $y - 0$, respectively.) Therefore, linearization requires only that we neglect the terms in xy. The Lotka–Volterra equations in this neighborhood become

$$\frac{dx}{dt} = \beta x,$$
$$\frac{dy}{dt} = -\delta y. \tag{S40}$$

According to the first equation in (S40), x increases exponentially. Therefore, the equilibrium point (0,0) is unstable.

(c)* We attempt to solve the Lotka–Volterra equations (1.43) and (1.44) by the method of separation of variables, which yields the equations

$$dt = \frac{dx}{x(\beta - \alpha y)} = \frac{dy}{y(-\delta + f\alpha x)}.$$

The last equality can be put into the separable form,

$$\frac{(-\delta + f\alpha x)dx}{x} = \frac{(\beta - \alpha y)dy}{y},$$

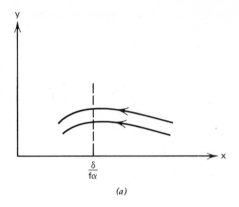

(a)

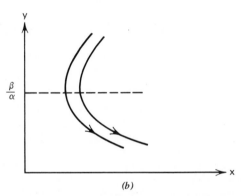

(b)

Figure S1.16. (a) Trajectories in the xy plane of the host–parasite problem when the birth-rate equals the death-rate in the host population. The parasite population has a maximum when dy/dx equals zero at $x = \delta/f\alpha$. (b) The trajectories for the case when none of the parasites' eggs hatch ($f = 0$). dy/dx is infinite at $y = \beta/\alpha$, at which point the host population is at a minimum.

and integrated. The result is

$$f\alpha x - \delta \log x = \beta \log y - \alpha y + c',$$

with c' a constant of integration. By applying the exponential operation to both sides of this equation, we can write it in the form

$$x^{-\delta}e^{f\alpha x} = cy^{\beta}e^{-\alpha y}, \tag{S41}$$

where c is a constant, determined by the initial condition. Equation (S41) describes a family of closed trajectories in the xy plane, a particular curve being determined by a particular initial condition.

1.17.* Let S and R denote the number of smooth and rough colony forming cells, respectively, and let A denote the amount of alanine present in the broth, at time t. It seems reasonable to suppose that S and R cells, without interaction, obey logistic growth laws. In addition, because of alanine production, S cells die at a rate proportional to AS. Furthermore, R cells arise originally by mutation from S cells, so that there is an additional production of R cells at a rate proportional to S. Neglect the effect of backmutations on the growth of S cells. Finally, assume alanine is produced at a rate which is proportional to S. Hence, we assume the following growth equations,

$$\frac{dS}{dt} = k_1 S - \beta_1 S^2 - \alpha S A, \tag{S42a}$$

$$\frac{dR}{dt} = k_2 R - \beta_2 R^2 + \mu S, \tag{S42b}$$

$$\frac{dA}{dt} = \rho S, \tag{S42c}$$

where k_1, k_2, β_1, β_2, α, μ, and ρ are constants. The initial condition is $(S,R,A) = (S_0,0,0)$. We can eliminate A from this set of equations by integrating equation (S42c) and substituting it into (S42a). The latter then becomes

$$\frac{dS}{dt} = k_1 S - \beta_1 S^2 - \alpha_1 S \int_0^t S(\tau)d\tau, \tag{S43}$$

where $\alpha_1 = \alpha\rho$. The equation system is then equations (S43) and (S42b).

The equilibrium states are determined by the conditions

$$0 = S\left[k_1 - \beta_1 S - \alpha_1 \int_0^t S(\tau)d\tau\right], \tag{S44a}$$

$$0 = k_2 R - \beta_2 R^2 + \mu S. \tag{S44b}$$

To satisfy the first equation, there are two possibilities, $S = 0$, or the quantity in the square brackets vanishes. Consider first the latter alternative. Since S is nonnegative, the integral will increase with time, so there is no fixed value of S that makes this quantity vanish. Thus, the only solution to equation (S44a) is $S = 0$. Considering now equation (S44b), we see that when $S = 0$, there are two roots of this equation, $R = 0$, and $R = k_2/\beta_2$. Hence, there are two stationary points, $(S,R) = (0,0)$, and $(S,R) = (0,k_2/\beta_2)$. The latter is stable and the former is unstable.

It appears plausible that, with a proper choice of parameter values, the above model can successfully represent the growth curve of *Brucella abortus* cells shown in Figure 1.20.

1.18.   (a)   Let n_1 and n_2 be the number of smooth type cells and rough type cells, respectively. Because cells from smooth shaped colonies do not give rise to rough shaped colonies following replating, the mutation rate μ in equation (1.65b) is zero. To analyze the given data, we can utilize the theory presented in Section 1.9, and assume that $k_1 = k_2 + \beta$. Then equations (1.69) and those that follow are applicable. According to the information provided,

$$\psi_\infty = 1, \alpha = 0.059/\text{hr}, \text{ and } \lambda = 1.30/\text{hr},$$

in the notation of Section 1.9. Hence, from equations (1.76), we deduce that

$$k_1 = 1.30/\text{hr}, k_2 = 1.24/\text{hr}, \beta = 0.059/\text{hr}. \tag{S45}$$

(b)*   The solution of equation (1.65b) with $\mu = 0$, subject to the initial condition that $n_2(0) = n_{20}$, is

$$n_2 = n_{20}e^{k_2 t}. \tag{S46}$$

Substituting this result into equation (1.65a) makes it of the form of equation (1.20). We find with the aid of equation (1.21) that its solution is

$$n_1 = \left(n_{10} - \frac{\beta n_{20}}{k_2 - k_1}\right) e^{k_1 t} + \frac{\beta n_{20}}{k_2 - k_1} e^{k_2 t}, \tag{S47}$$

assuming $k_1 \neq k_2$ and $n_1(0) = n_{10}$. From equations (S46) and (S47), it follows that the ratio $\varphi(t)$ of smooth shaped colony forming cells to rough shaped colony forming cells is

$$\varphi(t) \equiv \frac{n_1}{n_2} = \left(\frac{n_{10}}{n_{20}} - \frac{\beta}{k_2 - k_1}\right) e^{(k_1 - k_2)t} + \frac{\beta}{k_2 - k_1}. \tag{S48}$$

If $k_1 > k_2$, n_1/n_2 approaches infinity with time, as for growth on broth [see (a)]. From the fact that n_1/n_2 approaches the finite limit $\varphi_\infty = 0.17/0.83 = 0.20$, we infer that $k_2 > k_1$, and that

$$\frac{\beta}{k_2 - k_1} = \varphi_\infty = 0.20. \tag{S49}$$

Equation (S48) can be expressed with the aid of (S49) as

$$\log[\varphi(t) - \varphi_\infty] = \log\left[\frac{n_{10}}{n_{20}} - \frac{\beta}{k_2 - k_1}\right] - (k_2 - k_1)t. \tag{S50}$$

Hence, the observed decay rate of $\varphi(t) - \varphi_\infty$ determines the quantity $k_2 - k_1$ to be

$$k_2 - k_1 = 0.061/\text{hr}. \tag{S51}$$

The total population of the culture increases with time as follows:

$$n_1 + n_2 = \left(n_{10} - \frac{\beta n_{20}}{k_2 - k_1}\right) e^{k_1 t} + n_{20}\left(1 + \frac{\beta}{k_2 - k_1}\right) e^{k_2 t}$$

$$= n_{20}\left(1 + \frac{\beta}{k_2 - k_1}\right) e^{k_2 t}\left[1 + \frac{n_{10} - \beta n_{20}/(k_2 - k_1)}{n_{20}(1 + \beta/(k_2 - k_1))} e^{-(k_2 - k_1)t}\right]. \tag{S52}$$

Therefore,

$$\log(n_1 + n_2) = \log\left[n_{20}\left(1 + \frac{\beta}{k_2 - k_1}\right)\right] + k_2 t + \log[\cdots], \tag{S53}$$

where the last term in equation (S53) becomes vanishing small as t increases. Hence, the observed growth rate of the total population is

$$k_2 = 1.22/\text{hr}. \tag{S54}$$

From equations (S49), (S51), and (S54), we determine the quantities

$$k_1 = 1.16/\text{hr}, \beta = 0.0012/\text{hr} = 0.00069 \text{ per generation time}.$$

In the last equation, we have assumed that the generation time equals the doubling time (log 2)/$k_2 = 0.57$ hr.

2.1. Equations (2.2c) and (2.2d) must be replaced by the following two equations, which include the effect of the back-reaction:

$$\frac{dc}{dt} = k_{+1}es - (k_{-1} + k_{+2})c + k_{-2}pe, \tag{S55}$$

$$\frac{dp}{dt} = k_{+2}c - k_{-2}pe. \tag{S56}$$

When equation (2.4) is utilized to eliminate e from equation (S55), it becomes

$$\frac{dc}{dt} = k_{+1}e_0s - (k_{+1}s + k_{-2}p + k_{-1} + k_{+2})c + k_{-2}pe_0. \tag{S57}$$

Assuming the complex is a quasi-steady state, we infer from equation (S57) by setting $dc/dt = 0$ that

$$c = e_0 \frac{k_{+1}s + k_{-2}p}{k_{+1}s + k_{-2}p + k_{-1} + k_{+2}}. \tag{S58}$$

By substituting equations (2.4) and (S58) into equation (2.5a), we obtain the differential equation satisfied by $s(t)$ as

$$\frac{ds}{dt} = -k_{+2}c + k_{-2}pe = e_0 \frac{-k_{+1}k_{+2}s + k_{-1}k_{-2}p}{k_{+1}s + k_{-2}p + k_{-1} + k_{+2}}. \tag{S59}$$

To solve this equation, it must be supplemented by the differential equation for p. It is simpler to use instead the expression for the conservation of substrate molecules in the reactions,

$$s + c + p = s_0. \tag{S60}$$

By eliminating c between equations (S58) and (S60), we can express p as an explicit function of s. We shall not discuss the solution of equation (S59) here. From equation (S59) we infer that the reaction velocity at $t = 0$ is again given by the Michaelis–Menton equation (remember $p = 0$ at $t = 0$).

Haldane has also pointed out the desirability of including in the theory a product–enzyme complex C', so that the reaction scheme reads $S + E \rightleftharpoons C \rightleftharpoons C' \rightleftharpoons P + E$.

When equilibrium is established, $ds/dt = de/dt = dc/dt = dp/dt = 0$. In particular, from equations (2.2a) and (S56), we obtain

$$\begin{aligned} 0 &= -k_{+1}\bar{e}\bar{s} + k_{-1}\bar{c}, \\ 0 &= k_{+2}\bar{c} - k_{-2}\bar{p}\bar{e}, \end{aligned} \tag{S61}$$

where the equilibrium concentrations are denoted with an overbar. By eliminating $\bar{e}$ from these two equations, Holdane's relation equation (2.115) results.

2.2. By eliminating t_1 from equation (2.116) and equation (2.22a) with $t = t_1$ and $s = s_1$, we calculate $1/w_0$ to be

$$\frac{1}{w_0} = \frac{t_1}{s_0 - s_1} = \frac{1}{V}\left[1 - \frac{K_m}{s_0 - s_1}\log\left(\frac{s_1}{s_0}\right)\right]. \tag{S62}$$

Hence, with the aid of equation (2.15), the theoretical error is expressed as

$$\frac{1}{w_0} - \frac{1}{v_0} = -\frac{K_m}{s_0 V}\left[\frac{s_0}{s_0 - s_1}\log\left(\frac{s_1}{s_0}\right) + 1\right]. \tag{S63}$$

The power series for log(1 − ε), which can be generated by expanding the function in a Taylor series about ε = 0, is

$$\log(1 - \epsilon) = -\epsilon - \frac{\epsilon^2}{2} - \frac{\epsilon^3}{3} - \cdots.$$

Therefore,

$$\log\left(\frac{s_1}{s_0}\right) = \log\left(1 - \frac{s_0 - s_1}{s_0}\right) = -\frac{s_0 - s_1}{s_0} - \frac{1}{2}\left(\frac{s_0 - s_1}{s_0}\right)^2$$
$$- \frac{1}{3}\left(\frac{s_0 - s_1}{s_0}\right)^3 - \cdots,$$

(S64)

and the expression (S63) for $1/w_0 - 1/v_0$ becomes

$$\frac{1}{w_0} - \frac{1}{v_0} = \frac{K_m}{s_0 V}\left[\frac{1}{2}\left(\frac{s_0 - s_1}{s_0}\right) + \frac{1}{3}\left(\frac{s_0 - s_1}{s_0}\right)^2 + \cdots\right].$$

(S65)

The first term on the right above, of the order of $(s_0 - s_1)/s_0$, is the leading term in the error caused by using the secant approximation. Equation (S63), or equivalently equation (S65) if $(s_0 - s_1)/s_0$ is small, can be utilized to determine $1/v_0$ when $1/w_0$ is inferred from experimental data. Because the right-hand side of equation (S65) is positive, $1/w_0$ is an overestimate of the reciprocal reaction velocity $1/v_0$.

2.3. If we retain terms of order e_0/s_0, the inner solution for $s(t)$ is the solution of equation (2.18a), or its equivalent (2.5a), with s and c on the right-hand side being replaced by the zeroth order solution, equations (2.19) and (2.20), respectively. Thus,

$$\frac{ds}{dt} = -k_{+1}e_0 s_0 + (k_{+1}s_0 + k_{-1})\bar{c}[1 - e^{-(k_{+1}s_0 + k_{-1} + k_{+2})t}].$$

(S66)

By direct integration of the above equation for s, with $s = s_0$ at t = 0, we find that

$$s = s_0 + [-k_{+1}e_0 s_0 + (k_{+1}s_0 + k_{-1})\bar{c}]t$$
$$- \frac{(k_{+1}s_0 + k_{-1})\bar{c}}{k_{+1}s_0 + k_{-1} + k_{+2}}[1 - e^{-(k_{+1}s_0 + k_{-1} + k_{+2})t}].$$

If we make use of the relationships

$$\bar{c} = \frac{e_0 s_0}{s_0 + K_m},$$

$$\frac{k_{+1}s_0 + k_{-1}}{k_{+1}s_0 + k_{-1} + k_{+2}} = 1 - k_{+2}\tau_0,$$

then the desired expression (2.117) for $s(t)$ follows.

2.4. By the law of mass action, the equations for the rates of reaction of the concentrations s and p of substrate and product, respectively, are the following:

$$\frac{ds}{dt} = -ksp,$$

(S67a)

$$\frac{dp}{dt} = ksp,$$

(S67b)

where k is a rate constant. Initially, assume $s = s_0$ and $p = p_0$. By adding the above two equations, it follows that $s + p$ is a constant, and, in view of the initial condition,

$$s + p = s_0 + p_0. \tag{S68}$$

Using equation (S68) to eliminate s from equation (S67b), we obtain the relation

$$\frac{dp}{dt} = k(s_0 + p_0 - p)p. \tag{S69}$$

We recognize this to be the logistic equation [see equation (1.22)] whose solution we know, from equation (1.30), to be

$$p = (s_0 + p_0) \frac{p_0}{p_0 + s_0 e^{-k(s_0 + p_0)t}}. \tag{S70}$$

Thus, p approaches $s_0 + p_0$ as $t \to \infty$, when all the substrate has been converted to product. By substitution of equation (S70) into (S68), it follows that

$$s = (s_0 + p_0) \frac{s_0 e^{-k(s_0 + p_0)t}}{p_0 + s_0 e^{-k(s_0 + p_0)t}}. \tag{S71}$$

2.5. (a) According to Figure 2.22, the rate equations for the concentrations of D, C_1, A, C_2, and R, denoted as usual by lower case letters, are, respectively,

$$\frac{dd}{dt} = -k_{+1}rd + k_{-1}c_1, \tag{S72a}$$

$$\frac{dc_1}{dt} = k_{+1}rd - k_{-1}c_1, \tag{S72b}$$

$$\frac{da}{dt} = -k_{+2}ar + k_{-2}c_2, \tag{S72c}$$

$$\frac{dc_2}{dt} = k_{+2}ar - k_{-2}c_2, \tag{S72d}$$

$$\frac{dr}{dt} = -k_{+1}rd + k_{-1}c_1 - k_{+2}ar + k_{-2}c_2. \tag{S72e}$$

If the production of A by some other reaction is to be mathematically represented, then an additional positive term is needed on the right-hand side of the equation for a. Initially, assume that only A, D, and R are present, so that

$$(a, d, r, c_1, c_2) = (a_0, d_0, r_0, 0, 0) \text{ at } t = 0, \tag{S73}$$

where a_0, d_0, and r_0 are constants.

By adding equations (S72a) and (S72b), it follows that

$$d + c_1 = d_0, \tag{S74}$$

which represents the conservation of DNA molecules. Similarly by adding equations (S72c) and (S72d), the conservation of anti-repressor molecules is expressed as

$$a + c_2 = a_0. \tag{S75}$$

From the sum of equations (S72b), (S72d), and (S72e), we obtain the result

$$r + c_1 + c_2 = r_0, \tag{S76}$$

which represents the conservation of repressor molecules.

Using (S74)–(S76), we eliminate r, c_1, and c_2 from the equations (S72a) and (S72c) to obtain the following two coupled equations for d and a,

$$\frac{dd}{dt} = k_{-1}d_0 - [k_{-1} + k_{+1}(r_0 - d_0 - a_0)]d - k_{+1}(a + d)d,$$

$$\frac{da}{dt} = k_{-2}a_0 - [k_{-2} + k_{+2}(r_0 - d_0 - a_0)]a - k_{+2}(a + d)a. \tag{S77}$$

(b)   Now assume the reactions have reached equilibrium, so that the time derivatives are zero in equations (S77). With $K_1 \equiv k_{-1}/k_{+1}$, and $K_2 \equiv k_{-2}/k_{+2}$, equations (S77) become

$$0 = K_1 d_0 - [K_1 + r_0 - d_0 - a_0]d - (a + d)d, \tag{S78a}$$
$$0 = K_2 a_0 - [K_2 + r_0 - d_0 - a_0]a - (a + d)a. \tag{S78b}$$

If we subtract d^{-1} times (S78a) from a^{-1} times (S78b), the result is the following relation between the equilibrium values of a and d:

$$a = \frac{K_2 a_0 d}{K_1 d_0 + (K_2 - K_1)d}. \tag{S79}$$

Substituting this relation back into equation (S78a), we obtain the following cubic equation for the equilibrium value of d:

$$0 = K_1^2 d_0^2 + K_1 d_0(K_2 + a_0 + d_0 - r_0 - 2K_1)d + [K_2 d_0 - K_1(a_0 + 2d_0) \\ - (K_2 - K_1)(K_1 + r_0)]d^2 - (K_2 - K_1)d^3. \tag{S80}$$

(c)*   To solve equation (S80) approximately, it is easier to treat the nondimensional ratio $d/d_0 \equiv x$ as unknown. Thus, after division by d_0^4, equation (S80) becomes

$$0 = \left(\frac{K_1}{d_0}\right)^2 + \left(\frac{K_1}{d_0}\right)\left(1 + \frac{K_2}{d_0} + \frac{a_0}{d_0} - \frac{r_0}{d_0} - 2\frac{K_1}{d_0}\right)x + \left[\frac{K_2}{d_0} - 2\frac{K_1}{d_0}\right. \\ \left. - \frac{K_1 a_0}{d_0^2} + \left(\frac{K_1}{d_0} - \frac{K_2}{d_0}\right)\left(\frac{K_1}{d_0} + \frac{r_0}{d_0}\right)\right]x^2 + \left(\frac{K_1}{d_0} - \frac{K_2}{d_0}\right)x^3. \tag{S81}$$

Because $d < d_0$, we are interested in the root of this equation that lies in the interval $0 \le x \le 1$. Furthermore, we are given as reasonable estimates that $K_1/d_0 \approx 10^{-10}$ and $K_2/d_0 \approx 10^{-13}$. Hence $K_1 \approx 10^3 K_2$, which implies that the affinity of repressor molecules for antirepressor molecules is considerably greater than that of DNA.

Let us examine equation (S81) when it is simplified by the neglect of the terms in K_2/d_0. Then it becomes

$$0 = \frac{K_1}{d_0} + \left(1 + \frac{a_0}{d_0} - \frac{r_0}{d_0} - \frac{2K_1}{d_0}\right)x + \left(\frac{K_1}{d_0} + \frac{r_0}{d_0} - \frac{a_0}{d_0} - 2\right)x^2 + x^3. \tag{S82}$$

This equation has two roots: the root $x = 1$, and an approximate root for which x is small. We obtain the latter by neglecting the terms in x^2 and x^3 (remember that K_1/d_0 is very small), and the term in K_1/d_0 in the coefficient of x in comparison with 1, with the result

$$x = \frac{K_1}{r_0 - a_0 - d_0}, \text{ or } d = \frac{K_1 d_0}{r_0 - a_0 - d_0}. \tag{S83}$$

We note that this root is small and positive so long as a_0 is less than (and not too close to) $r_0 - d_0$.

We return to equation (S81) with the knowledge that it possesses a root in the neighborhood of $x = 1$ and/or an approximate root given by the expression (S83). We examine the possibility of a root in the neighborhood of $x = 1$ more closely by setting $x = 1 - \xi$, and retaining terms up to order ξ, inclusive. The equation for ξ is

$$-K_2 r_0 + [K_1(a_0 - r_0) + K_2(K_1 + d_0 + 2r_0)]\xi = 0, \tag{S84}$$

where the term in K_2 in the square brackets can be neglected in comparison to the first term. Hence, the approximate root of equation (S82) near $x = 1$ is

$$x = 1 - \frac{K_2 r_0}{K_1(a_0 - r_0)}, \text{ or } d = d_0 - \frac{d_0 K_2 r_0}{K_1(a_0 - r_0)}. \tag{S85}$$

This root is less than 1, and therefore meaningful, when a_0 is greater than r_0 (but not too close to it).

Now imagine that a_0 is increasing quasistatically during cell growth, because of an independent production process. Initially, $a_0 < r_0$, and the equilibrium value of d is determined by (S83). This value, which represents the amount of "free" DNA, is of the order of K_1 and is essentially zero, that is, the DNA molecule exists in its complexed form C_1. (Remember that $K_1 = 10^{-10} d_0$ and d_0 equals one molecule per cell.) As antirepressor molecules slowly accumulate in the cell, their number attains and surpasses the critical value $a_0 \approx r_0 - d_0$. The equilibrium value of d then shifts to the value given by the root (S85), or $d \approx d_0$. In this state, the DNA is free of repressor which is now complexed with antirepressor in the form C_2. The biochemical reason for this transition is, of course, that the affinity of repressor for antirepressor is considerably greater than for DNA. Thus, when the number of molecules of antirepressor becomes equal to or greater than the number of repressor molecules, all the repressor molecules bind with antirepressor, leaving the DNA in a "derepressed" state. This transition occurs abruptly at the critical value of a_0.

To find the equilibrium value of antirepressor concentration corresponding to the equilibrium value of d given by either equation (S83) or (S85), we combine these two equations with (S79). Thus, if (S83) is applicable so that $d \sim 0$, then according to (S79), the equilibrium value of antirepressor concentration is given approximately as

$$a = a_0 \frac{K_2}{r_0 - a_0 - d_0}, a_0 < r_0 - d_0. \tag{S86}$$

In obtaining this expression, we have neglected a term in the denominator of order $(K_2 - K_1)$. Equation (S86) states that there is virtually no free antirepressor, because $K_2/d_0 \ll 1$, it being bound to repressor in the form C_2. When equation (S85) is applicable, the equilibrium value of antirepressor is given approximately with the aid of (S79) as

$$a = a_0 - r_0, a_0 > r_0. \tag{S87}$$

In obtaining this result, we have neglected terms of order K_2/K_1. Equation (S87) states that the amount of free antirepressor is just the excess of antirepressor molecules over repressor molecules.

The above remarks are illustrated in a concrete fashion by the numerical solution of equation (S81) for the assumed values of K_1/d_0 and K_2/d_0, a hypothetical value of ten repressor molecules in the cell so that $r_0/d_0 = 10$, and the antirepressor molecule number a_0/d_0 considered as a parameter. The equilibrium value of d/d_0 is shown as a function of the ratio a_0/d_0 in Figure S2.5.

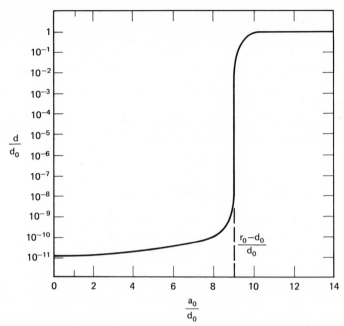

Figure S2.5. The equilibrium value of free DNA in a cell is shown as a function of the anti-repressor molecule number a_0/d_0, based on equation (S81), with $K_1/d_0 = 10^{-10}$, $K_2/d_0 = 10^{-13}$, and $r_0/d_0 = 10$. The ordinate scale is logarithmic, while the scale of the abscissa is linear.

2.6.* Setting $x = \xi - 1$ in equation (2.70), we find that the saturation function is expressible as

$$Y = \frac{\xi^n - \xi^{n-1}}{L + \xi^n}. \tag{S88}$$

The condition for a point of inflection is that $d^2 Y/d\xi^2$ is null. We find, by explicit differentiation, that the desired condition is

$$(L + \xi^n)^3 \frac{d^2 Y}{d\xi^2} = L^2[n(n - 1)\xi^{n-2} - (n - 1)(n - 2)\xi^{n-3}]$$
$$- L[n(n + 1)\xi^{2n-2} - (n^2 + 3n - 4)\xi^{2n-3}] - 2n\xi^{3n-3} = 0. \tag{S89}$$

We shall search for an inflection point when $L \gg 1$.

To find a real root of equation (S89) approximately, let us first retain only the term in L^2. Then it follows that $\xi = (n - 2)/n$ or $\xi = 0$. However, a meaningful solution requires that $\xi \geq 1$ because $x \geq 0$, so that these solutions are outside the physical range of interest.

Let us next search for a solution for which $\xi \gg 1$, and in fact varies as some positive power of L. Then it follows that the term in ξ^{n-3} is negligible in comparison with ξ^{n-2}, and the term in ξ^{2n-3} is negligible in comparison with ξ^{2n-2}. Thus, equation (S89) simplifies to

$$L^2(n - 1)\xi^{n-2} - L(n + 1)\xi^{2n-2} - 2\xi^{3n-3} = 0. \tag{S90}$$

Because the first term above is positive and the last two are negative, we can assume that the leading or most important terms, as a function of L, are either the first and second, or the first and third. Therefore, equation (S90) is tentatively replaced by either

or
$$L^2(n - 1)\xi^{n-2} - L(n + 1)\xi^{2n-2} = 0, \tag{S91a}$$
$$L^2(n - 1)\xi^{n-2} - 2\xi^{3n-3} = 0. \tag{S91b}$$

In the second case above, we find that

$$\xi \sim L^{2/(2n-1)}. \tag{S92}$$

Therefore, the term neglected in equation (S90) is

$$(n + 1)L\xi^{2n-2} \sim L^{3 - 2/(2n-1)}.$$

The terms retained in (S91b) are of order $L^2\xi^{n-2} \sim L^{3 - 3/(2n-1)}$. We see that the neglected term is of higher order than the retained terms, and therefore we were not justified in neglecting it. Consequently, (S92) is not an approximate solution of equation (S90).

Hence, we examine instead equation (S91a). Its solution is

$$\xi = \left(\frac{n - 1}{n + 1} L\right)^{1/n}. \tag{S93}$$

The terms in equation (S91a) are both of order $L^{3 - 2/n}$, as is readily verified. The neglected term is of order $L^{3 - 3/n}$ which is indeed of lower order than the retained terms for all positive values of n. We are therefore justified in neglecting it.

From (S93), it follows that the inflection point of $Y(x)$ is given approximately when $L \gg 1$ as

$$x = \left(\frac{n - 1}{n + 1} L\right)^{1/n} - 1. \tag{S94}$$

When $L = 1000$ and $n = 4$, we find, from the above formula, that $x = 3.95$. That is approximately where the inflection point is, according to the curve for these parameter values and $c = 0$, shown in Figure 2.11.

2.7. According to hypothesis, it follows as for equations (2.35) and (2.82), with $K = k_{-1}/k_{+1}$ and $\lambda_+/\lambda_- = 1 + L$, that

$$h_1 = 4\left(\frac{s_0}{K}\right) h_0,$$

$$h_2 = \frac{3}{2}\left(\frac{s_0}{K}\right) h_1 = 6\left(\frac{s_0}{K}\right)^2 h_0,$$

$$h_3 = \frac{2}{3}\left(\frac{s_0}{K}\right) h_2 = 4\left(\frac{s_0}{K}\right)^3 h_0, \tag{S95}$$

$$h_4 = \frac{1}{4}(1 + L)\left(\frac{s_0}{K}\right) h_3 = (1 + L)\left(\frac{s_0}{K}\right)^4 h_0.$$

Hence, by substituting these relations into the expression for the saturation function, equation (2.83), we obtain the result

$$Y(s_0) = \frac{s_0/K + 3(s_0/K)^2 + 3(s_0/K)^3 + (1 + L)(s_0/K)^4}{1 + 4s_0/K + 6(s_0/K)^2 + 4(s_0/K)^3 + (1 + L)(s_0/K)^4}. \tag{S96}$$

The numerical coefficients appearing above are binomial coefficients, and therefore,

$Y(x)$ is expressible, as in equation (2.41), in the form of equation (2.119) (Margaria, 1963), where $x = s_0/K$. From (S95), it is apparent, with K_1 to K_4 defined as in equation (2.81), that

$$(K_1, K_2, K_3, K_4) = K\left(\frac{1}{4}, \frac{2}{3}, \frac{3}{2}, \frac{4}{1+L}\right). \tag{S97}$$

Hence (Forbes and Roughton, 1931)

$$(K_1^{-1}, K_2^{-1}, K_3^{-1}, K_4^{-1}) = 4K^{-1}\left(1, \frac{3}{8}, \frac{1}{6}, \frac{1+L}{16}\right). \tag{S98}$$

2.8. The graph for a fully competitive reaction is given in Figure S2.8. The basic determinants

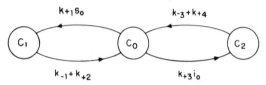

Figure S2.8. Graph for a fully competitive reaction.

are readily found from it to be

$$\begin{aligned}
D_0 &= (k_{-1} + k_{+2})(k_{-3} + k_{+4}), \\
D_1 &= (k_{-3} + k_{+4})k_{+1}s_0, \\
D_2 &= (k_{-1} + k_{+2})k_{+3}i_0.
\end{aligned} \tag{S99}$$

Hence, from the above equations and equation (2.95), the velocity v_0^s of the reaction producing the product P_1 is given as

$$\begin{aligned}
v_0^s = k_{+2}c_1 &= \frac{k_{+2}e_0(k_{-3} + k_{+4})k_{+1}s_0}{(k_{-1} + k_{+2})(k_{-3} + k_{+4} + k_{+3}i_0) + (k_{-3} + k_{+4})k_{+1}s_0} \\
&= \frac{V^s s_0 K_m^i}{K_m^s K_m^i + K_m^s i_0 + K_m^s s_0},
\end{aligned} \tag{S100}$$

in agreement with equation (2.30).

Similarly, the velocity of the reaction producing the product P_2 is

$$\begin{aligned}
v_0^i = k_{+4}c_2 &= \frac{k_{+4}e_0(k_{-1} + k_{+2})k_{+3}i_0}{(k_{-1} + k_{+2})(k_{-3} + k_{+4} + k_{+3}i_0) + (k_{-3} + k_{+4})k_{+1}s_0} \\
&= \frac{V^i i_0 K_m^s}{K_m^s K_m^i + K_m^s i_0 + K_m^s s_0},
\end{aligned} \tag{S101}$$

which likewise agrees with equation (2.30), when the substrate and inhibitor are interchanged.

2.9. The graph for noncompetitive inhibition representing the reactions (2.120) is shown in

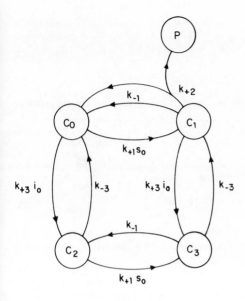

Figure S2.9(a). Graph for an enzyme reacting with two substrates in noncompetitive inhibition.

Figure S2.9(a). The basic determinants corresponding to the nodes c_0, c_1, c_2, and c_3 each consist of four trees, and are readily found to be given, respectively, by the equations

$$
\begin{aligned}
D_0 &= k_{-3}k_{-1}k_{+3}i_0 + k_{-3}k_{-1}(k_{-1} + k_{+2}) + k_{-3}^2(k_{-1} + k_{+2}) \\
&\quad + k_{-3}(k_{-1} + k_{+2})k_{+1}s_0, \\
D_1 &= k_{-3}k_{-1}k_{+1}s_0 + k_{-3}^2k_{+1}s_0 + k_{-3}k_{+1}^2s_0^2 + k_{-3}k_{+3}i_0k_{+1}s_0, \\
D_2 &= k_{-3}(k_{-1} + k_{+2})k_{+3}i_0 + k_{-1}(k_{-1} + k_{+2})k_{+3}i_0 + k_{-1}k_{+3}^2i_0^2 \\
&\quad + k_{-1}k_{+3}i_0k_{+1}s_0, \\
D_3 &= k_{-3}k_{+3}i_0k_{+1}s_0 + k_{+3}i_0k_{+1}^2s_0^2 + k_{+3}^2i_0^2k_{+1}s_0 \\
&\quad + (k_{-1} + k_{+2})k_{+3}i_0k_{+1}s_0.
\end{aligned}
\tag{S102}
$$

Let

$$
K_1 \equiv \frac{k_{-1}}{k_{+1}}, \; K_3 \equiv \frac{k_{-3}}{k_{+3}}, \; K_m = \frac{k_{-1} + k_{+2}}{k_{+1}}, \; \beta = \frac{k_{-3}}{k_{-1}}, \; V = k_{+2}e_0.
$$

Then (S102) can be rewritten as

$$
\begin{aligned}
D_0 &= k_{+1}^2k_{+3}[\beta K_1^2i_0 + K_1K_3K_m + \beta K_1K_3K_m + K_3K_ms_0], \\
D_1 &= k_{+1}^2k_{+3}s_0[K_1K_3 + \beta K_1K_3 + K_3s_0 + \beta K_1i_0], \\
D_2 &= k_{+1}^2k_{+3}i_0[\beta K_3K_m + K_3K_m + \beta K_1i_0 + K_3s_0], \\
D_3 &= k_{+1}^2k_{+3}s_0i_0[\beta K_1 + s_0 + \beta K_1i_0/K_3 + K_m].
\end{aligned}
\tag{S103}
$$

Hence, from equations (2.95) and (2.96),

$$
v = k_{+2}c_1 = V\frac{D_1}{D_0 + D_1 + D_2 + D_3},
$$

or

$$v = V \frac{s_0}{K_1} \left(1 + \beta + \frac{s_0}{K_1} + \beta \frac{i_0}{K_3} \right) \left\{ \left(1 + \frac{i_0}{K_3} \right) \left[(1 + \beta) \frac{K_m}{K_1} + \beta \frac{i_0}{K_3} \right. \right.$$
$$\left. \left. + \frac{s_0}{K_1} \left(1 + \beta + \frac{K_m}{K_1} + \beta \frac{i_0}{K_3} \right) + \frac{s_0^2}{K_1^2} \right] \right\}^{-1}. \tag{S104}$$

If we divide the numerator of (S104) into the denominator, we can rewrite (S104) as

$$v = V s_0 \left(1 + \frac{i_0}{K_3} \right)^{-1} \left[s_0 + K_m + \beta \frac{i_0}{K_3} \frac{K_1 - K_m}{1 + \beta + s_0/K_1 + \beta i_0/K_3} \right]^{-1}. \tag{S105}$$

If $k_{+2} \ll k_{-1}$, then $K_m \approx K_1$, and the last term in the square brackets is negligible. The expression for v^{-1} becomes the following:

$$\frac{1}{v} = \frac{1}{V} \left(1 + \frac{i_0}{K_3} \right) \left(1 + \frac{K_1}{s_0} \right), \tag{S106}$$

which is the result usually quoted. The double reciprocal plot of $1/v$ versus $1/s_0$ for various values of i_0, based on equation (S106), is shown in Figure S2.9(b). The resulting

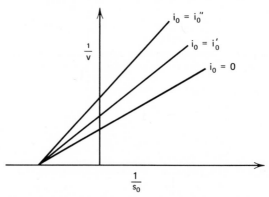

Figure S2.9(b).   Lineweaver–Burk plot for a pure noncompetitive enzyme–substrate–inhibitor system, based on equation (S106), for several values of the inhibitor concentration i_0.

curves, which have a common intercept on the axis of the abscissa, are distinguished from the corresponding curves in Figure 2.4 for a fully competitive system, which have a common intercept on the ordinate axis.

If $k_{+4} \neq 0$, then we must add to v the contribution v', where, from (2.95) and (2.96),

$$v' = k_{+4} c_3 = k_{+4} e_0 \frac{D_3}{D_0 + D_1 + D_2 + D_3}$$
$$= V' s_0 i_0 (K_3 + i_0)^{-1} \left[s_0 + K_1 + \frac{\beta(K_m - K_1)}{\beta + s_0/K_1 + \beta i_0/K_3 + K_m/K_1} \right]^{-1}, \tag{S107}$$

and $V' \equiv k_{+4} e_0$.

The total velocity of the reaction v_T is given as the sum

$$v_T = v + v', \tag{S108}$$

where v and v' are given by equation (S105) and (S107), respectively. If $k_{+2} \ll k_{-1}$, then $K_m \approx K_1$, and equation (S107) simplifies to

$$v' = \frac{V' i_0 s_0}{(i_0 + K_3)(s_0 + K_1)}. \tag{S109}$$

From equations (S106), (S108), and (S109), we find in this case of apparent competitive inhibition that the reciprocal reaction velocity is given as

$$\frac{1}{v_T} = \frac{1}{V} \frac{(1 + i_0/K_3)}{(1 + V' i_0/V K_3)} \left(1 + \frac{K_1}{s_0}\right). \tag{S110}$$

This expression is similar to equation (S106) if the double reciprocal plot of $1/v_T$ versus $1/s_0$ is made: A family of straight lines is obtained for different values of i_0, similar to those shown in Figure S2.9(b). However, the difference between the two results is more apparent if we examine the dependence on the inhibitor concentration of the apparent velocity maximum, determined from the intercepts of the straight lines in the Lineweaver–Burk plot. In the case of equation (S106), the dependence of the reciprocal velocity maximum on i_0 is linear, so that V and K_3 are easily determined. In the case of equation (S110), the relationship is not linear, and in fact we find that the apparent velocity maximum V_T is given as $V_T = V_T(i_0) = (V + V' i_0/K_3)/(1 + i_0/K_3)$. Nevertheless, we can determine the parameters V, V', and K_3 from this relationship and the observed values of V_T by noting that $V_T(0) = V$ and $V_T(\infty) = V'$. The quantity K_3 can be found from intermediate values of V_T by solving the above equation for i_0/K_3 and utilizing the predetermined values of V and V'. Then it is seen that K_3 is the slope of the straight line relating i_0 and the quantity $(V_T - V)/(V' - V_T)$.

If, in addition, $k_{+2} = k_{+4}$, so that $V = V'$, (S110) further simplifies to the usual Michaelis–Menten expression

$$\frac{1}{v_T} = \frac{1}{V} \left(1 + \frac{K_1}{s_0}\right). \tag{S111}$$

There is then no apparent effect of competition.

3.1. The rate of influx of dye mass into the reservoir is J_0, and the efflux rate is Qc. Hence, the rate equation governing $c = c(t)$ is the following:

$$\frac{d(Vc)}{dt} = -Qc + J_0, \tag{S112}$$

where Vc is the mass of dye in the reservoir at time t. The initial condition is $c(0) = 0$. Let

$$c = c_\infty + x(t),$$

$$c_\infty = \frac{J_0}{Q}. \tag{S113}$$

Here c_∞ is the steady-state solution, obtained from equation (S112) by setting $dc/dt = 0$. Substituting c as given by (S113) into (S112), we obtain the equation for x (remember V is constant),

$$\frac{dx}{dt} = -\frac{Q}{V} x, \tag{S114}$$

where $x(0) = -c_\infty$, according to (S113).

We find readily that $x(t) = -c_\infty e^{-Qt/V}$, so that

$$c(t) = c_\infty (1 - e^{-Qt/V}). \tag{S115}$$

Hence, $c(t)$ has the form of the curve shown in Figure 1.3(a), where $\bar{c}$ in the figure represents c_∞ in equation (S115).

The area S bounded by the curve c, the asymptote c_∞, and the ordinate axis is finite, and is easily measured with the aid of a planimeter. Its theoretical value is determined with the aid of equation (S115) as

$$S = \int_0^\infty (c_\infty - c)dt = c_\infty \int_0^\infty e^{-Qt/V}\, dt = \frac{c_\infty V}{Q}. \tag{S116}$$

By observing c_∞ and knowing J_0, we determine Q from the second of equations (S113) to be

$$Q = \frac{J_0}{c_\infty}. \tag{S117}$$

With Q and equation (S116), we infer V to be

$$V = \frac{QS}{c_\infty}. \tag{S118}$$

3.2. Equation (3.58a) with $\rho_1 = V_1 x_1$ yields the following equation for x_1,

$$\frac{dx_1}{dt} = -\frac{K_{11}}{V_1}x_1 + \frac{K_{12}}{V_1}x_2. \tag{S119}$$

In the limit as $V_1 \to \infty$, the right-hand side vanishes, and dx_1/dt is null. Hence

$$x_1 = x_{10}, \text{ a constant.}$$

Because the compartment system is closed, the mass flux leaving compartment 1 is the same as the mass flux leaving compartment 1 to enter compartment 2, or

$$K_{21} = K_{11}. \tag{S120}$$

Therefore, the term $K_{21}x_1$ appearing in (3.58b) represents a steady infusion of labeled mass at a constant rate J_1, where

$$J_1 = K_{21}x_1 = K_{11}x_{10}. \tag{S121}$$

Hence, equations (3.58) reduce to the form given by equations (3.119).

3.3.* The compartment system is represented by the graph shown in Figure 3.5. In addition, there is a steady flux of labeled mass per unit volume at a rate j_0 into compartment 1. Let

$$\begin{aligned} x_1(t) &= \bar{x}_1 + \xi_1(t), \\ x_2(t) &= \bar{x}_2 + \xi_2(t), \end{aligned} \tag{S122}$$

where $\bar{x}_1$ and $\bar{x}_2$ represent the equilibrium concentrations, determined as the solution to the compartmental equations when $dx_1/dt = dx_2/dt = 0$, or,

$$0 = -L_{11}\bar{x}_1 + M_{12}\bar{x}_2 + j_0, \tag{S123a}$$
$$0 = M_{21}\bar{x}_1 - L_{22}\bar{x}_2. \tag{S123b}$$

Because all the (nonlabeled) mass leaving compartment 2 comes from compartment 1, $K_{22} = K_{21}$, or,

$$L_{22} = V_2^{-1}L_{21}V_1 = M_{21}. \tag{S124}$$

Hence, from (S123b) and (S124), we find that

$$\bar{x}_1 = \bar{x}_2. \tag{S125}$$

Substituting this result into equation (S123a), we obtain the relation

$$\bar{x}_1 = \bar{x}_2 = \frac{j_0}{L_{11} - M_{12}}. \tag{S126}$$

Note that, because compartment 1 is leaky, $V_1 L_{11} > L_{12} V_2$, and therefore the denominator in equation (S126) is a positive quantity.

Substituting equations (S122) into the compartmental equations transforms them to the following:

$$\frac{d\xi_1}{dt} = -L_{11}\xi_1 + M_{12}\xi_2, \tag{S127a}$$

$$\frac{d\xi_2}{dt} = M_{21}\xi_1 - L_{22}\xi_2, \tag{S127b}$$

where, from (S122) and (S125),

$$\xi_1(0) = \xi_2(0) = -\bar{x}_1. \tag{S128}$$

The equilibration between coronary sinus blood and the extracellular compartment is assumed to be so rapid, that observations of the coronary blood effectively represents the extracellular compartment. Hence, we identify the observations given by equations (3.120) and (3.121) as representing $x_1(t)$. From this identification and equations (S122) and (S128), $\bar{x}_1$ and $\xi_1(t)$ are determined to be

$$\bar{x}_1 = A_{10}, \tag{S129a}$$
$$\xi_1(t) = A_{11}e^{\lambda_1 t} + A_{12}e^{\lambda_2 t}, \tag{S129b}$$

where $A_{11} + A_{12} = -A_{10}$.

Equations (S127)–(S129) represent a standard two-compartment problem with $\xi_1(t)$ known from observation. However, both $\xi_1(0)$ and $\xi_2(0)$ are nonzero, so that equation (3.48) is not applicable. Instead, we substituting equation (S129b) into (S127a), set $t = 0$, and using equation (S128), find that

$$-L_{11}\xi_1(0) + M_{12}\xi_2(0) = (L_{11} - M_{12})\bar{x}_1 = \lambda_1 A_{11} + \lambda_2 A_{12}. \tag{S130}$$

Furthermore, we know that the clearance rates satisfy equations (3.40). Equations (S130) and (3.40), together with (S124), constitute four equations for the four unknown flux coefficients L_{11}, L_{22}, M_{12}, and M_{21}. Their solution is found in a straightforward manner, for the given values of $A_{10}, A_{11}, A_{12}, \lambda_1$, and λ_2, to be

$$L_{11} = 4.92/\text{min},$$
$$M_{21} = L_{22} = 0.0440/\text{min}, \tag{S131}$$
$$M_{12} = 3.19/\text{min}.$$

The intracellular compartment 2 is leakproof. Hence, $L_{12} = L_{22}$. Because $M_{12} = V_1^{-1} L_{12} V_2$, it follows that

$$\frac{V_2}{V_1} = \frac{M_{12}}{L_{12}} = \frac{M_{12}}{L_{22}} = 72.5. \tag{S132}$$

From equations (S126), (S129a), and (S131), we find that

$$j_0 = 99.6 \ \mu\text{Ci/liter/min.} \tag{S133}$$

Alternatively, j_0 can be obtained by substituting the expression for $x_1(t)$ into the differential equation for $x_1(t)$ and evaluating it at $t = 0$. It is found then that

$$j_0 = A_{11}\lambda_1 + A_{12}\lambda_2. \tag{S134}$$

If we multiply the compartment equation for $x_1(t)$ by the amount V_1 of extracellular potassium, and compare it with the compartment equation (3.119a) of problem 2, we see with the aid of (S121) that $V_1 j_0 / \bar{x}_1$ represents the steady-state flux K_{10} of unlabeled material from the exterior (blood plasma) to compartment 1 (extracellular pool). From the observed mean value of $V_2 = 83.8$ mEq/kg and equation (S132), we infer that $V_1 = 1.16$ mEq/kg. Hence, with the aid of (S129a), (S133), and (3.121),

$$K_{10} = 2.01 \ \text{mEq/min/kg}, \tag{S135}$$

and the unlabeled flux rate between the extracellular pool and the intracellular pool is

$$K_{12} = K_{21} = L_{12}V_2 = L_{21}V_1 = 3.70 \ \text{mEq/min/kg}. \tag{S136}$$

Conn and Robertson (1955) determined the above quantities for each value of V_2 observed in six dogs, and then determined their averages. They found in this manner the average values $K_{10} = 1.93$ mEq/min/kg, and $K_{12} = K_{21} = 4.06$ mEq/min/kg.

3.4. The initial condition satisfied by the concentrations of the compartments of the mixing-cell model when $n = 3$ is assumed to be that

$$(c_1(t_d), c_2(t_d), c_3(t_d)) = \left(\frac{\rho_0}{V}, 0, 0\right). \tag{S137}$$

We solve equation (3.123a) first, and find readily that

$$c_1 = \frac{\rho_0}{V} e^{-(K/V)(t - t_d)}, \qquad t \geq t_d. \tag{S138}$$

Substituting this result into equation (3.123b) with $j = 2$, we find that $c_2(t)$ satisfies a first order inhomogeneous differential equation of the form of equation (1.20). Its solution, according to equation (1.21), is

$$c_2 = \frac{\rho_0}{V} (t - t_d) e^{-(K/V)(t - t_d)}, \qquad t \geq t_d. \tag{S139}$$

In an exactly similar manner, we find that the solution of equation (3.123b) with $j = 3$ is

$$c_3 = \frac{\rho_0}{V} \frac{(t - t_d)^2}{2} e^{-(K/V)(t - t_d)}, \qquad t \geq t_d. \tag{S140}$$

If we continue in this manner for a system of n compartments, we find that

$$c_4 = \frac{\rho_0}{V} \frac{(t - t_d)^3}{2 \times 3} e^{-(K/V)(t - t_d)}, \qquad t \geq t_d,$$

and, for any value of n,

$$c_n = \frac{\rho_0}{V} \frac{(t - t_d)^n}{(n - 1)!} e^{-(K/V)(t - t_d)}, \qquad t \geq t_d. \tag{S141}$$

3.5. Mathematically, this problem is similar to the example of creatinine clearance discussed in Section 3.8. The compartment system is represented by the graph of Figure 3.5. Thus, compartment 2 is leakproof, so that L_{02} is zero. With this equation and equations (3.40), (3.48), and (3.125), we compute the clearance rates to be as follows,

$$L_{11} = 0.093/\text{day},$$
$$L_{12} = L_{22} = 0.046/\text{day}, \tag{S142}$$
$$L_{21} = 0.054/\text{day}.$$

The size of compartment 1 (remember 1 μCi = 3.7 × 10^4 dis/sec) is given by equation (3.46a) and the given data as

$$V_1 = \frac{\rho_{10}}{A_{11} + A_{12}} = \frac{30 \times 3.7 \times 10^4 \times 60}{2650} \text{ mg} = 25 \text{ g}. \tag{S143}$$

As in equation (3.60), the steady state fluxes of unlabeled cholesterol into and out of the system are represented by the equations

$$L_{11}V_1 = L_{12}V_2 + J_1,$$
$$L_{22}V_2 = L_{21}V_1 + J_2. \tag{S144}$$

Here, J_1 represents the steady-state rate of influx of unlabeled cholesterol into compartment 1, due to ingestion and production, and J_2 represents the steady-state rate of influx of unlabeled cholesterol into compartment 2. If we add the above two equations, the compartment size V_2 (which we do not know) is eliminated because $L_{12} = L_{22}$, with the result that

$$J_1 + J_2 = (L_{11} - L_{21})V_1 = L_{01}V_1. \tag{S145}$$

In words, the total rate of influx of unlabeled cholesterol into the system, due to production and consumption, equals the total rate of efflux of unlabeled cholesterol from the system, which is the same as the excretion rate from compartment 1. With the aid of equations (S142) and (S143), we readily compute this flux rate to be

$$J_1 + J_2 = 1.0 \text{ g/day}. \tag{S146}$$

With a mean rate of ingestion of cholesterol estimated to be 0.2 g/day, we infer that the mean production rate of cholesterol is 0.8 g/day.

3.6. By integrating equation (3.126a) for $\xi = \xi(t)$, subject to the initial condition that $\xi(0) = \xi_0 \equiv 47$ μm, we find readily, as for equation (1.17), that

$$\xi = L - (L - \xi_0)e^{-k_1 t}. \tag{S147}$$

If we substitute this result into equation (3.126b) for x, it becomes

$$\frac{dx}{dt} = -(k_1 + k_2)x + (k_1 + k_2)L - k_2(L - \xi_0)e^{-k_1 t}. \tag{S148}$$

We recognize the above equation to be of the form of equation (1.20). Its solution is given by the straightforward application of equation (1.21) as

$$x = L - (L - \xi_0)e^{-k_1 t} + [x_0 - L + (L - \xi_0)e^{-k_1 t_0}]e^{-(k_1 + k_2)(t - t_0)}, t \geqq t_0, \tag{S149}$$

where $x = x_0$ when $t = t_0$.

When t is large, only the first two terms on the right-hand side of (S149) survive, and

$$L - x \sim (L - \xi_0)e^{-k_1 t} = (L - \xi_0)e^{-k_1 t_0 - k_1(t - t_0)} = (L - \xi). \tag{S150}$$

Comparison of this result with the observed behavior of $L - x$ for large times, equation (3.127a), yields

$$(L - \xi_0)e^{-k_1 t_0} = A, \tag{S151a}$$

$$k_1 = -\lambda_1. \tag{S151b}$$

By subtracting $(L - \xi)$ given by equation (S147) from $L - x$, we obtain

$$L - x - (L - \xi) = -[x_0 - L + (L - \xi_0)e^{-k_1 t_0}]e^{-(k_1 + k_2)(t - t_0)}. \tag{S152}$$

Hence, by comparison of this result with equation (3.217b), we infer that

$$x_0 - L + (L - \xi_0)e^{-k_1 t_0} = -B,$$

$$k_1 + k_2 = -\lambda_2. \tag{S153}$$

Solving equations (S151) and (S153) for k_1, k_2, x_0, and ξ_0, with the given values of λ_1, λ_2, A, B, L, and t_0, we obtain the results

$$k_1 = 0.087/\text{hr}, \ x_0 = 0,$$

$$k_2 = 0.53/\text{hr}, \ \ \xi_0 = 43 \ \mu\text{m}. \tag{S154}$$

From equations (S147) and (S151a), we deduce that

$$\xi(t_0) = L - A = 46 \ \mu\text{m}. \tag{S155}$$

3.7. Equations (3.68) and (3.72) constitute three equations for the four desired quantities. We require one additional equation, and either of the following equations:

$$\rho_1(0) = B_{11} + B_{12},$$

$$\rho_2(0) = B_{21} + B_{22}, \tag{S156}$$

which follow directly from equation (3.67) with $t = 0$, is satisfactory. Here, $\rho_1(0)$ and $\rho_2(0)$ are given in terms of observed quantities by equations (3.76). We note that it would be unsatisfactory to use as the additional equation, the relation

$$B_{21}\lambda_1 + B_{22}\lambda_2 = \lambda_1 A + \lambda_2 B,$$

which follows from equation (3.73) and the equation proceeding it, because this relation is not linearly independent of equations (3.68) and (3.72) (that is, it can be derived from the latter equations).

By solving equations (3.68), (3.72), and either of equations (S156), we find that

$$B_{11} = -\frac{(\lambda_1 - \lambda_2)\lambda_2 AB}{\lambda_1^2 A + \lambda_2^2 B}, B_{21} = \frac{\lambda_1 A(\lambda_1 A + \lambda_2 B)}{\lambda_1^2 A + \lambda_2^2 B},$$

$$B_{12} = \frac{(\lambda_1 - \lambda_2)\lambda_1 AB}{\lambda_1^2 A + \lambda_2^2 B}, \quad B_{22} = \frac{\lambda_2 B(\lambda_1 A + \lambda_2 B)}{\lambda_1^2 A + \lambda_2^2 B}. \tag{S157}$$

3.8. The compartmental equations representing the specific activities of the precursor x_1 and two successors x_2 and x_3, are assumed to be the following, respectively:

$$\frac{dx_1}{dt} = -L_{11}x_1, \tag{S158a}$$

$$\frac{dx_2}{dt} = V_2^{-1}L_{21}V_1 x_1 - L_{22}x_2, \tag{S158b}$$

$$\frac{dx_3}{dt} = V_3^{-1}L_{32}V_2 x_2 - L_{33}x_3. \tag{S158c}$$

Assume $(x_1, x_2, x_3) = (x_{10}, 0, 0)$ at $t = 0$. Equation (S158a) indicates that dx_1/dt is always negative or zero, so that x_1 is a monotonic decreasing function of the time. From equation (S158b), we see that dx_2/dt is initially positive. Because x_1 and x_2 both ultimately go to zero, it follows that x_2 will at first increase and later decrease with time. Hence, the curve for $x_2 = x_2(t)$ possesses a maximum, which occurs at the particular time $t = \tau$ when $dx_2/dt = 0$, or, from equation (S151b), when

$$L_{21} V_1 x_1(\tau) = L_{22} V_2 x_2(\tau). \tag{S159}$$

Furthermore, because the form of the graph of the reaction system is as shown in Figure 3.11, it follows that $L_{02} = 0$. In addition, the mass flux J_2 from the exterior into compartment 2 is also zero, because the quantity 1 is assumed to be the only precursor of the successor 2. Hence conservation of mass flux through compartment 2 requires that $L_{21} V_1 = L_{22} V_2$. Therefore, equation (S159) states that

$$x_1(\tau) = x_2(\tau), \tag{S160}$$

or the maximum in the curve for the product $x_2 = x_2(t)$ occurs when the activity of the product equals the activity in the precursor. This property is called the precursor–product relation (Zilversmit, Entenman, and Fishler, 1943).

By a similar argument, it follows that the successor $x_3(t)$ possesses a maximum when $x_2 = x_3$. These remarks are illustrated in Figure 3.13, which shows the concentrations in the plasma of injected triglyceride and the products "resynthesized triglyceride," and "lipoprotein." The form of the curves and their crossings indicate that the metabolic sequence in vivo is as follows,

exogenous triglyceride → resynthesized triglyceride → lipoprotein.

3.9. Because the system is closed, there is no flux from any given compartment to the exterior, or $L_{0i} = 0$, $i = 1, 2, \ldots, n$. In addition, there can be no nonlabeled flux from the exterior into any compartment, so that $J_i = 0$, $i = 1, 2, \ldots, n$. Hence, the equality of the mass flux entering and leaving the ith compartment is expressed as

$$K_{21} = K_{12},$$
$$K_{i, i-1} + K_{i, i+1} = K_{i-1, i} + K_{i+1, i}, \, i = 2, 3, \ldots, n - 1, \tag{S161}$$
$$K_{n, n-1} = K_{n-1, n}.$$

By solving equations (S161) in sequence, for $i = 2, 3, \ldots, n - 1$, the desired relation that $K_{i, i+1} = K_{i+1, i}$ is obtained.

By definition, $K_{ij} = L_{ij} V_j$, and $M_{ij} = V_i^{-1} L_{ij} V_j$. Therefore, $K_{ij} = K_{ji}$ implies that

$$L_{ij} = V_j^{-1} L_{ji} V_i = M_{ji}. \tag{S162}$$

3.10. Let n and p represent labeled thymine in the DNA and precursor compartments, respectively, at any time. Because the culture medium is maintained at a constant concentration of labeled thymine (only a negligibly small amount of it enters the cells), we assume that there is a constant influx j of labeled thymine into the direct precursor compartment. Because the labeled thymine accumulates with time in the DNA, we assume that the DNA compartment contains no exit flux. Hence, we assume (Rubinow and Yen, 1972) that n and p satisfy the equations

$$\frac{dp}{dt} = -\alpha p + j, \tag{S163a}$$

$$\frac{dn}{dt} = \beta p, \tag{S163b}$$

where α and β are rate constants. The solution to equation (S163a) is

$$p = p_\infty(1 - e^{-\alpha t}),$$

$$p_\infty = \frac{j}{\alpha}. \tag{S164}$$

Substituting this result into equation (S163b), it follows with the aid of equation (1.21) that

$$n = c(\alpha t - 1 + e^{-\alpha t}). \tag{S165}$$

If the above theory is a satisfactory representation of the quoted experimental results of Werner (1971), then the experimental observations shown in Figure 3.14 can be fitted by the functional form of equation (S165). The fitting is readily performed in two stages: first fitting the asymptote $c(\alpha t - 1)$ to observation, and then, after subtraction, fitting a function $c\, e^{-\alpha' t}$. If equation (S165) is valid, then the values of α and α' should be sensibly the same. In fact, it was found they were, with a common value $\alpha = 0.27/\text{min}$.

Furthermore, the true direct precursor should be describable by equation (S164), with the same common value of α. It was found, by fitting the data shown in Figure 3.15, that the candidate precursors could be represented in a reasonable manner by equation (S164), with the fitted values of the decay constant α determined to be as follows:

Precursor Candidate:	TMP	TDP	TTP
α:	1.76/min	0.35/min	0.43/min.

By comparing these results with the characteristic value of α for DNA, it may be inferred that, allowing for unknown experimental error, TDP is the most likely direct precursor of the three candidates. Alternatively, it can be concluded that an unknown molecular species is the direct thymine bearing precursor, possessing a characteristic decay constant $\alpha = 0.27/\text{min}$ that is associated with its incorporation into DNA.

3.11.* As $t \to \infty$, the steady-state solution ρ_0 is approached asymptotically. Setting $d\rho/dt = 0$ in equation (3.129), we find that ρ_0 is given as

$$\rho_0 = -L^{-1}J, \tag{S166}$$

assuming the matrix L^{-1} exists. To reduce the equation to standard compartmental form, let

$$\rho = \rho_0 + \xi. \tag{S167}$$

Substituting equation (S167) into the equation for ρ, we obtain the following equation for $\xi = \xi(t)$:

$$\frac{d\xi}{dt} = L\xi. \tag{S168}$$

Because the matrix L is of order n, a generic component $\xi_i(t)$ of the vector ξ contains in general n exponential terms. Therefore, from equation (S167), the corresponding component $\rho_i(t)$ contains $n + 1$ terms: the n terms from ξ_i and the constant term ρ_{0i}, the ith component of ρ_0.

3.12.* In the notation of the problem, equations (3.48) and (3.49) are expressed as follows:

$$L_{11} = -f_1\lambda_1 - f_2\lambda_2, \tag{S169a}$$
$$L_{22} = -f_2\lambda_1 - f_1\lambda_2. \tag{S169b}$$

Substituting these equations into equation (3.40b), with $M_{12}M_{21}$ replaced by $L_{12}L_{21}$, yields the result

$$L_{12}L_{21} = f_1 f_2 (\lambda_1 - \lambda_2)^2. \tag{S170}$$

From equations (3.33) and (S169), it follows that

$$L_{21} \leq L_{11} = -f_1\lambda_1 - f_2\lambda_2, \tag{S171a}$$
$$L_{12} \leq L_{22} = -f_2\lambda_1 - f_1\lambda_2. \tag{S171b}$$

From the first inequality above, it follows by transposition that

$$\frac{1}{L_{11}} \leq \frac{1}{L_{21}}. \tag{S172}$$

Now multiply both sides of this inequality by $L_{12}L_{21}$, and substitute equations (S169a) and (S170) into it on the left-hand side only. Hence, we find (Rubinow and Winzer, 1971), in combination with equation (S171b), that

$$\frac{f_1 f_2 (\lambda_1 - \lambda_2)^2}{-f_1\lambda_1 - f_2\lambda_2} \leq L_{12} \leq -f_2\lambda_1 - f_1\lambda_2. \tag{S173}$$

Similarly, it follows from equation (S171b) that $1/L_{22} \leq 1/L_{12}$. Multiplying this equation by $L_{12}L_{21}$ and substituting equations (S169b) and (S170) into it on the left-hand side only, we find in analogous fashion that

$$\frac{f_1 f_2 (\lambda_1 - \lambda_2)^2}{-f_2\lambda_1 - f_1\lambda_2} \leq L_{21} \leq -f_1\lambda_1 - f_2\lambda_2. \tag{S174}$$

From equations (3.60) and (3.32), it follows that

$$L_{12}V_2 \leq L_{11}V_1,$$
$$L_{21}V_1 \leq L_{22}V_2. \tag{S175}$$

The first equation above states that the mass flux leaving compartment 2 and entering compartment 1 is less than or equal the total mass efflux from compartment 1. The second equation makes a similar statement with regard to the mass flux entering and leaving compartment 2. Dividing the first equation by $L_{12}V_1$ and the second equation by $L_{22}V_1$, we infer that

$$\frac{L_{21}}{L_{22}} \leq \frac{V_2}{V_1} \leq \frac{L_{11}}{L_{12}}. \tag{S176}$$

These inequalities certainly remain valid if we substitute for L_{21} and L_{12} their minimum values $(L_{12})_{min}$ and $(L_{21})_{min}$ given by the left-hand sides of equations (S173) and (S174), respectively, because $L_{21}/L_{22} \geq (L_{21})_{min}/L_{22}$, and $L_{11}/L_{12} \leq L_{11}/(L_{12})_{min}$. Hence, with the aid of equations (S169), equation (S176) can be written as

$$\frac{f_1 f_2 (\lambda_1 - \lambda_2)^2}{(-f_2\lambda_1 - f_1\lambda_2)^2} \leq \frac{V_2}{V_1} \leq \frac{(-f_1\lambda_1 - f_2\lambda_2)^2}{f_1 f_2 (\lambda_1 - \lambda_2)^2}. \tag{S177}$$

3.13.   According to the method of small perturbations, we set $(c,N) = (\bar{c} + \xi, \bar{N} + \eta)$ in equations (1.34) and (1.36), and linearize these equations. That is to say, we expand the functions of c and N appearing on the right-hand sides in Taylor series about the stationary state $(\bar{c},\bar{N})$, just as we did in Section 1.8, and neglect all powers in $c - \bar{c} = \xi$

and $N - \bar{N} = \eta$ greater than the first power. For example, from equations (1.35) and (1.37a),

$$k(c) = k(\bar{c}) + k_1 \xi + \cdots,$$
$$k(\bar{c}) = q,$$
$$k_1 = \left(\frac{dk}{dc}\right)_{c=\bar{c}} = \frac{k_m K}{(K + \bar{c})^2}. \tag{S178}$$

The linearized equations satisfied by $\xi = \xi(t)$ and $\eta = \eta(t)$ are the following:

$$\frac{d\eta}{dt} = \bar{N} k_1 \xi,$$
$$\frac{d\xi}{dt} = -\left(q + \frac{\bar{N} k_1}{y}\right)\xi - \frac{q}{y}\eta. \tag{S179}$$

These equations are of the form of equations (3.35). As in equations (3.37) and those that follow, we search for a solution of the form $\xi = \xi_0 \exp(\lambda t)$, $\eta = \eta_0 \exp(\lambda t)$, where ξ_0 and η_0 are constants. By substituting these expressions into equation (S179), we find that

$$e^{\lambda t}[-\lambda \eta_0 + \bar{N} k_1 \xi_0] = 0,$$
$$e^{\lambda t}\left[-\frac{q}{y}\eta_0 - \left(\lambda + q + \frac{\bar{N} k_1}{y}\right)\xi_0\right] = 0. \tag{S180}$$

The coefficients of $\exp(\lambda t)$ must each vanish, if these equations are to hold true for any value of t. A nontrivial solution of the resulting algebraic equations for ξ_0 and η_0 requires that

$$\lambda^2 + \lambda\left(q + \frac{\bar{N} k_1}{y}\right) + \frac{\bar{N} k_1 q}{y} = 0. \tag{S181}$$

The permissible values of λ are the roots $\lambda_{1,2}$ of this equation, given by inspection or by the quadratic formula as

$$\lambda_1 = -q,$$
$$\lambda_2 = -\frac{\bar{N} k_1}{y}. \tag{S182}$$

We see that these two values of λ are both negative. Therefore, the perturbations ξ and η decay to zero with time, and the equilibrium state $(\bar{c}, \bar{N})$ is stable, by definition.

3.14.* (a) With the aid of the graph shown in Figure 3.17, we can write down the rate equations satisfied by e, c, and c' as follows:

$$\frac{ds}{dt} = -k_{+1}se + k_{-1}c,$$

$$\frac{de}{dt} = -k_{+1}se + k_{-1}c,$$

$$\frac{dc}{dt} = k_{+1}se - (k_{-1} + k_{+2})c + k_{-2}c', \tag{S183}$$

$$\frac{dc'}{dt} = k_{+2}c - k_{-2}c'.$$

Conservation of substrate and enzyme molecules is expressed as follows:

$$s + c + c' = s_0, \tag{S184a}$$

$$e + c + c' = e_0, \tag{S184b}$$

where s_0 and e_0 are the concentrations of substrate and enzyme, respectively, before the reaction occurs. By eliminating s and c from equations (S183) by means of (S184), they become

$$\frac{de}{dt} = k_{-1}e_0 - [k_{+1}(s_0 + e - e_0) + k_{-1}]e - k_{-1}c',$$
$$\frac{dc'}{dt} = k_{+2}e_0 - k_{+2}e - (k_{+2} + k_{-2})c'. \tag{S185}$$

The equilibrium values of e and c', denoted by $\bar{e}$ and $\bar{c}'$, satisfy equations (S185) with the time derivatives set equal to zero, or

$$0 = k_{-1}e_0 - [k_{+1}(s_0 + \bar{e} - e_0) + k_{-1}]\bar{e} - k_{-1}\bar{c}',$$
$$0 = k_{+2}e_0 - k_{+2}\bar{e} - (k_{+2} + k_{-2})\bar{c}'. \tag{S186}$$

Utilizing the method of small perturbations, we set $(e,c') = (\bar{e} + x, \bar{c}' + y)$ in equations (S185), where x and y represent the perturbations from equilibrium induced by the sudden change in an external parameter. Retaining only the linear terms in the perturbations, we find the equations for x and y with the aid of equations (S186) to be

$$\frac{dx}{dt} = -[k_{+1}(\bar{s} + \bar{e}) + k_{-1}]x - k_{-1}y, \tag{S187a}$$

$$\frac{dy}{dt} = -k_{+2}x - (k_{+2} + k_{-2})y. \tag{S187b}$$

In deriving equation (S187a), we have replaced $s_0 + \bar{e} - e_0$ by $\bar{s}$, which follows from the subtraction of equation (S184b) from (S184a).
(b)   Setting $(x,y) = (x_0,y_0) \exp(-\lambda t)$, where (x_0,y_0) are the perturbation amplitudes at $t = 0$, we find (as in equations (3.37) and problem 3.13, but note the change in sign convention) that the decay constant λ must satisfy the characteristic equation

$$\lambda^2 - [k_{+2} + k_{-2} + k_{-1} + k_{+1}(\bar{s} + \bar{e})]\lambda + k_{+1}(\bar{s} + \bar{e})(k_{+2} + k_{-2}) + k_{-1}k_{-2} = 0. \tag{S188}$$

Let us denote the roots of equation (S188) by λ_1 and λ_2. Then, as in the derivation of equations (3.40) from (3.39), it follows from equation (S188) that

$$\frac{1}{\tau_1} + \frac{1}{\tau_2} = \lambda_1 + \lambda_2 = k_{+2} + k_{-2} + k_{-1} + k_{+1}(\bar{s} + \bar{e}),$$
$$\frac{1}{\tau_1\tau_2} = \lambda_1\lambda_2 = k_{+1}(k_{+2} + k_{-2})(\bar{s} + \bar{e}) + k_{-1}k_{-2}. \tag{S189}$$

According to equations (S189), the sum and product of the reciprocal relaxation times are linearly dependent on the quantity $\bar{s} + \bar{e}$. By identifying $\bar{s}$ and $\bar{e}$ as the concentrations proflavin ($\overline{PF}$) and chymotripin ($\overline{CT}$) of Figure 3.16, we can determine from the latter the slopes and intercepts of the two straight lines representing the data, and compare these with equations (S189) to infer the values of the four rate constants k_{+1}, k_{-1}, k_{+2}, and k_{-2}. In this manner the values recorded in the following table were found (Havsteen, 1967):

pH	k_{+1} (liter M^{-1} sec^{-1})	k_{-1} (sec^{-1})	k_{+2} (sec^{-1})	k_{-2} (sec^{-1})
8.42	$(0.63 \pm 0.03) \times 10^8$	1230 ± 640	525 ± 350	7045 ± 350
9.18	$(1.14 \pm 0.02) \times 10^8$	2150 ± 250	7300 ± 700	2000 ± 300

The error terms quoted above represent $\pm$ standard deviations.

4.1. The mean velocity of flow v through the aorta equals, by definition, the cardiac output K divided by the cross-sectional area of the aorta πR^2, where R is the mean radius of the aorta, namely,

$$v = \frac{K}{\pi R^2}. \tag{S190}$$

With the given values of $K = 5.5$ liter/min and $R = 1.1$ cm, we calculate from this equation

$$v = 24 \text{ cm/sec.}$$

Assuming Poiseuille's law is applicable to the mean flow of blood through a capillary, we infer from equations (4.15) and (S190) that

$$v = \frac{Q}{\pi a^2} = \frac{p_1 - p_2}{8\mu L} a^2, \tag{S191}$$

where a and L are the radius and length of the capillary, respectively, v, Q, and μ are the mean flow velocity, volume flux, and viscosity of the blood, respectively, and p_1 and p_2 are pressures at the arterial end and venous end of the capillary, respectively. With $p_1 - p_2 = 15$ mm Hg ($= 15 \times 1333$ dyne/cm^2), $L = 0.075$ cm, $\mu = 0.012$ P, and $a = 3 \times 10^{-4}$ cm for a systemic capillary, we find that an estimate of the mean velocity in a capillary is

$$v = 2.5 \text{ mm/sec.} \tag{S192}$$

This velocity is the order of magnitude of the mean velocity of red cells observed in capillaries of the cat, which may be taken to represent the mean plasma velocity. Such red cell velocities are unsteady and nonuniform, and range from 0.2 to 2.8 mm/sec (Johnson and Wayland, 1967).

Assume, for definiteness, that the mean capillary radius in the entire body is 3.5 μm, with the other parameters appearing on the right-hand side of equation (S191) as given above. Then, from (S191), the mean blood flux through a capillary is $Q = 0.13 \times 10^{-6}$ ml/sec. The cardiac output K equals the mean flux Q through one capillary times the mean number of open capillaries, because the circulatory flow is in a steady state. (The oscillations in the flow are assumed to be averaged out.) Let N equal the total number of capillaries in the body, and let f be the fraction of them that are open, $0 \leq f \leq 1$. Then

$$K = fNQ. \tag{S193}$$

With Q as just calculated, K as given above, and $f = 0.7$, it follows from (S193) that $N = 1.0 \times 10^9$. This estimate agrees with other such estimates (for example, see Green, 1944) in order of magnitude.

4.2. Denote the pressures at the inlet, the position x, and the outlet by p_1, p_x, and p_2, respectively. The dissipated power P through all the branches of the system is given by the expression

$$P = Q_1(p_1 - p_x) + 2Q_2(p_x - p_2),$$ (S194)

where Q_1 and Q_2 represents the blood fluxes in the upstream branch and each of the downstream branches, respectively. By conservation of flux,

$$Q_1 = 2Q_2.$$ (S195)

Eliminating Q_2 between the above two equations leads to the simpler expression for the dissipated power,

$$P = Q_1(p_1 - p_2).$$ (S196)

The power P depends on the position x, because the flux through the upstream branch Q_1 as given by Poiseuille's law is

$$Q_1 = \frac{\pi}{8\mu} \frac{a_1^4(p_1 - p_x)}{x},$$ (S197)

where μ is the blood viscosity.

However, p_x is not yet known explicitly as a function of x. To find it, we make use of the following explicit expression for Q_2, in accordance with Poiseuille's law,

$$Q_2 = \frac{\pi}{8\mu} a_2^4 \frac{(p_x - p_2)}{\sqrt{d^2 + (L - x)^2}}.$$ (S198)

Substituting equations (S197) and (S198) into (S195) and solving for p_x, we find that

$$p_x = \frac{p_1 a_1^4 \sqrt{d^2 + (L - x)^2} + 2p_2 a_2^4 x}{a_1^4 \sqrt{d^2 + (L - x)^2} + 2a_2^4 x}.$$ (S199)

Substituting equations (S197) and (S199) into (S196) yields the dissipated power as an explicit function of x,

$$P = \frac{16\mu(p_1 - p_2)^2}{\pi[a_1^4 \sqrt{d^2 + (L - x)^2} + 2a_2^4 x]}.$$ (S200)

To minimize P with respect to x, we set $dP/dx = 0$, obtaining the condition

$$-\frac{a_1^4(L - x)}{\sqrt{d^2 + (L - x)^2}} + 2a_2^4 = 0.$$ (S201)

Solving for x, we find that the bifurcation point x is given as

$$x = L - \frac{2a_2^4 d}{[a_1^8 - 4a_2^8]^{1/2}}.$$ (S202)

(The fact that this value of x minimizes P can be verified by showing that d^2P/dx^2 evaluated at this value of x is positive.) From Figure 4.15, we see that $\cos(\theta/2) = (L - x)/[d^2 + (L - x)^2]^{1/2}$, so that equation (S201) or (S202) is equivalent to the condition

$$\frac{\theta}{2} = \cos^{-1}\left[2\left(\frac{a_2}{a_1}\right)^4\right].$$ (S203)

Because x is restricted to the range $0 \leqq x \leqq L$, the permitted range of values of the half bifurcation angle $\theta/2$ is $\cos^{-1}(L/[d^2 + L^2]^{1/2}) \leqq \theta/2 \leqq \pi/2$. Hence, by combining this lower bound for the half bifurcation angle with equation (S203), we infer that a minimum in P is realized only so long as

$$2 \left(\frac{a_2}{a_1}\right)^4 \geqq \frac{L}{\sqrt{d^2 + L^2}}. \tag{S204}$$

When $a_2 = 0$, $x = L$ and the bifurcation angle θ is π. As a_2 is increased, the bifurcation point x moves to the left in Figure 4.15 and the bifurcation angle θ decreases until $x = 0$ when, from equation (S202),

$$a_2 = a_1 \left[\frac{L^2}{4(d^2 + L^2)}\right]^{1/8} \tag{S205}$$

and

$$\theta = 2 \cos^{-1} \frac{L}{\sqrt{d^2 + L^2}}.$$

4.3. (a) For an arbitrarily shaped particle falling under the influence of gravity through a viscous fluid, the drag force F_D is given by the generalization of equation (4.16) as

$$F_D = f v_s, \tag{S206}$$

where f is the frictional coefficient for the particle and v_s is the particle's settling or terminal velocity. For an equivalent sphere of radius r_0 falling through the medium which experiences the same drag force,

$$F_D = f_0 v_0, \tag{S207}$$

where $f_0 = 6\pi\mu r_0$ and v_0 is the terminal velocity of the sphere, given by equation (4.18). Equating the right-hand sides of equations (S206) and (S207) yields the desired equation (4.72).
(b) From the given values of the semi-major axis a and volume V, we calculate with the aid of equation (4.71) the following additional quantities,

$$r_0 = \left(\frac{3\pi V}{4}\right)^{1/3} = 1.95 \ \mu m, \quad \frac{a}{b} = 1.88,$$

$$b = \frac{3\pi V}{4a^2} = 1.28 \ \mu m, \quad \frac{f}{f_0} = 0.966.$$

The significance of the third figure in the above numbers is in doubt, but has been given for the purpose of reducing round-off error. With the above quantities and the given values of μ, ρ, ρ_c, and $g = 980 \ cm/sec^2$, we compute from equation (4.72) that

$$v_s = \begin{cases} 1.3 \ mm/hr, \ T = 4°C, \\ 2.2 \ mm/hr, \ T = 22°C. \end{cases}$$

These theoretical velocities are in reasonable agreement with the observed velocities quoted in the problem.

4.4 (a) Assume that the carotid artery of the sheep is a thin elastic cylindrical tube with thickness h_1, radius r_1, and tension per unit thickness t_1. Then, by equation (4.20), the law of Young and Laplace for mechanical equilibrium,

$$t_1 h_1 = (p_1 - p_0)r_1, \tag{S208}$$

where p_1 is the internal pressure in the artery, and p_0 is the external pressure. Treating the carotid artery of the ox in a similar manner, with the subscript 2 denoting the ox, we have

$$t_2 h_2 = (p_2 - p_0)r_2. \tag{S209}$$

It seems reasonable to suppose that the tension in a muscle fiber of an ox is the same as in a muscle fiber of a sheep, so that $t_1 = t_2$. Solving equations (S208) and (S209) for t_1 and t_2, respectively, and equating them, we obtain

$$\frac{h_2}{h_1} = \frac{p_2 - p_0}{p_1 - p_0} \frac{r_2}{r_1}. \tag{S210}$$

With $h_1 = 0.616$ mm, $p_1 - p_0 = 40$ mm Hg, $r_1 = 1.5$ mm, $p_2 - p_0 = 60$ mm Hg, and $r_2 = 3$ mm, we compute

$$h_2 = 1.85 \text{ mm}.$$

The observed value of the thickness is reported to be $h_2 = 1.74$ mm (Feldman, 1935). (b) According to equation (4.20),

$$T = (p - p_0)r, \tag{S211}$$

where T is the tension in the capillary wall, r is the capillary radius, and $p - p_0$ is the transmural pressure difference. With $p - p_0 = 30$ mm Hg $= 30 \times 1333$ dyne/cm^2, and $r = 3$ μm, we obtain for the capillary wall tension the value $T = 0.12$ dyne/cm.

4.5. Change the integration variable in equation (4.26) from p' to r. It becomes

$$Q = \frac{\pi}{8\,\mu L} \int_{r_2}^{r_1} r^4 \frac{dp}{dr}\, dr, \tag{S212}$$

where the limits of integration are determined by setting $r = r_1$ when $p = p_1$, and $r = r_2$ when $p = p_2$, in the relation of Young and Laplace,

$$\frac{T(r)}{r} = p - p_0. \tag{S213}$$

Here, the function $T(r)$ is given by equation (4.73). From equation (S213), we calculate

$$\frac{dp}{dr} = \frac{1}{r}\frac{dT(r)}{dr} - \frac{T(r)}{r^2}. \tag{S214}$$

Substituting equations (4.73) and (S214) into (S212), we obtain for Q the following integral expression:

$$Q = \frac{\pi}{8\,\mu L} \int_{r_2}^{r_1} \left\{ r^3 \left[\frac{t_1}{r_0} + 5\frac{t_2}{r_0}\left(\frac{r}{r_0} - 1\right)^4 \right] - r^2\left[t_1\left(\frac{r}{r_0} - 1\right) + t_2\left(\frac{r}{r_0} - 1\right)^5 \right] \right\} dr, \tag{S215}$$

which can be easily integrated.

4.6. Let us imagine that Figure 4.5 represents a cross section of a hemisphere of radius r. Mechanical equilibrium of this hemisphere requires that the net downward force due to tension $th2\pi r$ equals the pressure difference $p - p_0$ times the cross-sectional area πr^2, or

$$th = \frac{(p - p_0)r}{2}. \tag{S216}$$

We combine this with Hooke's law, equation (4.21) with E constant, which is assumed to hold true for all values of r between the equilibrium radius r_0 and the yield stress radius $r = r_{max}$. By eliminating t between these two equations, we obtain the following pressure-radius relation, in analogy with equation (4.23) for a cylindrical elastic tube,

$$p - p_0 = \frac{2Eh}{r_0}\left(1 - \frac{r_0}{r}\right). \tag{S217}$$

Now let us take into account the fact that h changes with radius. Assuming the elastic wall is incompressible, its volume when the wall has a radius r equals its volume when the wall radius equals the value r_0. Because the wall is thin, this volume is expressible as

$$4\pi r^2 h = 4\pi r_0^2 h_0, \tag{S218}$$

where h_0 is the wall thickness when $r = r_0$. Solving this equation for h and substituting it into equation (S217), we obtain the explicit dependence of $p - p_0$ on r, namely,

$$p - p_0 = \frac{2Eh_0}{r_0}\left(\frac{r_0^2}{r^2} - \frac{r_0^3}{r^3}\right). \tag{S219}$$

For $r > r_0$, $p - p_0$ is always positive because the first term in the parentheses is larger than the second. Since $p - p_0 = 0$ when $r = r_0$ and $r = \infty$, it follows that $p - p_0$ possesses a maximum for $r > r_0$, determined by the condition $dp/dr = 0$. By differentiation of equation (S219), we find that this maximum occurs when $r = 3r_0/2$, and $p - p_0 = 8Eh_0/27r_0$. Assuming that $r_{max} > 3r_0/2$, we infer that the transmural pressure difference at rupture is exceeded prior to rupture, and in fact $p - p_0$ is decreasing as the elastic limit is approached. These conclusions are compatible with the observation that growth and maturation of the ovarian follicle proceeds without a measurable increase in the internal pressure. Furthermore, the rapid increase in size of the follicle immediately prior to rupture is accompanied by a decrease in the internal pressure. Figure S4.6a, taken from Rodbard (1968), illustrates the pressure–radius dependence of equation (S219). Figure S4.6b shows, for comparison, the curve based on actual measurements of the pressure inside a rubber latex balloon as a function of the balloon radius, in static equilibrium.

4.7.* When the fluid is compressible, there is a contribution to the rate of change of momentum of an infinitesimal fluid volume element due to the variability of the density. The equation of motion for an incompressible fluid, equation (4.29), is replaced by the following:

$$A\left[\frac{\partial}{\partial t}(\rho u) + u\frac{\partial}{\partial x}(\rho u)\right] = -\frac{\partial}{\partial x}[(p - p_0)A]. \tag{S220}$$

Furthermore, the equation of continuity for a compressible fluid reads, in place of (4.30),

$$\frac{\partial}{\partial t}(\rho A) + \frac{\partial}{\partial x}(\rho A u) = 0. \tag{S221}$$

In addition, it is assumed that p and A are related by equation (4.31), and p and ρ are related by equation (4.74).

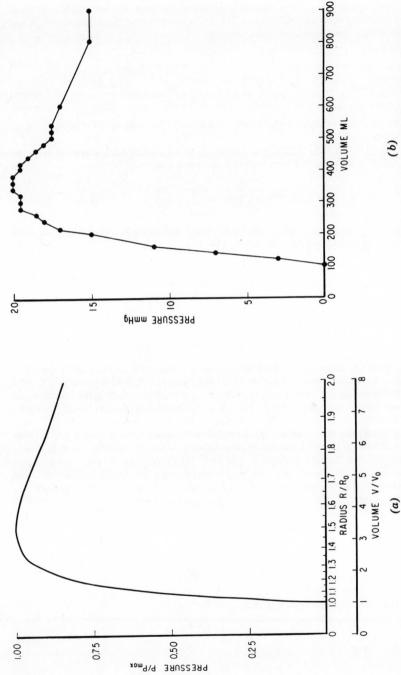

Figure S4.6. (a) The theoretical pressure–radius and pressure–volume dependence of an inflated spherical elastic membrane is shown based on equation (S219). (b) The observed pressure–volume dependence of an inflated latex rubber balloon. From Rodbard (1968), with permission.

Assume u, $p - p_0$, $A - A_0$, and $\rho - \rho_0$, together with their derivatives, are small. Linearizing equations (S220) and (S221) in the same manner that equations (4.29) and (4.30) were linearized leads to the equations

$$\rho_0 \frac{\partial u}{\partial t} = -\frac{\partial p}{\partial x}, \tag{S222a}$$

$$\rho_0 \frac{\partial A}{\partial t} + A_0 \frac{\partial \rho}{\partial t} + \rho_0 A_0 \frac{\partial u}{\partial x} = 0. \tag{S222b}$$

Linearization of equation (4.31) leads to equation (4.34), while equation (4.74) is already in linearized form.

By differentiating (S222a) with respect to x and (S222b) with respect to t, we can eliminate the terms in u from these two equations, obtaining

$$\rho_0 \frac{\partial^2 A}{\partial t^2} + A_0 \frac{\partial^2 \rho}{\partial t^2} = A_0 \frac{\partial^2 p}{\partial x^2}. \tag{S223}$$

Finally, substituting ρ from equation (4.74) and A from equation (4.34) into the above equations, we obtain the wave equation for p,

$$\frac{1}{c^2} \frac{\partial^2 p}{\partial t^2} = \frac{\partial^2 p}{\partial x^2}, \tag{S224}$$

where

$$\frac{1}{c^2} = \frac{2\rho_0 r_0}{Eh} + \frac{1}{c_0^2}. \tag{S225}$$

This result, due to Korteweg (1878), generalizes Young's formula when the fluid medium is compressible. The velocity of sound in horse blood at $37°C$ is 1571 m/sec (Urick, 1947). Consequently, the term c_0^{-2} on the right above is small compared to the first term, and may sensibly be neglected. When this is done, equation (S225) reduces to Young's formula. According to the formula, if $E \rightarrow 0$, so that the blood flow becomes a free jet, then $c \rightarrow 0$, and the pulse wave vanishes. This conclusion illustrates how the elasticity of the tube wall is essential for the existence of the pulse wave. At the other extreme, the limit $E \rightarrow \infty$ implies that the tube becomes a rigid wall. Then $c \rightarrow c_0$, or the pulse wave velocity approaches the sound wave velocity. Furthermore, we infer from equation (S225) that the pulse wave velocity can never exceed the velocity of sound in the fluid medium.

4.8. If we define $c_y^2/c^2 \equiv f(m)$, then according to equation (4.75)

$$f(m) = \frac{1}{2} + \frac{m}{4}\left(1 + \frac{1}{K_0^2}\right) + \left(\left[\frac{1}{2} + \frac{m}{4}\left(1 + \frac{1}{K_0^2}\right)\right]^2\right.$$
$$\left. - \frac{m}{4}\left[2(1 - \sigma^2) + \frac{m}{K_0^2}\right]\right)^{1/2}. \tag{S226}$$

The linear approximation to $f(m)$ leads to the formula

$$\frac{c_y^2}{c^2} = f(0) + \left(\frac{df}{dm}\right)_{m=0} m, \tag{S227}$$

which provides a simple correction to Young's formula when m is small. From equation (S226), we calculate

$$
\frac{df}{dm} = \frac{1}{4}\left(1 + \frac{1}{K_0^2}\right) + \left(\left[\frac{1}{2} + \frac{m}{4}\left(1 + \frac{1}{K_0^2}\right)\right]^2 - \frac{m}{4}\left[2(1 - \sigma^2) + \frac{m}{K_0^2}\right]\right)^{-1/2}
$$
$$
\times \frac{1}{4}\left(\left[\frac{1}{2} + \frac{m}{4}\left(1 + \frac{1}{K_0^2}\right)\right]\left(1 + \frac{1}{K_0^2}\right) - (1 - \sigma^2) - \frac{m}{K_0^2}\right).
$$

(S228)

Hence, from equations (S226) and (S228),

$$
f(0) = 1,
$$
$$
\left(\frac{df}{dm}\right)_{m=0} = \frac{1}{2K_0^2} + \frac{\sigma^2}{2}.
$$

(S229)

Substituting equations (S229) into (S227) we find that when m is small,

$$
\frac{1}{c^2} = \frac{1}{c_y^2}\left(1 + \frac{m}{2K_0^2} + \frac{\sigma^2 m}{2}\right).
$$

(S230)

Expressing this result in terms of dimensional parameters, it becomes

$$
\frac{1}{c^2} = \frac{1}{c_y^2} + \frac{1}{c_0^2} + \frac{2\sigma^2 \rho_1}{E}.
$$

(S231)

The second term on the right is the correction to Young's formula due to the compressibility of the blood, already discussed in problem 4.7. The last term on the right is an additional correction term which arises as a result of the inertia of the tube wall. We note that it depends on wall properties only. To estimate the relative importance of the two correction terms, it is more convenient to utilize equation (S230). We note that

$$
\frac{m}{2K_0^2} = \frac{E}{2\rho_1 c_0^2}\frac{\rho_1 h}{\rho_0 r_0} = \frac{c_y^2}{c_0^2}.
$$

We saw in Section 4.5 that for a mammalian aorta, $c_y \sim 5$ m/sec, while from problem 4.7, $c_0 \sim 1600$ m/sec. Thus $m/2K_0^2 \sim 10^{-5}$. Similarly, we estimate, with $\rho_1 \sim \rho_0$, $h/a \sim 0.1$, and $\sigma \sim 1/2$, that $\sigma^2 m/2 \sim 0.013$. Hence, from equation (S230), we infer that the multiplicative correction to c/c_y is $(1 + 10^{-5} + 0.013)^{-1/2} = 0.994$. This correction represents a decrease of 0.6% due to arterial wall inertia, and a negligible decrease due to the compressibility of blood.

4.9.* (a) Figure 4.16 is a representation of the function $E = E(p)$, where E is the tangent modulus of the aorta at a given radius, and p represents the pressure in the aorta in excess of the external pressure p_0. After dividing equation (4.76) by $r^3 E$, we obtain by integration the relation

$$
\int_{r_0}^r \frac{dr'}{r'^3} = \frac{(1 - \sigma^2)}{h_0 r_0} \int_0^p \frac{dp'}{E(p')}.
$$

(S232)

Assuming E is given empirically for the thoracic aorta by equation (4.77), then we can estimate approximately, from Figure 4.16, that $E_0 = 0.6 \times 10^6$ dyne/cm^2, and $A = 3.7 \times 10^2$ dyne/cm^2 (mm Hg)^2 $= 2.1 \times 10^{-4}$ cm^2/dyne. The alternative to a curve-fitting procedure is to perform the integration indicated on the right hand side of equation (S232) numerically.

By substituting equation (4.77) into (S232) and integrating, we obtain the pressure–radius relation:

$$\frac{r_0^2}{r^2} = 1 - \frac{2r_0(1 - \sigma^2)}{h_0(E_0 A)^{1/2}} \tan^{-1}\left(p \sqrt{\frac{A}{E_0}}\right). \tag{S233}$$

The last term in (S233) is obtained from the integral on the right side of equation (S232) either by changing the integration variable p' to $p' = (E_0/A)^{1/2} \tan \xi$, or by consulting a table of integrals. The quantity r_0/h_0 appearing on the right-hand side of the above equation is not yet known. To determine it, we make use of equation (S233) with $p = p_1 \equiv 100$ mm Hg, and the fact that the aortic wall is essentially incompressible, so that the wall volume in one stretched state is the same as in another. This conservation of aortic wall volume is expressed mathematically as

$$2\pi r h = 2\pi r_0 h_0. \tag{S234}$$

In particular, by setting $h/r = 0.105$ and $p = 100$ mm Hg in equations (S233) and (S234), and eliminating h_0 between them, we find that

$$\frac{r_1^2}{r_0^2} = 1 + \frac{2r_1(1 - \sigma^2)}{h_1(E_0 A)^{1/2}} \tan^{-1}\left(p_1 \sqrt{\frac{A}{E_0}}\right), \tag{S235}$$

where r_1/h_1 is the aortic radius–thickness ratio at the pressure $p_1 = 100$ mm Hg. Substituting the values of the parameters appearing on the right above, we find that $r_1^2/r_0^2 = 2.52$. Hence, from equation (S234) evaluated at $r = r_1$,

$$\frac{h_0}{r_0} = \frac{r_1^2}{r_0^2}\frac{h_1}{r_1} = 0.265. \tag{S236}$$

(b) We assume that Young's formula is applicable to the observations of Figure 4.17, where now E represents the tangent modulus at a given pressure, and h/r is the thickness–radius ratio at the same pressure. Thus, we can write Young's formula in the present notation, utilizing equation (S234), as

$$c = \left(\frac{Eh}{2\rho r}\right)^{1/2} = \frac{r_0}{r}\left(\frac{Eh_0}{2\rho r_0}\right)^{1/2}. \tag{S237}$$

With $\rho = 1.06$ g/cm^2 and the aid of equations (4.76), (S233), and (S236), we construct the following table of the theoretical wave velocity c as a function of the excess pressure p.

p (mm Hg)	20	60	100	140	180	220
E (10^6 dyne/cm^2)	0.75	1.9	4.3	7.9	12.6	18.5
r_0/r	0.88	0.70	0.63	0.59	0.56	0.55
c (m/sec)	2.7	3.5	4.6	5.8	7.1	8.0

Comparison of the calculated values of the wave velocity, listed in the last line of the table, with the observed values shown in Figure 4.17 suggests a qualitative agreement between theory and experiment.

4.10.* The solution is based on the work of Karreman (1952).

(a) Because the fluid velocity u in the arteries also satisfies the wave equation (see Section 4.5), we express the incident, reflected and transmitted velocities, denoted by

u_i, u_r, and u_t, respectively, as

$$u_i = A' \sin \omega \left(t - \frac{x}{c_1} \right), \qquad x < 0,$$

$$u_r = B' \sin \omega \left(t + \frac{x}{c_1} \right), \qquad x < 0, \qquad \text{(S238)}$$

$$u_t = C' \sin \omega \left(t - \frac{x}{c_2} \right), \qquad x > 0,$$

where A', B', and C' are amplitudes to be determined. Substituting equations (S238) and (4.78) into equation (4.32), we find that the primed amplitudes are expressed in terms of the unprimed amplitudes as follows:

$$A' = \frac{A}{\rho c_1},$$

$$B' = -\frac{B}{\rho c_1}, \qquad \text{(S239)}$$

$$C' = \frac{C}{\rho c_2},$$

where ρ is the density of the blood.

We shall enforce the physical requirement that the pressure and flux of the blood at the junction $x = 0$ are continuous. Continuity of the pressure at the junction for all values of the time requires that

$$p_i + p_r = p_t, \qquad x = 0,$$

or

$$A + B = C. \qquad \text{(S240)}$$

The instantaneous flux of blood at a given station in the tube is expressed as the cross-sectional area times the velocity at that station. Hence, continuity of the instantaneous blood flux at the junction is expressed as

$$\pi r_1^2 (u_i + u_r) = \pi r_2^2 u_t, \qquad x = 0. \qquad \text{(S241)}$$

By substituting (S238) and (S239) into the above equation, we obtain the relation

$$A - B = \zeta C,$$

$$\zeta = \frac{r_2^2 c_1}{r_1^2 c_2}. \qquad \text{(S242)}$$

By dividing equations (S240) and (S242) by A, we can solve for the reflection and transmission coefficients R and T defined in equations (4.79), and express them in terms of the single parameter ζ as follows:

$$R = \frac{1 - \zeta}{1 + \zeta},$$

$$T = \frac{2}{1 + \zeta}. \qquad \text{(S243)}$$

If we use Young's formula, equation (4.37), for the velocities c_1 and c_2, then the parameter ζ can be written as

$$\zeta = \left(\frac{r_2}{r_1} \right)^{5/2} \left(\frac{h_1 E_1}{h_2 E_2} \right)^{1/2}, \qquad \text{(S244)}$$

where the subscripts 1 and 2 refer to the arterial sections upstream and downstream of the junction, respectively. When $h_1 E_1 = h_2 E_2$, ζ depends only on r_2/r_1, which may be called the "coarctation ratio," because this ratio is a measure of the extent to which an artery is **coarcted** or narrowed. Note that if $r_2 \ll r_1$, then $\zeta \sim 0$, $R \sim 1$, and $T = 2$, which means that the incident wave is totally reflected at the junction, and the transmitted wave has an amplitude about twice that of the incident wave. If $r_2 = r_1$, then $\zeta = 1$ and $R = 0$ and $T = 1$, which means that the wave is totally transmitted at the junction, and there is no reflected wave. However, if $h_1 E_1 \neq h_2 E_2$, then even if $r_1 = r_2$, there does exist a reflected wave in the artery upstream of the junction.

(b)   When there is a bifurcation at the junction, as in Figure 4.15, let us denote the two downstream branches by the subscripts 2 and 3, respectively. Thus, the transmitted pressures in the two branches are denoted by p_2 and p_3, respectively, where

$$p_2 = D \sin \omega \left(t - \frac{\xi_2}{c_2} \right), \qquad \xi_2 > 0,$$

$$p_3 = E \sin \omega \left(t - \frac{\xi_3}{c_3} \right), \qquad \xi_3 > 0. \tag{S245}$$

Here ξ_2 and ξ_3 represent distance along the axes of the arteries 2 and 3, respectively, measured from the junction. The associated tube velocities are given, as in equations (S238), by the expressions

$$u_2 = \frac{D}{\rho c_2} \sin \omega \left(t - \frac{\xi_2}{c_2} \right),$$

$$u_3 = \frac{E}{\rho c_3} \sin \omega \left(t - \frac{\xi_3}{c_3} \right), \tag{S246}$$

where we have required u_2, u_3, p_2, and p_3 to satisfy equation (4.32).

   Continuity of pressure and blood flux at the junction requires that at $x = \xi_2 = \xi_3 = 0$,

$$p_i + p_r = p_2 = p_3,$$
$$\pi r_1^2 (u_i + u_r) = \pi r_2^2 u_2 + \pi r_3^2 u_3. \tag{S247}$$

Substitution of equations (4.78), (S238), (S245), and (S246) into (S247) yields the relations

$$A + B = D = E,$$
$$A - B = \zeta D + \zeta' E, \tag{S248}$$

where, if Young's formula is applicable, ζ is defined as before and given by equation (S244), and

$$\zeta' = \frac{r_3^2 c_1}{r_1^2 c_3} = \left(\frac{r_3}{r_1} \right)^{5/2} \left(\frac{h_1 E_1}{h_3 E_3} \right)^{1/2}. \tag{S249}$$

Hence, the transmission coefficients in tubes 2 and 3, defined as

$$T_2 = \frac{D}{A},$$

$$T_3 = \frac{E}{A}, \tag{S250}$$

respectively, are given by equations (S248) as

$$T_2 = T_3 = \frac{2}{1 + \zeta + \zeta'}. \tag{S251}$$

Similarly, the reflection coefficient, defined as $R = B/A$, is given as

$$R = T_2 - 1 = T_3 - 1 = \frac{1 - \zeta - \zeta'}{1 + \zeta + \zeta'}. \tag{S252}$$

When the downstream branches are identical, equations (S251) and (S252) reduce to

$$T_2 = T_3 = \frac{2}{1 + 2\zeta},$$
$$R = \frac{1 - 2\zeta}{1 + 2\zeta}. \tag{S253}$$

4.11. We assume that the head has a propulsive horizontal velocity u but no motion in the vertical direction. Thus, it contributes a drag force to the organism of magnitude $6\pi\mu R_H u$, with a direction opposite to that of u. Here R_H is the radius of the head. The total force on the spermatozoon (head plus flagellum) in the direction of translation is expressed as

$$F - 6\pi\mu R_H u = 0, \tag{S254}$$

where F is given by equation (4.51). Substituting equation (4.51) into (S254) and solving for u, we find that

$$u = \left[\left(\frac{C_N}{C_L} - 1\right)\frac{I}{J}\right]\left[1 + \frac{C_N}{C_L}\frac{I}{J} + \frac{6\pi\mu R_H}{\gamma C_L J}\right]^{-1} c. \tag{S255}$$

To make use of this formula, let us assume that I and J are given by equations (4.56), that (4.54) holds, and that C_L is given by equation (4.70a). By means of these assumptions and equation (4.50), we can express (S255) as

$$u = \frac{(k\eta)^2}{2}\left\{1 + (k\eta)^2 + \frac{3R_H}{2b}\left[1 + \frac{1}{2}(k\eta)^2\right]^{1/2}\left[\log\frac{4b}{a} - \frac{3}{2}\right]\right\}^{-1} c. \tag{S256}$$

Gray (1955) found that $R_H = 0.5$ μm for spermatozoa of the sea urchin *P. miliaris*. Combining this result with the results of his other observations given in equations (4.62), we calculate from equation (S256) that

$$u = 193 \text{ μm/sec},$$

which is in excellent agreement with the observed velocity of 191 μm/sec (Gray and Hancock, 1955).

4.12.* Substituting equation (4.42) into equation (4.52a), we find that the integral I is given as

$$I = \int_0^{2\pi/k} \frac{(k\eta)^2 \cos^2 k(x + ct)dx}{[1 + (k\eta)^2 \cos^2 k(x + ct)]^{1/2}}. \tag{S257}$$

Change the integration variable from x to $kx + kct = \xi$. Then

$$I = \int_{kct}^{2\pi+kct} \frac{(k\eta)^2 \cos^2 \xi \, d\xi}{[1 + (k\eta)^2 \cos^2 \xi]^{1/2}} = \int_0^{2\pi} \frac{(k\eta)^2 \cos^2 \xi \, d\xi}{[1 + (k\eta)^2 \cos^2 \xi]^{1/2}}. \tag{S258}$$

The last integral follows from the previous one because the integrand is periodic with period 2π, so that the integral over a period is independent of the initiation point of the integration process. By adding and subtracting unity from the numerator of the integrand, we can rewrite equation (S258) as

$$I = \int_0^{2\pi} [1 + (k\eta)^2 \cos^2 \xi]^{1/2} \, d\xi - \int_0^{2\pi} \frac{d\xi}{[1 + (k\eta)^2 \cos^2 \xi]^{1/2}}. \quad \text{(S259)}$$

Finally, setting $\cos^2 \xi = 1 - \sin^2 \xi$ above, we find that

$$I = [1 + (k\eta)^2]^{1/2} \int_0^{2\pi} (1 - m \sin^2 \xi)^{1/2} \, d\xi$$
$$- \frac{1}{[1 + (k\eta)^2]^{1/2}} \int_0^{2\pi} \frac{d\xi}{(1 - m \sin^2 \xi)^{1/2}}, \quad \text{(S260)}$$

$$m \equiv \frac{(k\eta)^2}{1 + (k\eta)^2}. \quad \text{(S261)}$$

Hence, from equations (4.80),

$$I = 4[1 + (k\eta)^2]^{1/2} E(m) - 4[1 + (k\eta)^2]^{-1/2} K(m). \quad \text{(S262)}$$

Since $(k\eta)^2 = m/(1 - m)$, it is possible to express equation (S262) equivalently as

$$I = 4(1 - m)^{-1/2} E(m) - 4(1 - m)^{1/2} K(m). \quad \text{(S263)}$$

In a similar manner, we find that

$$J = 4[1 + (k\eta)^2]^{-1/2} K(m) = 4(1 - m)^{1/2} K(m). \quad \text{(S264)}$$

Substituting equations (S262) and (S264) into equation (4.55), it becomes

$$u = \frac{[1 + (k\eta)^2]^{1/2} E(m) - [1 + (k\eta)^2]^{-1/2} K(m)}{2[1 + (k\eta)^2]^{1/2} E(m) - [1 + (k\eta)^2]^{-1/2} K(m)} c, \quad \text{(S265)}$$

with m defined by equation (S261). We note that the first two terms in the power series representation of $K(m)$ and $E(m)$ are

$$K(m) = \frac{\pi}{2} \left(1 + \frac{m}{4} + \cdots \right),$$
$$E(m) = \frac{\pi}{2} \left(1 - \frac{m}{4} - \cdots \right). \quad \text{(S266)}$$

If we treat $(k\eta)^2$ as a small quantity, then, to order $(k\eta)^2$, $m = (k\eta)^2$ and equations (S266) become

$$K(m) = \frac{\pi}{2} \left[1 + \frac{(k\eta)^2}{4} + \cdots \right],$$
$$E(m) = \frac{\pi}{2} \left[1 - \frac{(k\eta)^2}{4} - \cdots \right]. \quad \text{(S267)}$$

Substituting (S267) into (S265) reduces the latter to the expression

$$u = \frac{1}{2} \frac{(k\eta)^2}{1 + 5(k\eta)^2/4} c, \quad \text{(S268)}$$

which is very nearly the same as equation (4.57).

5.1. By explicit differentiation of equation (5.196), we calculate that

$$\frac{\partial c}{\partial r} = -\frac{2r}{4Dt} c,$$

$$\frac{\partial}{\partial r}\left(r^2 \frac{\partial c}{\partial r}\right) = -\frac{2}{4Dt} \frac{\partial}{\partial r}(r^3 c) = -\frac{2}{4Dt}\left[3r^2 c - \frac{2r^4}{4Dt} c\right],$$

$$\frac{\partial c}{\partial t} = -\frac{3}{2t} c + \frac{r^2}{4Dt^2} c.$$

Substituting these equations into the diffusion equation (5.195), we verify that $c = c(r,t)$ is a solution.

Substituting (5.196) into (5.197) determines the desired integral with the aid of equation (5.36) as

$$\frac{4\pi m}{(4\pi Dt)^{3/2}} \int_0^\infty r^2 e^{-r^2/4Dt}\, dr = \frac{4\pi m}{(4\pi Dt)^{3/2}}\left[\frac{\pi^{1/2}(4Dt)^{3/2}}{4}\right] = m. \qquad (S269)$$

An alternate procedure for evaluating the integral on the left-hand side of (S269) is to integrate by parts, setting $u = r$, and $dv = \exp[-r^2/4Dt]r\, dr$. Then, with the aid of equation (5.17), it follows that

$$\frac{4\pi m}{(4\pi Dt)^{3/2}} \int_0^\infty r^2 e^{-r^2/4Dt}\, dr = \frac{4\pi m}{(4\pi Dt)^{3/2}}\left\{-Dtr e^{-r^2/4Dt}\Big|_0^\infty + 2Dt \int_0^\infty e^{-r^2/4Dt}\, dr\right\}$$

$$= \frac{4\pi m}{(4\pi Dt)^{3/2}}\, 2\, Dt(\pi Dt)^{1/2} = m. \qquad (S270)$$

5.2. A maximum in the function $j(x,t)$ at a given time t requires that $\partial j/\partial x$ vanishes, or by definition,

$$\frac{\partial}{\partial x}\left(-D\frac{\partial c}{\partial x}\right) = 0. \qquad (S271)$$

Here $c = c(x,t)$ is given by the one-dimensional unit source solution, equation (5.15). Because c satisfies the one-dimensional diffusion equation (5.7), (S271) is equivalent to the condition

$$\frac{\partial c}{\partial t} = 0, \qquad (S272)$$

which requires less differentiation of the function c to calculate, than does the condition (S271). Substituting (5.15) into (S272), we find that the equation determining $x = \xi(t)$ is

$$\frac{\partial}{\partial t}(t^{-1/2}e^{-x^2/4Dt}) = \left(-\frac{1}{2t} + \frac{x^2}{4Dt^2}\right)t^{-1/2}e^{-x^2/4Dt} = 0. \qquad (S273)$$

Because the exponential factor vanishes only when x is infinite for t finite, the desired solution of equation (S273) is given by the vanishing of the quantity in parentheses as

$$x = \xi(t) = (2Dt)^{1/2}.$$

This expression for ξ is exactly the same as the r.m.s. displacement $\tilde{x}$, equation (5.38).

5.3.   The r.m.s. displacement $\tilde{r}$ with respect to the unit three-dimensional source solution, equation (5.196) with $m = 1$, is defined by the equation

$$\tilde{r}^2 = \left[\frac{1}{(4\pi Dt)^{3/2}} \int_0^\infty r^2 e^{-r^2/4Dt} 4\pi r^2 \, dr \right] \left[\frac{1}{(4\pi Dt)^{3/2}} \int_0^\infty e^{-r^2/4Dt} 4\pi r^2 \, dr \right]^{-1}. \quad \text{(S274)}$$

Here $4\pi r^2 \, dr$ represents the volume element. We know from equation (5.197) that the last factor on the right above is unity. To evaluate the first factor, we can either use integration by parts, or make use of the evaluation method utilized in equation (5.36). Using the latter method and setting $\alpha = 1/4Dt$, equation (S274) becomes

$$\tilde{r}^2 = \left(\frac{\alpha}{\pi} \right)^{3/2} 4\pi \int_0^\infty e^{-\alpha r^2} r^4 \, dr = \left(\frac{\alpha}{\pi} \right)^{3/2} 4\pi \frac{d^2}{d\alpha^2} \int_0^\infty e^{-\alpha r^2} \, dr$$

$$= \left(\frac{\alpha}{\pi} \right)^{3/2} 4\pi \frac{d^2}{d\alpha^2} \left(\frac{\pi}{4\alpha} \right)^{1/2} = \frac{3}{2\alpha} = 6Dt,$$

which is the desired result, equation (5.40).

5.4.   Using Einstein's relation, equation (5.41), with $T = 293°K$, $\mu = 0.01$ P, and the given values of a and k, we calculate the diffusion constant of hemoglobin to be

$$D = \frac{kT}{6\pi\mu a} = \frac{1.38 \times 293 \times 10^{-16}}{6\pi \times 10^{-2} \times 30 \times 10^{-8}} \text{ cm}^2/\text{sec} = 7 \times 10^{-7} \text{ cm}^2/\text{sec}.$$

The experimentally observed value of the diffusion coefficient of hemoglobin at low concentration is about 6.3×10^{-7} cm^2/sec (Wittenberg, 1970).

5.5.   We noted in Section 5.4 that when the half-infinite plane $x < 0$ is initially at concentration c_0, and the half-infinite plane $x > 0$ is initially at zero concentration, then the concentration at $x = 0$ is maintained for all positive times at the value $c_0/2$ (see equation (5.32a)). This observation permits us to infer the solution to the present problem. Thus, with a little thought, we recognize that this solution is

$$c(x,t) = c_0 \, \text{erfc}(z), \quad \text{(S275)}$$

where $z = x/(4Dt)^{1/2}$. It is readily verified that the expression (S275) satisfies the one-dimensional diffusion equation subject to the boundary conditions $c(0,t) = c_0$, $c(\infty,t) = 0$, and the initial condition $c(x,0) = 0$ for $x > 0$.

5.6.   (a)   A unit point mass of solute placed at the position x_0 at time $t = 0$ is represented by the source solution

$$\frac{1}{(4\pi Dt)^{1/2}} e^{-(x-x_0)^2/4Dt},$$

according to equation (5.22). This expression is a solution to the one-dimensional diffusion equation and satisfies the boundary condition at $x = \infty$ as well as the unit source condition at $x = x_0$ and $t = 0$, but it does not satisfy the boundary condition at $x = 0$. To satisfy this latter condition, we add to the above particular solution a positive unit source at the image point $x = -x_0$. Thus, let

$$c(x,t) = \frac{1}{(4\pi Dt)^{1/2}} e^{-(x-x_0)^2/4Dt} + \frac{1}{(4\pi Dt)^{1/2}} e^{-(x+x_0)^2/4Dt}. \quad \text{(S276)}$$

The vanishing of the current density at the wall requires that $\partial c/\partial x$ vanish at $x = 0$. From equation (S276), we calculate that

$$\frac{\partial c}{\partial x} = (4\pi Dt)^{-1/2}\left[-\frac{2(x - x_0)}{4Dt}e^{-(x-x_0)^2/4Dt} - \frac{2(x + x_0)}{4Dt}e^{-(x+x_0)^2/4Dt}\right],$$

which vanishes when $x = 0$. Hence, the differential equation, the initial condition, and all the boundary conditions are satisfied by $c(x,t)$.

(b) Instead of a positive unit source at the image point $x = -x_0$, we place a negative unit source there. Thus, let

$$c(x,t) = \frac{1}{(4\pi Dt)^{1/2}}e^{-(x-x_0)^2/4Dt} - \frac{1}{(4\pi Dt)^{1/2}}e^{-(x+x_0)^2/4Dt}. \tag{S277}$$

We see immediately that $c(0,t) = 0$, so that $c(x,t)$ as given by equation (S277) is the required solution now. From (S277), we calculate the mass flux per unit area $j(t)$ incident on the wall at time t to be

$$j(t) = D\left(\frac{\partial c}{\partial x}\right)_{x=0} = \frac{D}{(4\pi Dt)^{1/2}}\left[\frac{-2(x - x_0)}{4Dt}e^{-(x-x_0)^2/4Dt} + \frac{2(x + x_0)}{4Dt}e^{-(x+x_0)^2/4Dt}\right]_{x=0}$$

$$= \frac{x_0}{(4\pi D)^{1/2}t^{3/2}}e^{-x_0^2/4Dt}. \tag{S278}$$

Note that there is no minus sign preceding D in the definition of $j(t)$ because the mass flux incident on the wall is in the negative x direction.

The total amount of solute mass $m(t)$ per unit area that adheres to the wall up to a time t is given with the aid of equation (S278) as

$$m(t) = \int_0^t j(\tau)d\tau = \frac{x_0}{(4\pi D)^{1/2}}\int_0^t \frac{1}{\tau^{3/2}}e^{-x_0^2/4D\tau}d\tau. \tag{S279}$$

Change the integration variable, setting $x_0^2/4D\tau = \xi^2$, so that $-x_0^2\,d\tau/4D\tau^2 = 2\xi\,d\xi$. Then

$$m(t) = \frac{2}{\pi^{1/2}}\int_{x_0/(4Dt)^{1/2}}^{\infty} e^{-\xi^2}\,d\xi = \text{erfc}\left[\frac{x_0}{(4Dt)^{1/2}}\right], \tag{S280}$$

from equation (5.33). Of course $m(0) = 0$ and $m(\infty) = 1$. The error function complement is shown plotted as a function of the time t for fixed x_0 and D in Figure S5.6.

5.7. According to equations (5.23) and (5.24), the concentration $c(x,t)$ is given with the aid of the integration variable transformation $(x - \xi)/(4Dt)^{1/2} = \eta$ as

$$c(x,t) = \frac{c_0}{(4\pi Dt)^{1/2}}\int_{-a}^{a} e^{-(x-\xi)^2/4Dt}\,d\xi = \frac{c_0}{\pi^{1/2}}\int_{(x-a)/(4Dt)^{1/2}}^{(x+a)/(4Dt)^{1/2}} e^{-\eta^2}\,d\eta$$

$$= \frac{c_0}{\pi^{1/2}}\left\{\int_0^{(x+a)/(4Dt)^{1/2}} e^{-\eta^2}\,d\eta + \int_{(x-a)/(4Dt)^{1/2}}^{0} e^{-\eta^2}\,d\eta\right\}$$

$$= \frac{c_0}{2}\left(\text{erf}\left[\frac{x + a}{(4Dt)^{1/2}}\right] - \text{erf}\left[\frac{x - a}{(4Dt)^{1/2}}\right]\right),$$

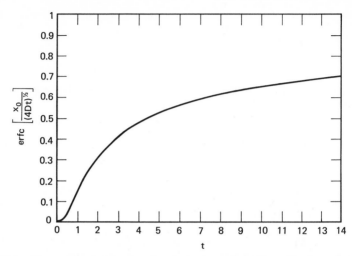

Figure S5.6. The complementary error function representing either $m(t)$ of equation (S280), $c(x_0, t)/c_0$ of equation (S275), or $2c(x_0, t)/c_0$ of equation (5.32a), is shown as a function of the time t for a fixed value of the distance x_0 and of the diffusion constant D. The unit of time is $x_0/(4D)^{1/2}$.

or

$$c(x,t) = \frac{c_0}{2}\left(\text{erfc}\left[\frac{x-a}{(4Dt)^{1/2}}\right] - \text{erfc}\left[\frac{x+a}{(4Dt)^{1/2}}\right]\right). \tag{S281}$$

The first term on the right above represents the concentration due to an initial concentration distribution $c(x,0) = c_0$ for $x < a$, and $c(x,0) = 0$ for $x > a$, as we learned in Section 5.4. Similarly, we see that the second term represents the concentration due to an initial concentration $c(x,0) = -c_0$, for $x < -a$, and $c(x,0) = 0$ for $x > -a$. The given initial concentration distribution, equation (5.198), is the sum of these two initial concentration distributions. In fact, the given problem could have been decomposed into two separate problems, as suggested by the above observation, and solved independently. The fundamental reason why such a procedure is feasible is that the diffusion equation (with D constant) is linear, so that solutions to it are additive.

5.8.   (a)   The solution to the steady-state diffusion equation (5.200), which satisfies the given boundary conditions, is

$$c = c_0\left(1 - \frac{x}{\xi}\right). \tag{S282}$$

We substitute this result into the flux condition (5.199) at the interface between the oxygen zone and the lactic acid zone to obtain the following equation for ξ:

$$\frac{Dc_0}{\xi} = \lambda\frac{d\xi}{dt}. \tag{S283}$$

The solution to this differential equation which satisfies the initial condition $\xi(0) = 0$

is easily found to be

$$\xi = \left(\frac{2Dc_0 t}{\lambda}\right)^{1/2}.$$ (S284)

Hence the velocity of the advancing front, which decreases with time, is given by the expression

$$\frac{d\xi}{dt} = \left(\frac{Dc_0}{2\lambda t}\right)^{1/2}.$$ (S285)

(b)   Instead of equation (5.200), $c(x,t)$ must satisfy the time-dependent diffusion equation (5.7). Our experience with this equation and its solutions suggests that we attempt to find a solution of the form

$$c(x,t) = c_0 - \frac{2A}{\pi^{1/2}} \int_0^{x/(4Dt)^{1/2}} e^{-\eta^2}\, d\eta = c_0 - A\, \text{erf}\left[\frac{x}{(4Dt)^{1/2}}\right],$$ (S286)

where A is a constant, to be determined. The above expression clearly satisfies the diffusion equation, as well as the boundary condition at the muscle boundary, $c(0,t) = c_0$. To find A, we enforce the boundary condition $c(\xi,t) = 0$, so that, from (S286),

$$c_0 - A\, \text{erf}\left[\frac{\xi}{(4Dt)^{1/2}}\right] = 0.$$ (S287)

Because this equation must be true for all values of the time t, it follows that the argument of the error function must be a constant, which we denote by α:

$$\frac{\xi}{(4Dt)^{1/2}} = \alpha.$$ (S288)

Hence, from equations (S287) and (S288), the constant A is given as

$$A = \frac{c_0}{\text{erf}(\alpha)}.$$ (S289)

Now we impose the flux condition (5.199) on the intended solution (S286), keeping in mind that $(d/dz)\text{erf}(z) = 2\exp(-z)^2/\pi^{1/2}$. The result is

$$\frac{2DA}{(4\pi Dt)^{1/2}} e^{-\xi^2/4Dt} = \lambda\alpha\left(\frac{D}{t}\right)^{1/2},$$

or, with the aid of (S288) and (S289),

$$\frac{c_0}{\pi^{1/2}\lambda} = \alpha e^{\alpha^2}\, \text{erf}(\alpha).$$ (S290)

The above equation is a transcendental equation for the determination of α, for a given value of the quantity appearing on the left-hand side. The function $\alpha \exp(\alpha^2)\text{erf}(\alpha)$ can be found plotted as a function of α in Carslaw and Jaeger (1959).

  The solutions given in both parts a and b above are taken from Hill (1928). The mathematical problem posed in part b, which involves a moving boundary point, arises in the consideration of the melting of ice and the freezing of water. The problem and its solution was first given by Stefan (1891) and is known as the **Stefan problem**.

5.9. The solutions to the various parts of this problem are to be found in Bossert and Wilson (1963).

(a) We calculate, from equation (5.201), that

$$\frac{\partial c}{\partial r} = \int_0^t \frac{2\eta(t_0)(-2r)e^{-r^2/4D(t-t_0)}}{\pi^{3/2}[4D(t-t_0)]^{5/2}}\, dt_0,$$

$$\frac{\partial^2 c}{\partial r^2} = \int_0^t \frac{2\eta(t_0)}{\pi^{3/2}}\left\{\frac{-2}{[4D(t-t_0)]^{5/2}} + \frac{4r^2}{[4D(t-t_0)]^{7/2}}\right\}e^{-r^2/4D(t-t_0)}\, dt_0, \quad \text{(S291)}$$

$$\frac{\partial c}{\partial t} = \int_0^t \frac{2\eta(t_0)}{(4\pi D)^{3/2}}\left\{\frac{-3}{2(t-t_0)^{5/2}} + \frac{r^2}{4D(t-t_0)^{7/2}}\right\}e^{-r^2/4D(t-t_0)}\, dt_0.$$

Note that there is no contribution to $\partial c/\partial t$ arising from the differentiation of the upper limit of integration because the integrand vanishes when $t_0 = t$. Equation (5.195) can be expressed as

$$\frac{\partial c}{\partial t} = D\left(\frac{2}{r}\frac{\partial c}{\partial r} + \frac{\partial^2 c}{\partial r^2}\right). \quad \text{(S292)}$$

By substituting (S291) into (S292), we verify that $c(r,t)$ as given by (5.201) is a solution of the diffusion equation (5.195).

(b) Setting $\eta(t_0) = \eta_0$ in (5.201) and changing the integration variable from t_0 to $r/[4D(t-t_0)]^{1/2} = \xi$, we find that $c(r,t)$ is expressible as

$$c(r,t) = \frac{\eta_0}{\pi^{3/2}Dr}\int_{r/(4Dt)^{1/2}}^{\infty} e^{-\xi^2}\, d\xi, \quad \text{(S293)}$$

which we recognize with the aid of (5.33) to be the same as equation (5.202).

(c) As t approaches infinity, the argument of the error function complement in equation (5.202) approaches zero, when $\text{erfc}(0) = 1$ and $c(r,\infty) = \eta_0/2\pi Dr$.

The radius of influence is determined as that distance $\bar{r}$ for which the concentration $c(\bar{r},t)$ just equals the threshold level K, or, from (5.202),

$$\frac{\eta_0}{2\pi D\bar{r}}\,\text{erfc}\left[\frac{\bar{r}}{(4Dt)^{1/2}}\right] = K. \quad \text{(S294)}$$

This equation determines $\bar{r} = \bar{r}(t)$. For large times, the radius of influence is essentially $\bar{r}(\infty)$, and the argument of the error function complement is essentially zero. Hence, from (S294),

$$\bar{r} \sim \bar{r}(\infty) = \frac{\eta_0}{2\pi DK}. \quad \text{(S295)}$$

(d) The quantity $\eta(t_0)\Delta t_0$ represents the number of pheromone molecules emitted by ants in the time interval t_0 to $t_0 + \Delta t_0$. Let one ant emit N_0 molecules at each emission. According to the given information, when $\Delta t_0 = 5$ sec, there is an emission from one ant (of N_0 molecules), or

$$\eta(t_0)\Delta t_0 = \eta_0 \cdot 5 \text{ sec} = N_0,$$

so that $\eta_0 = (N_0/5) \text{ sec}^{-1}$. Utilizing this expression together with the values of N_0/K and D given in equations (5.54), we calculate from equation (S295) that

$$\bar{r} \sim \frac{1400}{2\pi \times 5 \times 0.4} \frac{\text{cm}^3 \text{ sec}^{-1}}{\text{cm}^2 \text{ sec}^{-1}} = 1 \text{ m.}$$

5.10. (*a*) There is a discontinuity in the diffusion constant at the interface of the two slabs, but no discontinuity in the concentration there, which we denote by c. The mass current through each slab, which is the same, is denoted by J. By considering either slab, J is expressible by means of equation (5.203) in the alternative manner as

$$J = \frac{AD_1}{L_1}(c - c_1) = \frac{AD_2}{L_2}(c_2 - c). \qquad (S296)$$

Solving the last equality above for c, we find that

$$c = \frac{D_2 L_1 c_2 + D_1 L_2 c_1}{D_2 L_1 + D_1 L_2}. \qquad (S297)$$

Substituting (S297) back into (S296), we obtain

$$J = A \frac{D_1 D_2}{D_1 L_2 + D_2 L_1}(c_2 - c_1). \qquad (S298)$$

The diffusional resistance R of the two slabs in series is accordingly defined as

$$R \equiv \frac{c_2 - c_1}{J} = \frac{D_2 L_1 + D_1 L_2}{AD_1 D_2} = \frac{L_1}{AD_1} + \frac{L_2}{AD_2} = R_1 + R_2, \qquad (S299)$$

where $R_1 = L_1/AD_1$ and $R_2 = L_2/AD_2$ are the diffusional resistances for each slab.
(*b*) For the idealized cell of Figure 5.22, which is considered as three slabs in series, application of equation (S299) to the steady-state diffusion through it leads to the following expression for the cell resistance R',

$$R' = \frac{1}{AP} + \frac{L}{AD} + \frac{1}{AP}. \qquad (S300)$$

Here A is the cross-sectional area of the idealized cell, and the permeability constant P associated with the membrane has replaced D_m/δ, where D_m is the diffusion constant of the solute in the membrane. If the cell were a single homogeneous slab of width L and effective diffusion constant D', then the resistance R' would be would be expressed as

$$R' = \frac{L}{AD'}. \qquad (S301)$$

By equating the expressions for R' in (S300) and (S301), we see that equation (5.205) follows directly.

5.11. We follow the method of solution by separation of variables, discussed in Section 5.7. Thus, commencing with the general form of the solution given by equations (5.75), we impose the boundary conditions of equations (5.206) on equation (5.75a). Then $A = 0$, $B = c_1$, and $c(x,t)$ assumes the form

$$c(x,t) = c_1 + (C \sin \lambda x + E \cos \lambda x)e^{-\lambda^2 t}. \qquad (S302)$$

The boundary condition (5.206a) applied to the above expression requires that

$$\left(\frac{\partial c}{\partial x}\right)_{x=0} = \lambda C e^{-\lambda^2 t} = 0,$$

or $C = 0$. The boundary condition (5.206b) applied to (S302) requires that

$$E \cos(\lambda L)e^{-\lambda^2 t} = 0.$$

Therefore, λL must be an odd multiple of $\pi/2$, or

$$\lambda = \lambda_n = \frac{n}{L}\frac{\pi}{2}, n = 1, 3, 5, \ldots \tag{S303}$$

Hence, the most general form of the solution compatible with the boundary conditions is

$$c(x,t) = c_1 + \sum_{n=1, 3, \ldots}^{\infty} E_n \cos \lambda_n x e^{-\lambda_n^2 t}. \tag{S304}$$

Now we apply the initial condition (5.206c) to the above solution with the result

$$c_0 - c_1 = \sum_{n=1, 3, \ldots}^{\infty} E_n \cos \lambda_n x. \tag{S305}$$

Multiplying both sides of this equation by $\cos \lambda_m x$, and integrating from $x = 0$ to $x = L$, we obtain the equation

$$(c_0 - c_1) \int_0^L \cos \lambda_m x \, dx = \sum_{n=0}^{\infty} E_n \int_0^L \cos \lambda_m x \cos \lambda_n x \, dx, m = 1, 3, \ldots. \tag{S306}$$

The integral appearing on the right-hand side can be expressed with the aid of the change of variable $x = L\xi/\pi$ and equations (5.208) as

$$\int_0^L \cos \lambda_m x \cos \lambda_n x \, dx = \frac{2L}{\pi} \int_0^{\pi/2} \cos n\xi \cos m\xi \, d\xi$$

$$= \frac{L}{\pi} \int_0^{\pi} \cos n\xi \cos m\xi \, d\xi = \frac{L}{2} \delta_{nm}. \tag{S307}$$

The integral on the left-hand side of (S306) is readily evaluated to be

$$\lambda_m^{-1} \sin \lambda_m L = \lambda_m^{-1} \sin m\frac{\pi}{2} = \pm \frac{2L}{m\pi} \begin{cases} + \text{ if } \dfrac{m+1}{2} \text{ is odd,} \\[2mm] - \text{ if } \dfrac{m+1}{2} \text{ is even.} \end{cases}$$

A concise way to express this result is to let $m = 2j + 1$, so that the integral on the left side of (S306) becomes

$$\int_0^L \cos \lambda_m x \, dx = \int_0^L \cos \lambda_{2j+1} x \, dx = \frac{2L}{(2j+1)\pi} (-)^j, j = 0, 1, 2, \ldots. \tag{S308}$$

Substituting (S307) and (S308) into (S306), we find that the unknown coefficients are given as

$$E_{2j+1} = \frac{4(c_0 - c_1)}{(2j+1)\pi} (-)^j, \quad j = 0, 1, 2, \ldots. \tag{S309}$$

By substituting (S309) and (S303) into equation (S304), we obtain the concentration distribution as

$$c(x,t) = c_1 + \frac{4(c_0 - c_1)}{\pi} \sum_{j=0}^{\infty} \frac{(-)^j}{(2j+1)} \cos \left[\frac{(2j+1)\pi}{2L} x \right] e^{-(2j+1)^2\pi^2 Dt/4L^2}. \tag{S310}$$

From the solution, we easily calculate the desired expression

$$\frac{c(0,t) - c_1}{c_0 - c_1} = \frac{4}{\pi} \sum_{j=0}^{\infty} \frac{(-)^j}{(2j + 1)} e^{-(2j+1)^2\pi^2 Dt/4L^2}. \tag{S311}$$

The solid line of Figure 5.23 is based on the above formula. When the time t is large compared to L^2/D, the leading term in the above sum is the first term, or

$$\frac{c(0,t) - c_1}{c_0 - c_1} \sim \frac{4}{\pi} e^{-\pi^2 Dt/4L^2}. \tag{S312}$$

Hence, a semilog plot of the quantity on the left versus time yields a straight line, as illustrated in Figure 5.23. According to equations (S312) and (5.207), a tenfold decrease in the magnitude of the quantity $(p_b(t) - p_1)/(p_0 - p_1)$ along the rectilinear segment of the curve of Figure 5.23 occurs when

$$\frac{\pi^2 D \, \Delta t}{4L^2} = \log 10, \text{ or } D = \frac{0.9332 \, L^2}{\Delta t}, \tag{S313}$$

where Δt is the time interval during which the tenfold decrease occurs. For the curve shown in Figure 5.23, $L = 0.0961$ cm, and $\Delta t = 411$ sec. The inferred value of D at $T = 23.3°C$ (Goldstick and Fatt, 1970) was

$$D = 2.19 \times 10^{-5} \text{ cm}^2/\text{sec}.$$

The method described herein has also been utilized to determine the diffusion constant of dissolved oxygen in other solutions, such as an aqueous solution of human oxyhemoglobin.

5.12. Setting $c = c_0 + u/r$ in equation (5.195), we find by explicit differentiation that $u = u(r,t)$ satisfies the differential equation

$$\frac{\partial u}{\partial t} = D \frac{\partial^2 u}{\partial r^2}, \quad r > a. \tag{S314}$$

The boundary conditions satisfied by $c(r,t)$ are that $c(a,t) = 0$, and $c(\infty,t) = c_0$. Therefore, the boundary conditions to be satisfied by u are as follows:

$$u(a,t) = -ac_0,$$
$$\frac{u(r,t)}{r} \to 0 \text{ as } r \to \infty. \tag{S315}$$

The initial condition that u must satisfy is that

$$u(r,0) = 0, r > a. \tag{S316}$$

The problem posed is very similar to the one-dimensional diffusion problem posed in problem 5.5, and the one-dimensional diffusion problem of a column of solution of uniform concentration initially, in contact with a column of pure solvent, discussed in Section 5.4. In those problems the position of the discontinuity in concentration was the origin, whereas it is the position $r = a$ in the present problem. At this value of r, the function u must be maintained at the value $-ac_0$, whereas it was maintained at the value $c_0/2$ or c_0, previously. Hence we infer from either equation (5.32) or (S275) that

$$u(r,t) = -ac_0 \text{ erfc} \left[\frac{r - a}{(4Dt)^{1/2}} \right], \quad r > a. \tag{S317}$$

The concentration $c(r,t)$ is then given (Carlson, 1962) as

$$c(r,t) = c_0 \left\{ 1 - \frac{a}{r} \operatorname{erfc} \left[\frac{r - a}{(4Dt)^{1/2}} \right] \right\}, \qquad r > a. \tag{S318}$$

The mass flux J of nutrient into the spherical cell is the surface area of the cell times the negative of the radial current density j_r (because the flux into the cell is in the inward radial direction), or

$$J = -4\pi a^2 j_r = 4\pi a^2 D \left(\frac{\partial c}{\partial r} \right)_{r=a}$$

$$= 4\pi a^2 D c_0 \left\{ \frac{a}{r^2} \operatorname{erfc} \left[\frac{r - a}{(4Dt)^{1/2}} \right] + \frac{a}{r(4Dt)^{1/2}} e^{-(r-a)^2/4Dt} \right\}_{r=a} \tag{S319}$$

$$= 4\pi a D c_0 \left\{ 1 + \frac{a}{(4Dt)^{1/2}} \right\}.$$

When $t \to \infty$, J approaches the steady-state value

$$\bar{J} = 4\pi a D c_0. \tag{S320}$$

The result (S319) also suggests why it is desirable for microorganisms to possess flagella or cilia (see problem 1.3). By being able to swim, the microorganisms can move to a region where the local value of c_0 is greater. An alternative explanation of the utility of flagella is that they make it possible for the microorganism to stir up the medium, and therefore in effect decrease the value of t in equation (S319), so that J is increased in comparison with its steady-state value.

5.13. (a) By setting $c_1 = c_0, c_2 = 0$, and $\delta = L$ in equation (5.87), we obtain the concentration $c(x,t)$ in the line of cells of width L as

$$c(x,t) = \frac{c_0(L - x)}{L} - \frac{2c_0}{\pi} \sum_{n=1}^{\infty} \frac{1}{n} \sin \frac{n\pi x}{L} e^{-n^2 \pi^2 Dt/L^2}. \tag{S321}$$

The first term on the right-hand side is seen to be the final steady-state concentration distribution $\bar{c}(x) = c_0(L - x)/L$. Define the error Δc as the difference between $c(x,t)$ and $\bar{c}(x)$, that is, $\Delta c \equiv c(x,t) - \bar{c}(x)$. The fractional error $|\Delta c/\bar{c}|$ at the position x at the time t is given with the aid of (S321) as

$$\left| \frac{\Delta c}{\bar{c}} \right| = \frac{2L}{\pi(L - x)} \left| \sum_{n=1}^{\infty} \frac{1}{n} \sin \frac{n\pi x}{L} e^{-n^2 \pi^2 Dt/L^2} \right|. \tag{S322}$$

If the value of this fractional error evaluated at the midpoint of the line of cells $x = L/2$ is specified, then the above relation becomes a transcendental equation for the dimensionless quantity Dt/L^2. Under these circumstances, we can write the solution as

$$t = \gamma \frac{L^2}{D}, \tag{S323}$$

where γ is a pure number. This expression represents the time needed to establish the gradient at the given position. When c at the midpoint attains its final steady-state value within 1%, $|\Delta c/c|_{x=L/2} = 0.01$, and if we retain for the sake of simplicity only the first term in the summation in equation (S322), then the quantity γ is determined explicitly (Munro and Crick, 1971) to be

$$\gamma = -\frac{1}{\pi^2} \log \left(\frac{\pi}{4} \times 0.01 \right) \doteq 0.49. \tag{S324}$$

(b)   From Table 5.1, the diffusion constant in water at 20°C of a molecular species with $M = 500$ is about $D_0 = 5 \times 10^{-6}$ cm²/sec. With the viscosity μ for cytoplasm about six times the value μ_0 for water, the estimate of the diffusion constant D of the morphogen in cytoplasm is, according to equation (5.41),

$$D = \frac{\mu_0}{\mu} D_0 = \frac{1}{6} \times 5 \times 10^{-6} \text{ cm}^2/\text{sec} \sim 0.8 \times 10^{-6} \text{ cm}^2/\text{sec.} \qquad \text{(S325)}$$

To evaluate the effective diffusion constant of the morphogen in the line of cells, it is sufficient to evaluate the effective diffusion of constant D' of the morphogen in a single cell, taking into account that the morphogen must diffuse through the membrane of each cell twice. Thus, using equations (S325) and (5.205) with L replaced by d, and the given estimates of P and d, we find that

$$D' = \frac{DPd}{2D + Pd} \sim \begin{cases} 1.8 \times 10^{-7} \text{ cm}^2/\text{sec}, d = 10 \ \mu\text{m}, \\ 3.7 \times 10^{-7} \text{ cm}^2/\text{sec}, d = 30 \ \mu\text{m}. \end{cases} \qquad \text{(S326)}$$

Now replace D by D' and set $L = nd$ in equation (S323), where n is the number of cells in the line. Then

$$n = \frac{1}{d}\left(\frac{D't}{\gamma}\right)^{1/2}, \qquad \text{(S327)}$$

where t is the time necessary to establish a gradient in the field. With the given estimates of d and t and the aid of equations (S324) and (S326), we find from equation (S327) that the estimated number of cells in an embryonic field (Crick, 1970) is

$$n \sim \begin{cases} 60, d = 10 \ \mu\text{m}. \\ 30, d = 30 \ \mu\text{m}. \end{cases}$$

The agreement in order of magnitude of these estimates with the observed size of embryonic fields suggests that diffusion is a possible mechanism for the establishment of morphogenetic gradients in embryonic development.

5.14.   In the steady state, the one-dimensional convective diffusion equation (5.99) applied to a pore of the membrane of length L, simplifies to the form

$$u\frac{dc}{dx} = D\frac{d^2c}{dx^2}, \qquad 0 < x < L. \qquad \text{(S328)}$$

Here $c = c(x)$ represents the steady-state concentration of tritium oxide in the water inside a single pore of the model membrane, and u is the mean velocity of the bulk flow of water in the pore. The concentration $c(x)$ satisfies the boundary conditions

$$\begin{aligned} c(0) &= c_1, \\ c(L) &= 0. \end{aligned} \qquad \text{(S329)}$$

It is convenient to introduce the dimensionless variable $\xi = x/L$ and the dimensionless parameter $\alpha = uL/D$. Then equations (S328) and (S329) become

$$\alpha\frac{dc}{d\xi} = \frac{d^2c}{d\xi^2}, \qquad 0 < \xi < 1, \qquad \text{(S330a)}$$

$$c(0) = c_1, \qquad \text{(S330b)}$$

$$c(1) = 0, \qquad \text{(S330c)}$$

where now $c = c(\xi)$. A first integration with respect to ξ of (S330a) yields the relation

$$\frac{dc}{d\xi} = Ae^{\alpha\xi},$$

where A is a constant of integration. A second integration yields the result

$$c = \frac{A}{\alpha} e^{\alpha\xi} + B, \tag{S331}$$

where B is likewise a constant. By applying the boundary conditions (S330b) and (S330c) to (S331), we determine A and B to be

$$A = -\frac{\alpha c_1}{e^{\alpha} - 1},$$
$$B = \frac{c_1 e^{\alpha}}{e^{\alpha} - 1}. \tag{S332}$$

Hence, the concentration c is

$$c = c_1 \frac{e^{\alpha} - e^{\alpha\xi}}{e^{\alpha} - 1}. \tag{S333}$$

From the above solution, we can calculate the current density j in the pore to be

$$j = uc - D\frac{dc}{dx} = \frac{D}{L}\left(\alpha c - \frac{dc}{d\xi}\right) = \frac{D}{L}\frac{\alpha c_1 e^{\alpha}}{e^{\alpha} - 1}. \tag{S334}$$

The total mass current J of tritiated water through a membrane consisting of N identical pores is then $J = N\pi a^2 j$, and the associated volume flux of water is

$$Q = \frac{J}{c_1} = N\pi a^2 \frac{D}{L}\frac{\alpha e^{\alpha}}{e^{\alpha} - 1}, \tag{S335}$$

where πa^2 is the cross-sectional area of a circular cylindrical pore of radius a.

If Poiseuille's law applies to the bulk flow through the pore, then the mean velocity of flow u is given by equation (S191), namely,

$$u = \frac{a^2 \Delta p}{8\mu L}. \tag{S191}$$

We see that u is proportional to Δp, so that α can be thought of as a dimensionless form of Δp:

$$\alpha = \frac{a^2 \Delta p}{8\mu D} = \frac{\Delta p}{\Delta p^*},$$
$$\Delta p^* \equiv \frac{8\mu D}{a^2}. \tag{S336}$$

The solid curve in Figure 5.24 is based on equation (S335) with α given by (S336).

The condition $|u| \ll D/L$ is equivalent to the condition $|\alpha| \ll 1$. A simple way to obtain the limiting form of equation (S335) under this condition is to replace e^{α} in it by its linear expansion in the neighborhood of $\alpha = 0$, namely $1 + \alpha$. The leading term in Q then becomes

$$Q = N\pi a^2 \frac{D}{L}. \tag{S337}$$

In this limit, the mass current $J = c_1 Q$ is seen to be entirely due to diffusion. This limiting value of Q is represented in Figure 5.24 by the point on the solid curve for which $\Delta p = 0$.

In the other extreme for which $|u| \gg D/L$ or $|\alpha| \gg 1$, $e^\alpha \to \infty$ if u is positive, and $e^\alpha \to 0$ if u is negative. Equation (S335) reduces to

$$Q = \begin{cases} N\pi a^2 u, & u \text{ positive,} \\ 0, & u \text{ negative.} \end{cases} \tag{S338}$$

In this limit, the bulk flow due to the applied pressure gradient is the controlling mechanism of transport. When $\Delta p \to \infty$, Q is asymptotically linear in Δp, as shown by equations (S338) and (S191).

The quantity a^2, or equivalently Δp^*, is the only unknown parameter appearing in equation (S335), so that a least-square procedure can be utilized to determine a unique best value for it, associated with the given data. By a procedure similar in spirit to this, it was found (Bean, 1972) that $a \sim 1500$ Å. The associated value of Δp^*, based on the values $\mu = 0.89$ cP and $D = 2.44 \times 10^{-5}$ cm^2/sec at 25°C, is indicated by the point on the abscissa labeled p^* in Figure 5.24.

5.15. According to the discussion following equation (5.102), the solution $c(x,t)$ to the one-dimensional convective diffusion equation (5.99), which satisfies the given initial condition, vanishes as $|x| \to \infty$, and satisfies the normalization condition

$$A \int_{-\infty}^{\infty} c(x,t)dx = \rho_0,$$

is found with the aid of equation (5.15) to be

$$c(x,t) = \frac{\rho_0}{A(4\pi Dt)^{1/2}} \exp\left[-\frac{(x - ut)^2}{4Dt}\right]. \tag{S339}$$

Hence the indicator concentration at position x_0, $c(t) \equiv c(x_0,t)$, can be written as

$$c(t) = \frac{\rho_0}{A(4\pi Dt)^{1/2}} \exp\left[-\frac{u^2}{4D}\left(t - \frac{2x_0}{u} + \frac{x_0^2}{u^2 t}\right)\right]. \tag{S340}$$

From the definition of cardiac output K and equation (5.210), it follows that the cardiac output is

$$K = uA = \left(\overline{\frac{x_0}{t}}\right)A = \frac{Ax_0 \displaystyle\int_0^\infty t^{-1}c(t)dt}{\displaystyle\int_0^\infty c(t)dt}. \tag{S341}$$

To evaluate the integral appearing in the numerator of the last expression on the right, we follow the hint, and change the integration variable to $t = (4D/u^2)\xi^2$. Then

$$\int_0^\infty \frac{c(t)}{t}\,dt = \frac{\rho_0 u}{2\pi^{1/2} AD}\int_0^\infty \frac{1}{\xi^2}\exp\left[-\xi^2 + 2\beta - \frac{\beta^2}{\xi^2}\right]d\xi, \tag{S342}$$

$$\beta = \frac{ux_0}{4D}.$$

The integral on the right can be written with the aid of equations (5.211) as

$$e^{2\beta}\int_0^\infty \frac{1}{\xi^2}\exp\left[-\xi^2 - \frac{\beta^2}{\xi^2}\right]d\xi = \frac{\sqrt{\pi}}{2\beta} = \frac{2\sqrt{\pi}\,D}{ux_0}. \tag{S343}$$

Substitution of equations (S342) and (S343) into (S341) yields Hamilton's formula, equation (5.209).

The theoretical curve (dashed line) of Figure 5.25 was obtained by fitting equation (S340) to the experimental observations (solid line), treating D and x_0/u as parameters. It was found (Norwich and Zelin, 1970) that $A^2D \sim 600 \; cm^6/sec$, and $\tau \equiv x_0/u \sim 5 \; sec$. The constant D appearing in equation (S340) should be thought of as an empirical parameter that can vary with the mean speed of blood flow. The reason for this attitude is that indicator is dispersed not only because of diffusion, but because of the (approximate) Poiseuille nature of the flow in the aorta. Thus, it has been shown (Taylor, 1953) that the indicator in a Poiseuille flow in a tube has an effective diffusion constant D_e given as

$$D_e = D + \frac{a^2 u}{48D}, \tag{S344}$$

where D is the actual diffusion constant of the indicator, a is the tube radius, and u is the mean flow velocity.

5.16. In the steady state, c, m, and u are time-independent and denoted by $\bar{c}(x)$, $\bar{m}(x)$, and $\bar{u}(x)$, respectively. These quantities satisfy the steady-state form of equations (5.212),

$$0 = D \frac{d^2\bar{c}}{dx^2} - k_+\bar{c}\bar{m} + k_-\bar{u}, \tag{S345a}$$

$$0 = D_m \frac{d^2\bar{m}}{dx^2} - k_+\bar{c}\bar{m} + k_-\bar{u}, \tag{S345b}$$

$$0 = D_m \frac{d^2\bar{u}}{dx^2} + k_+\bar{c}\bar{m} - k_-\bar{u}. \tag{S345c}$$

By adding the last two equations above, we see that

$$\frac{d^2}{dx^2}(\bar{m} + \bar{u}) = 0. \tag{S346}$$

By integration, it follows that

$$\frac{d}{dx}(\bar{m} + \bar{u}) = B, \tag{S347}$$

where B is a constant. A second integration yields the relation

$$\bar{m}(x) + \bar{u}(x) = A + Bx, \tag{S348}$$

where A is a constant. Because the myoglobin and oxymyoglobin molecules cannot cross the boundary of the slab, the diffusive flux density of these solute species vanishes at the boundary, or

$$\frac{d\bar{m}}{dx} = \frac{d\bar{u}}{dx} = 0, \qquad x = 0, L. \tag{S349}$$

Applying this condition to equation (S347), we see that $B = 0$. Assume that before the myoglobin solution is exposed to oxygen, there is only myoglobin present in the slab, at a uniform concentration m_0. From this consideration in conjunction with equation (S349) applied to (S348), we determine the constant A, namely,

$$\bar{m}(x) + \bar{u}(x) = m_0. \tag{S350}$$

By subtracting equation (S345a) from (S345b), we obtain

$$D_m \frac{d^2\bar{m}}{dx^2} - D\frac{d^2\bar{c}}{dx^2} = 0. \tag{S351}$$

Two successive integrations yield the relation

$$D_m\bar{m}(x) - D\bar{c}(x) = a + jx, \tag{S352}$$

where a and j are constants to be determined. We note that equation (S352) becomes, after differentiation,

$$D_m \frac{d\bar{m}}{dx} - D\frac{d\bar{c}}{dx} = j. \tag{S353}$$

The first term on the left represents the negative of the myoglobin current density at x. If we set $x = 0$ or $x = L$ in (S353), then the myoglobin current density vanishes, and we see that j represents the steady-state value of the current density of oxygen entering or leaving the slab.

By eliminating $\bar{m}(x)$ from equations (S350) and (S352), it follows that $\bar{u}(x)$ can be expressed in terms of $\bar{c}(x)$ as

$$\bar{u}(x) = m_0 - \frac{1}{D_m}[D\bar{c}(x) + a + jx]. \tag{S354}$$

Now we can use equations (S352) and (S354) to eliminate $\bar{m}(x)$ and $\bar{u}(x)$ from equation (S345a) and obtain the nonlinear ordinary differential equation (Wyman, 1966) satisfied by $\bar{c}(x)$,

$$0 = D\frac{d^2\bar{c}}{dx^2} + k_-m_0 - \frac{1}{D_m}(k_+\bar{c} + k_-)(D\bar{c} + a + jx). \tag{S355}$$

The solution to this differential equation for given values of $\bar{c}(0)$ and $\bar{c}(L)$, or equivalently, for given values of the oxygen partial pressure on both sides of the slab, determines the facilitated oxygen flux through the slab.

5.17. By integrating equation (5.136) with respect to time between the limits 0 and t, we find with the aid of equation (5.131) that

$$\int_0^t \frac{d}{dt'}\left[\int_{r_1}^{r_p} c(r,t')r\,dr\right]dt' = -s_0\omega^2 r_p^2 \int_0^t c_p(t')dt' = \frac{1}{2}r_p^2 \int_0^t \frac{dc_p(t')}{dt'}\,dt'.$$

Therefore, with the aid of the initial conditions (5.132) and $c_p(0) = c_0$, it follows that

$$\int_{r_1}^{r_p} c(r,t)r\,dr - \int_{r_1}^{r_p} c_0 r\,dr = \frac{1}{2}r_p^2[c_p(t) - c_0],$$

or

$$\int_{r_1}^{r_p} c(r,t)r\,dr = \frac{1}{2}r_p^2 c_p(t) - \frac{1}{2}r_1^2 c_0. \tag{S356}$$

From equations (5.143) and (5.130), the left-hand side above can also be expressed as

$$\int_{r_1}^{r_p} c(r,t)r\,dr = \frac{1}{2}r_p^2 c_p(t) - \frac{1}{2}r_1^2 c(r_1,t) - \frac{1}{2}\int_{r_1}^{r_p} \frac{\partial c}{\partial r}r^2\,dr. \tag{S357}$$

When the right-hand sides of equations (S356) and (S357) are equated, equation (5.214) follows directly.

5.18.   As we see from equation (5.162), the major contribution to $\partial c/\partial r$ arises from the neighborhood of $\log(r/r_0) = s_0\omega^2 t$. If $s_0\omega^2 t \ll 1$, then this neighborhood becomes the neighborhood of $r = r_0$, and we can sensibly represent $\log(r/r_0)$ by its linear approximation in the neighborhood of $r = r_0$, namely

$$\log\left(\frac{r}{r_0}\right) = \frac{1}{r_0}(r - r_0) + \cdots.$$

We make this replacement in equation (5.162) and substitute the resulting expression into the integrals appearing in the definition of $r_B(t)$ in the case when $c(r_1,t)$ is null, equation (5.147). Because of the Gaussian nature of $\partial c/\partial r$, a negligible error is made by replacing the limits r_1 and r_p appearing in equation (5.147) by $-\infty$ and $+\infty$, respectively, so long as r_1 and r_p are far from r_0. Then equation (5.147) becomes

$$r_B^2(t) = \left\{ \int_{-\infty}^{\infty} \exp\left[-\frac{(r_0 s_0 \omega^2 t + r_0 - r)^2}{4D_0 t}\right] r^2\, dr \right\}$$
$$\times \left\{ \int_{-\infty}^{\infty} \exp\left[-\frac{(r_0 s_0 \omega^2 t + r_0 - r)^2}{4D_0 t}\right] dr \right\}^{-1}. \qquad \text{(S358)}$$

Change the integration variable in these two integrals from r to ρ by setting

$$r = r_0 + r_0 s_0 \omega^2 t + (4D_0 t)^{1/2} \rho.$$

Then (S358) becomes

$$r_B^2(t) = \frac{4D_0 t \displaystyle\int_{-\infty}^{\infty} e^{-\rho^2}(\rho - \bar\rho)^2\, d\rho}{\displaystyle\int_{-\infty}^{\infty} e^{-\rho^2} d\rho}, \qquad \text{(S359a)}$$

$$\bar\rho = \frac{r_0 + r_0 s_0 \omega^2 t}{(4D_0 t)^{1/2}}. \qquad \text{(S359b)}$$

The integral in the numerator of equation (S359a) can be written as

$$\int_{-\infty}^{\infty} e^{-\rho^2}(\rho - \bar\rho)^2\, d\rho = \int_{-\infty}^{\infty} e^{-\rho^2}\rho^2\, d\rho - 2\bar\rho \int_{-\infty}^{\infty} e^{-\rho^2}\rho\, d\rho + \bar\rho^2 \int_{-\infty}^{\infty} e^{-\rho^2}\, d\rho. \qquad \text{(S360)}$$

From equation (5.36), the first integral appearing on the right above equal $\sqrt{\pi}/2$. The second integral is null because the integrand is an odd function. The last integral on the right equals $\sqrt{\pi}$ according to (5.17). Hence $r_B^2(t) = 4D_0 t(1/2 + \bar\rho^2)$, or

$$r_B(t) = (r_0 + r_0 s_0 \omega^2 t)\left[1 + \frac{2D_0 t}{(r_0 + r_0 s_0 \omega^2 t)^2}\right]^{1/2}. \qquad \text{(S361)}$$

An alternative way to derive equation (S361) is to recognize that $r_B^2(t)$ as given by equation (S358) is just the second moment about zero of the normal density function $f(r)$, namely

$$f(r) = \left\{ \exp\left[-\frac{(r_0 s_0 \omega^2 t + r_0 - r)^2}{4D_0 t}\right] \right\} \left\{ \int_{-\infty}^{\infty} \exp\left[-\frac{(r_0 s_0 \omega^2 t + r_0 - r)^2}{4D_0 t}\right] dr \right\}^{-1}. \qquad \text{(S362)}$$

Comparison of $f(r)$ with the defining equation (5.112) shows that $f(r)$ possesses the parameters $\mu = r_0 + r_0 s_0 \omega^2 t$ and $\sigma = (2D_0 t)^{1/2}$. According to equations (5.114)–(5.117), the second moment about zero of the normal density function is $\mu^2 + \sigma^2$. Therefore, $r_B^2(t) = \mu^2 + \sigma^2$, with μ and σ as given above, which is the result (S361).

The expression in the first bracket in equation (S361) is seen to be the linear expansion of $\bar{r}(t)$ as given by equation (5.163), valid for small values of $s_0 \omega^2 t$. The expression in the second bracket in equation (S361) is a multiplicative correction term, which differs very little from unity in practical circumstances. For example, if we evaluate it for the representative values of the parameters in a sedimentation velocity experiment, given by equations (5.168), we calculate with $t = 100$ min that

$$\left[1 + \frac{2D_0 t}{(r_0 + r_0 s_0 \omega^2 t)^2} \right]^{1/2} \sim [1 + 10^{-5}]^{1/2} \sim 1.$$

Hence $r_B(t)$ is virtually identical with $\bar{r}(t)$ when $s_0 \omega^2 t \ll 1$.

5.19.   According to the conservation of mass principle, the integral of the concentration over the volume of the cell is invariant in time, or

$$\int_{r_1}^{r_2} c(r,t) r \, dr = \text{constant}. \tag{S363}$$

By equating this integral evaluated at $t = 0$ (when $c(r,t) = c_0$) to its value when t is infinite and the steady state is established, so that $c(r,\infty)$ is given by equation (5.177), we infer that

$$\int_{r_1}^{r_2} c_0 r \, dr = c_1 \int_{r_1}^{r_2} \exp[AM(r^2 - r_1^2)] r \, dr. \tag{S364}$$

By carrying out the indicated integrations, we find that

$$c_1 = \frac{c_0(r_2^2 - r_1^2)AM}{\exp[AM(r_2^2 - r_1^2)] - 1}. \tag{S365}$$

5.20.   From equation (4.18), an *E. coli* cell, assumed to be a sphere of radius a, falls under the influence of gravity with a terminal velocity u given as

$$u = \frac{2}{9\mu} a^2 (\rho - \rho_0) g. \tag{S366}$$

Here μ is viscosity of water and g is the gravitational constant. We treat the *E. coli* cells as a solute with an effective diffusion constant M, and denote the volume concentration of cells in a column of water of height L as $c = c(x,t)$. Here t represents the time and x represents height, $x = 0$ being the bottom of the column, and $x = L$ being the top. The concentration c satisfies the one-dimensional convective diffusion equation, (5.99) with D replaced by M, and u there replaced by $-u$, because cells fall in the negative x direction. The boundary condition to be satisfied by c is that no cells can traverse the ends of the column, or

$$j(x,t) = -M \frac{\partial c}{\partial x} - uc = 0, \qquad x = 0, L. \tag{S367}$$

In the steady state, $c = \bar{c}(x)$, and equation (5.99) simplifies to

$$M \frac{d^2 \bar{c}}{dx^2} + u \frac{d\bar{c}}{dx} = 0. \tag{S368}$$

A first integration yields

$$M \frac{d\bar{c}}{dx} + u\bar{c} = 0, \tag{S369}$$

where the constant of integration has been taken to be zero to satisfy the boundary conditions (S367). A second integration yields the solution

$$\bar{c}(x) = c_0 e^{-ux/M}, \tag{S370}$$

where c_0 is a constant, representing the concentration of cells at the bottom of the column. It is determined by the condition that there are N cells in the column of water, so that

$$N = A \int_0^L \bar{c}(x)dx = \frac{Ac_0 M}{u} [1 - e^{-uL/M}],$$

where A is the cross-sectional area of the column. Solving for c_0, we obtain

$$c_0 = \frac{Nu}{AM[1 - e^{-uL/M}]}. \tag{S371}$$

According to equation (S370), the decay length δ associated with the concentration distribution $\bar{c}(x)$ is $\delta \equiv M/u$.

To determine the height to which the cells extend, we substitute equation (S370) into the extension criterion (5.215), and obtain the relation

$$c_0 e^{-ux_0/M} A \, \Delta x_0 = 1,$$

or

$$x_0 = \frac{M}{u} \log(c_0 A \, \Delta x_0). \tag{S372}$$

From equation (S366), we calculate with $a \sim 1 \ \mu m$ and $\mu = 0.01$ P that

$$u \sim 2 \times 10^{-5} \text{ cm/sec} = 0.2 \ \mu m/sec.$$

With this estimate, we find that $\delta \sim 3.4$ cm, and the value of c_0 from equation (S371) is

$$c_0 \sim 3 \times 10^7 \text{ cm}^{-3}.$$

Hence with the calculated or given values of M, u, c_0, A, and Δx_0, we infer from equation (S372) that the equilibrium distribution of cells extends to the height

$$x_0 \sim 35 \text{ cm}.$$

These considerations ignore any directed swimming response of the microorganisms to the effect of gravity, or **geotaxis**.

5.21.* (a)   In the steady state, equation (5.216) reduces to equation (5.184). Hence $\bar{c}(x)$ is given by $c(x)$ of equation (5.185), with c_T there replaced by $\bar{c}_T$. By substituting the steady-state functions $\bar{c}(x)$ and $\bar{c}_T$ into equation (5.217), we find that they are related by the expression

$$0 = \int_0^L [\bar{c}(x) - \bar{c}_T]dx,$$

or

$$0 = (c_A - \overline{c}_T) \int_0^L e^{-PSx/QL} \, dx. \tag{S373}$$

This equation can be satisfied only if the factor in parentheses is zero, so that the steady-state solution is given as

$$\overline{c}(x) = \overline{c}_T = c_A. \tag{S374}$$

(b)   We substitute equations (5.218) into (5.216) and (5.217), with the result that

$$-\lambda f(x) + u \frac{df}{dx} = -u \frac{PS}{QL} [f(x) + c_A],$$

$$\lambda c_A = \frac{PS}{V_T L} \int_0^L [f(x) + c_A] dx. \tag{S375}$$

Here, as for equation (5.184), $Q = \pi a^2 u$ and $S = 2\pi aL$. The solution to the first equation above, which satisfies the required boundary condition, is

$$f(x) = \frac{c_A}{QL\lambda/PSu - 1} \left\{ 1 - \exp\left[\left(\frac{\lambda}{u} - \frac{PS}{QL} \right) x \right] \right\}. \tag{S376}$$

By substituting this result into the second of equations (S375) and performing the indicated integration, we find that the decay rate λ satisfies the transcendental equation

$$\lambda = \frac{PS}{V_T} \left\{ 1 + \left(\frac{QL\lambda}{PSu} - 1 \right)^{-1} \left[1 - \left(\frac{\lambda L}{u} - \frac{PS}{Q} \right)^{-1} \left\{ \exp\left[\frac{\lambda L}{u} - \frac{PS}{Q} \right] - 1 \right\} \right] \right\}. \tag{S377}$$

(c)   When we seek an approximate solution of this equation, subject to the condition $\lambda \ll PSu/QL$, we note that all the terms in λ appearing on the right-hand side of the equation can be neglected, and equation (S377) reduces to (Johnson and Wilson, 1966)

$$\lambda = \frac{Q}{V_T} (1 - e^{-PS/Q}). \tag{S378}$$

If, in addition, $PS \gg Q$, the decay rate λ further simplifies to

$$\lambda = \frac{Q}{V_T}. \tag{S379}$$

This result must be consistent with the simplifying conditions on λ by means of which it was obtained, or $QL/uV_T \ll PS/Q$. Since $QL/u = \pi a^2 L \equiv V_C$, the capillary volume, the latter condition reads $V_C/V_T \ll PS/Q$, which is readily satisfied, since PS/Q has been assumed to be large.

At the other extreme, suppose $PS \ll Q$. By replacing the exponential in equation (S378) by its linear approximation, we find in this case that

$$\lambda = \frac{PS}{V_T}. \tag{S380}$$

The consistency condition on λ now reads $1/V_T \ll u/QL$, or $V_C/V_T \ll 1$. This condition is likewise satisfied under normal physiological circumstances, a typical value of V_C/V_T being 1/20.

REFERENCES

E. Ackerman, J. W. Rosevear, and W. F. McGuckin (1964), *Phys. Med. Biol.* **9**, 203.

C. S. Adair (1925), *J. Biol. Chem.* **63**, 529.

J. Adler and M. M. Dahl (1967), *J. Gen. Microbiol.* **46**, 161.

W. J. Archibald (1938), *Phys. Rev.* **53**, 746; **54**, 371.

W. J. Archibald (1947), *J. Phys. Chem.* **51**, 1204.

G. L. Atkins (1969), *Multicompartment Models for Biological Systems*, Methuen, London.

C. P. Bean (1972), in *Membranes*, Vol. I, G.Eisenman, ed., Dekker, New York, pp. 1–54.

D. H. Bergel (1961), *J. Physiol.* **156**, 445.

M. Berman and R. Schoenfeld (1956), *J. Appl. Phys.* **27**, 1361.

W. H. Bossert and E. O. Wilson (1963), *J. Theoret. Biol.* **5**, 443.

J. Botts and M. Morales (1953), *Trans. Faraday Soc.* **49**, 696.

G. A. Brecher (1952), *Amer. J. Physiol.* **169**, 423.

G. E. Briggs and J. B. S. Haldane (1925), *Biochem. J.* **19**, 338.

C. J. Brokaw (1970), *J. Exp. Biol.* **53**, 445.

R. Brown (1828), *Philos. Mag.* **4**, 161.

M. I. Bunting (1940), *J. Bacteriol.* **40**, 69.

F. D. Carlson (1959), in *Proc. First Nat. Biophys. Conf.*, H. Quastler and H. J. Morowitz, eds., Yale University Press, New Haven, Conn., pp. 443–449.

F. D. Carlson (1962), in *Spermatozoan Motility*, D. W. Bishop, ed., Amer. Assoc. Advan. Sci., Washington, D. C., pp. 137–146.

H. S. Carslaw and J. C. Jaeger (1959), *Conduction of Heat in Solids*, 2nd ed., Clarendon, Oxford, Chapter 11.

J. P. Changeux and M. M. Rubin (1968), *Biochemistry* **7**, 553.

H. Chick (1908), *J. Hyg.* **8**, 92.

C. M. Child (1941), *Patterns and Problems of Development*, Chicago University Press, Chicago.

W. W. Cleland (1963), *Biochim. Biophys. Acta* **67**, 104, 173.

W. W. Cleland (1967), *Ann. Rev. Biochem.* **36**, 77.

H. L. Conn, Jr., and J. S. Robertson (1955), *Amer. J. Physiol.* **181**, 319.

R. G. Cox (1970), *J. Fluid Mech.* **44**, 791.

342

F. Crick (1970), *Nature (London)* **225**, 420.

C. Crone (1963), *Acta Physiol. Scand.* **54**, 292.

H. Davson and J. F. Danielli (1952), *The Permeability of Natural Membranes*, 2nd ed., Cambridge Univ. Press, Cambridge.

M. Deskowitz and A. Shapiro (1935), *Proc. Soc. Exp. Biol. Med.* **32**, 573.

R. D. Dyson and I. Isenberg (1971), *Biochemistry* **10**, 3233.

M. Eigen (1954), *Discuss. Faraday Soc.* **17**, 194.

M. Eigen (1967), in *Fast Reactions and Primary Processes in Chemical Kinetics, Proc. Fifth Nobel Symposium, Södergarn*, S. Claesson, ed., Interscience, New York, pp. 333–369.

M. Eigen and L. De Maeyer (1963), in *Techniques of Organic Chemistry*, Vol. VIII, Part 2, 2nd ed., S. L. Friess, E. S. Lewis, and A. Weissberger, eds., Wiley, New York, pp. 895–1054.

M. Eigen and G. G. Hammes (1963), *Advan. Enzymol.* **25**, 1.

A. Einstein (1905), *Ann. Phys.* **17**, 549 (Engl. trans. by A. D. Cowper in A. Einstein, *Investigations on the Theory of the Brownian Movement*, R. Fürth, ed., Dover, New York, 1956).

A. Einstein (1906), *Ann. Phys.* **19**, 289 (Engl. trans. by A. D. Cowper in A. Einstein, *Investigations on the Theory of the Brownian Movement*, R. Fürth, ed., Dover, New York, 1956).

H. Faxén (1929), *Ark. Mat. Astron. Fys.* **21B**, No. 3.

W. M. Feldman (1935), *Biomathematics*, 2nd ed., Charles Griffin, London.

A. Fick (1855), *Ann. Phys. Chem.* **94**, 59; *Philos. Mag.* **10**, 30.

B. Folkow and E. Neil (1971), *Circulation*, Oxford Univ. Press, New York.

H. Fujita (1962), *Mathematical Theory of Sedimentation Analysis*, Academic, New York.

G. F. Gause (1934), *The Struggle for Existence*, Williams and Wilkins, Baltimore (reprinted by Hafner, New York, 1969).

G. F. Gause (1935), *J. Exp. Biol.* **12**, 44.

J. C. Gerhart (1970), *Curr. Top. Cell Reg.* **2**, 275.

J. C. Gerhart and H. K. Schachman (1965), *Biochemistry* **4**, 1054.

R. J. Goldberg (1953), *J. Phys. Chem.* **57**, 194.

T. K. Goldstick and I. Fatt (1970), in *Mass Transfer in Biological Systems*, A. L. Shrier and T. G. Kaufmann, eds., Amer. Inst. Chem. Eng., New York, pp. 101–113.

R. J. Goodlow, L. A. Mika, and W. Braun (1950), *J. Bacteriol.* **60**, 291.

D. S. Goodman and R. P. Noble (1968), *J. Clin. Invest.* **47**, 231.

J. Gray (1955), *J. Exp. Biol.* **32**, 775.

J. Gray and G. J. Hancock (1955), *J. Exp. Biol.* **32**, 802.

H. D. Green (1944), in Medical Physics, Vol. I, O. Glasser, ed., Year Book Publishers, Chicago, pp. 208–232.

H. Gutfreund (1972), *Enzymes: Physical Principles*, Wiley–Interscience, New York, p. 93.

J. B. S. Haldane (1930), *Enzymes*, 2nd ed., Longmans, Green, London, (reprinted by M.I.T. Press, Cambridge, 1965) Chapter 5.

D. Hallberg (1965), *Acta Physiol. Scand.* **64**, 306.

G. J. Hancock (1953), *Proc. Roy. Soc. London Ser. A* **217**, 96.

H. H. Hartridge and F. J. W. Roughton (1925), *Proc. Roy. Soc. London Ser. A* **107**, 654.

R. H. Haschemeyer and A. E. V. Haschemeyer (1973), *Proteins*, John Wiley, New York, Chapter 8.

B. H. Havsteen (1967), *J. Biol. Chem.* **242**, 769.

J. Z. Hearon, S. A. Bernhard, S. L. Friess, D. J. Botts, and M. F. Morales (1959), in *The Enzymes*, Vol. I, 2nd ed., P. D. Boyer, H. Lardy, and K. Myrbäck, eds., Academic, New York, pp. 49–142.

F. G. Heineken, H. M. Tsuchiya, and R. Aris (1967), *Math. Biosci.* **1**, 115.

E. A. Hemmingsen (1965), *Acta Physiol. Scand.* **64**, Suppl. 246, 1.

D. Herbert, R. Elsworth, and R. C. Telling (1956), *J. Gen. Microbiol.* **14**, 601.

A. V. Hill (1910), *J. Physiol.* **40**, iv.

A. V. Hill (1928), *Proc. Roy. Soc. London Ser. B* **104**, 39.

A. V. Hill (1938), *Proc. Roy. Soc. London Ser. B* **126**, 136.

M. B. Histand and M. Anliker (1973), *Circ. Res.* **32**, 524.

D. I. Hitchcock (1947), in *Physical Chemistry of Cells and Tissues*, R. Höber, ed., Churchill, London, pp. 1–91.

A. L. Hodgkin and A. F. Huxley (1952), *J. Physiol.* **117**, 500.

A. L. Hodgkin and R. D. Keynes (1953), *J. Physiol.* **119**, 513.

M. E. J. Holwill (1966), *Physiol. Rev.* **46**, 696.

B. L. Horecker, J. Thomas, and J. Monod (1960), *J. Biol. Chem.* **235**, 1580.

A. F. Huxley (1960), in *Mineral Metabolism*, Vol. I, C. L. Comar and F. Bronner, eds., Academic, New York, pp. 163–165.

J. S. Huxley (1932), *Problems of Relative Growth*, Dial, London.

F. Jacob and J. Monod (1961), *J. Mol. Biol.* **3**, 318.

G. R. Jacobson and G. R. Stark (1973), in *The Enzymes*, Vol. 9, 3rd ed., P. D. Boyer, ed., Academic, New York, pp. 225–308.

J. A. Johnson and T. A. Wilson (1966), *Amer. J. Physiol.* **210**, 1299.

P. C. Johnson and H. Wayland (1967), *Amer. J. Physiol.* **212**, 1405.

S. G. Jokipii and O. Turpeinen (1954), *J. Clin. Invest.* **34**, 452.

G. Karreman (1952), *Bull. Math. Biophys.* **14**, 327.

S. S. Kety and C. F. Schmidt (1948), *J. Clin. Invest.* **27**, 476.

E. L. King (1956), *J. Phys. Chem.* **60**, 1378.

E. L. King and C. Altman (1956), *J. Phys. Chem.* **60**, 1375.

J. M. Kinsman, J. W. Moore, and W. F. Hamilton, (1929), *Amer. J. Physiol.* **89**, 322.

S. M. Klainer and G. Kegeles (1955), *J. Phys. Chem.* **59**, 952.

D. J. Korteweg (1878), *Ann. Phys. Chem. Neue Folge* **5**, 525.

D. E. Koshland, Jr., G. Nemethy, and D. Filmer (1966), *Biochemistry* **5**, 365.

A. Krogh (1919), *J. Physiol.* **52**, 457.

A. K. Laird (1965), *Brit. J. Cancer* **19**, 278.

A. K. Laird, S. A. Tyler, and A. D. Barton (1965), *Growth* **29**, 233.

O. Lamm (1929), *Ark. Mat. Astron. Fys.* **21B**, 528.

O. Lamm and A. Polson (1936), *Biochem. J.* **30**, 528.

A. Levitzki and D. E. Koshland, Jr. (1969), *Proc. Nat. Acad. Sci. USA* **62**, 1121.

H. Lineweaver and D. Burk (1934), *J. Amer. Chem. Soc.* **56**, 658.

A. J. Lotka (1925), *Elements of Physical Biology*, Williams and Wilkins, Baltimore, pp. 88–94 (reprinted as *Elements of Mathematical Biology*, Dover, New York, 1956).

S. E. Luria and M. Delbruck (1943), *Genetics* **28**, 491.

K. E. Machin (1958), *J. Exp. Biol.* **35**, 796.

D. A. MacLulich (1937), *Fluctuations in the numbers of the varying hare (Lepus americanus).* Univ. Toronto studies, Biol. Ser., No. 43.

R. Margaria (1963), *Clin. Chem.* **9**, 745.

P. Martin and D. Yudilevich (1964), *Amer. J. Physiol.* **207**, 162.

D. A. McDonald (1968), *J. Appl. Physiol.* **24**, 73.

P. Meier and K. L. Zierler (1954), *J. Appl. Physiol.* **6**, 731.

M. Meselson and F. W. Stahl (1958), *Proc. Nat. Acad. Sci. USA* **44**, 671.

M. Meselson, F. W. Stahl, and J. Vinograd (1957), *Proc. Nat. Acad. Sci. USA* **43**, 581.

L. Michaelis and M. L. Menten (1913), *Biochem. Z.* **49**, 333.

H. K. Miller and M. E. Balis (1969), *Biochem. Pharmacol.* **18**, 2225.

R. G. Miller and R. A. Phillips (1969), *J. Cell Physiol.* **73**, 191.

J. Monod (1949), *Ann. Rev. Microbiol.* **3**, 371.

J. Monod (1950), *Ann. Inst. Pasteur* **79**, 390.

J. Monod, J. P. Changeux, and F. Jacob (1963), *J. Mol. Biol.* **6**, 306.

J. Monod, J. Wyman, and J. P. Changeux (1965), *J. Mol. Biol.* **12**, 88.

J. W. Moore, J. M. Kinsman, W. F. Hamilton, and R. G. Spurling (1929), *Amer. J. Physiol.* **89**, 331.

M. Munro and F. H. C. Crick (1971), *Symp. Soc. Exp. Biol.* **25**, 439.

C. D. Murray (1926), *J. Gen. Physiol.* **9**, 35.

J. Needham (1931), *Chemical Embryology,* Cambridge Univ. Press, Cambridge.

E. V. Newman, M. Merrell, A. Genecin, C. Monge, W. R. Milnor, and W. P. McKeever (1951), *Circulation* **4**, 735.

K. H. Norwich and S. Zelin (1970), *Bull. Math. Biophys.* **32**, 25.

A. Novick (1955), *Ann. Rev. Microbiol.* **9**, 97.

A. Novick and L. Szilard (1950), *Science* **112**, 715; *Proc. Nat. Acad. Sci. USA* **36**, 708.

W. J. V. Osterhout (1940), *Cold Spring Harbor Symp. Quant. Biol.* **8**, 51.

E. Overton (1899), *Vierteljahreschr. Naturforsch. Ges. Zurich* **44**, 88.

L. Pauling (1935), *Proc. Nat. Acad. Sci. USA* **21**, 186.

P. R. Payne and E. F. Wheeler (1967), *Nature (London)* **215**, 849.

W. Perl (1960), *Int. J. Appl. Rad. Isot.* **8**, 211.

J. Perrin (1910), *Brownian Motion and Molecular Reality,* Taylor and Francis, London.

M. F. Perutz, M. G. Rossman, A. F. Cullis, H. Muirhead, G. Will, and A. C. T. North (1960), *Nature (London)* **185**, 416.

E. M. Renkin (1959), *Amer. J. Physiol.* **197**, 1205.

E. M. Renkin (1967), in *Coronary Circulation and Energetics of the Myocardium,* G. Marchetti and B. Taccardi, eds., S. Karger AG, Basel, pp. 18–30.

E. M. Renkin, O. Hudlická, and R. M. Sheehan (1966), *Amer. J. Physiol.* **211**, 87.

E. M. Renkin and S. Rosell (1962), *Acta Physiol. Scand.* **54**, 223.

H. Résal (1876), *C. R. Acad. Sci.* **82**, 698.

M. R. Roach and A. C. Burton (1957), *Can. J. Biochem. Physiol.* **35**, 681.

D. Rodbard (1968), *J. Clin. Endocrin. Metabol.* **28**, 849.

E. Rogus and K. L. Zierler (1973), *J. Physiol.* **233**, 227.

B. H. Rosenberg, L. F. Cavalieri, and G. Ungers (1969), *Proc. Nat. Acad. Sci. USA* **63**, 1410.

J. P. Rosenbusch and K. Weber (1971), *J. Biol. Chem.* **246**, 1644.

L. Rothschild and K. W. Clelland (1952), *J. Exp. Biol.* **29**, 66.

F. J. W. Roughton, E. C. DeLand, J. C. Kernohan, and J. W. Severinghaus (1972), in *Oxygen Affinity of Hemoglobin and Red Cell Acid Base States*, M. Rorth and P. Astrup, eds., Academic, New York, pp. 73–83.

F. J. W. Roughton and R. L. J. Lyster (1965), *Hvalradets Skrifter Norske Vadingskaps–Akad. Oslo* **48**, 185.

S. I. Rubinow (1973), *Math. Biosci.* **18**, 245.

S. I. Rubinow and J. B. Keller (1971), *J. Acoust. Soc. Amer.* **50**, 198.

S. I. Rubinow and J. B. Keller (1972), *J. Theoret. Biol.* **35**, 299.

S. I. Rubinow and J. L. Lebowitz (1970), *J. Amer. Chem. Soc.* **92**, 3888.

S. I. Rubinow and A. Winzer (1971), *Math. Biosci.* **11**, 203.

S. I. Rubinow and A. Yen (1972), *Nature New Biol.* **239**, 73.

L. A. Sapirstein, D. G. Vidt, M. J. Mandel, and G. Hanusek (1955), *Amer. J. Physiol.* **181**, 330.

H. K. Schachman (1959), *Ultracentrifugation in Biochemistry*, Academic, New York.

A. Shapiro (1946), *Cold Spring Harbor Symp. Quant. Biol.* **11**, 228.

D. B. Smith, G. C. Wood, and P. A. Charlwood (1956), *Can. J. Chem.* **34**, 364.

R. E. Smith and M. F. Morales (1944), *Bull. Math. Biophys.* **6**, 133.

J. Stefan (1891), *Ann. Phys. Chem.* **42**, 269.

W. D. Stein (1962), in *Comprehensive Biochemistry*, Vol. II, M. Florkin and E. H. Stotz, eds., Elsevier, Amsterdam, Chapter 3.

W. D. Stein (1967), *The Movement of Molecules across Cell Membranes*, Academic, New York.

G. M. Stewart (1921), *Amer. J. Physiol.* **58**, 20.

G. G. Stokes (1851), *Cambridge Phil. Trans.* **9**, 8.

T. Svedberg and K. O. Pedersen (1940), *The Ultracentrifuge*, Oxford Univ. Press, London and New York.

S. L. Tamm (1967), *J. Exp. Zool.* **164**, 163.

C. Tanford (1961), *Physical Chemistry of Macromolecules*, Wiley, New York.

G. Taylor (1951), *Proc. Roy. Soc. London Ser. A* **209**, 447.

G. Taylor (1952), *Proc. Roy. Soc. London Ser. A* **211**, 225.

G. Taylor (1953), *Proc. Roy. Soc. London Ser. A* **219**, 186.

D'A. Thompson (1942), *On Growth and Form*, Vol. I, 2nd ed., Cambridge Univ. Press, Cambridge, pp. 205–212.

S. Thovert (1910), *Compt. Rend.* **150**, 270.

A. M. Turing (1952), *Philos. Trans. Roy. Soc. London* **237**, 37.

R. J. Urick (1947), *J. Appl. Phys.* **18**, 983.

K. E. Van Holde and R. L. Baldwin (1958), *J. Phys. Chem.* **62**, 734.

H. D. Van Liew (1967), *J. Theoret. Biol.* **16**, 43.

M. V. Volkenshtein (1969), *Enzyme Physics*, Plenum, New York.

M. V. Volkenstein and B. N. Goldstein (1966a), *Biochim. Biophys. Acta* **115**, 471.

M. V. Volkenstein and B. N. Goldstein (1966b), *Biochim. Biophys. Acta* **115**, 478.

V. Volterra (1926), *Memoria della R. Accademia Nazionale dei Lincei*, **2**, 31 (translated by

M. E. Wells in R. N. Chapman, *Animal Ecology*, McGraw-Hill, New York, pp. 409–448, 1931).

J. D. Watson and F. H. C. Crick (1953), *Nature (London)* **171**, 964.

K. Weber (1968), *Nature (London)* **218**, 1116.

W. Weber (1866), *Ber. Sachs. Ges. Wissenschaft. Math. Phys. Cl.* **18**, 353.

R. Werner (1971), *Nature New Biol.* **233**, 99.

E. O. Wilson (1958), *Psyche* **65**, 41.

J. B. Wittenberg (1966), *J. Biol. Chem.* **241**, 104.

J. B. Wittenberg (1970), *Physiol. Rev.* **50**, 559.

L. Wolpert (1969), *J. Theoret. Biol.* **25**, 1.

J. Wyman (1966), *J. Biol. Chem.* **241**, 115.

T. Young (1808), *Philos. Trans. Roy. Soc. London* **98**, 164.

D. B. Zilversmit, C. Entenman, and M. C. Fishler (1943), *J. Gen. Physiol.* **26**, 325.

APPENDIX A
BRIEF REVIEW

1. VARIABLES, FUNCTIONS

The idea of a variable is based on the idea of a class of numbers.

EXAMPLES OF CLASSES

1. The positive integers.
2. The real numbers.
3. $-1, 0, +1$.
4. The numbers >2 and <3.

A **class** of numbers is just the totality of numbers having some specified properties. The class may be finite or infinite. A **variable** is a symbol that can take for values any of the members of a pre-assigned class.

Let x be a variable which can assume any value in the number scale between the numbers a and b,

$$a \leqq x \leqq b.$$

Then x is called a **continuous variable** in the interval from a to b. If now, to each value of x, there corresponds a single value of the variable y, where x and y are related by any law whatsoever, we say that **y is a function of x**. This correspondence may be written in a variety of ways, namely,

$$y = y(x), \ y = f(x), \ y = g(x),$$

or some similar expression. In other words, $y(x)$ is a prescription for finding y, given x. We call x the **independent variable**, and y the **dependent variable**, or we call x the **argument** of the function y.

Functions are represented graphically by making use of a rectangular

348

coordinate system and marking for each value of the abscissa x, the corresponding ordinate $y = f(x)$ (see Figure 1). In the example illustrated by Figure 1, y is a **single-valued** function of x. The graph may also be viewed as presenting x as a function of y, which is called the **inverse function**. It is represented as $x = f^{-1}(y)$, or $f^{-1}(y) = f^{-1}(f(x)) = x$. In the graphical example, x is not a single valued function of y, but is rather **multiple-valued**. Thus, corresponding to the value $y = y_1$, x can assume any of three values, x_1, x_2, or x_3.

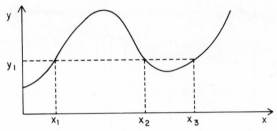

Figure 1. Graphical representation of the function $y = f(x)$. $x = f^{-1}(y)$ is multiple-valued.

EXAMPLES

$$y = x^2 \text{ is a single-valued function for all values of } x.$$

The inverse function $x = \sqrt{y}$ is single-valued provided x and $y > 0$.

The curve in Figure 1 exhibits a property which is of great importance in the study of functions, namely, the property of **continuity**. The intuitive concept of continuity is that a small change in x causes only a small change in y, and not a sudden jump in value.

2. THE CALCULUS

Historically, the invention of the calculus stems from two central problems of the 17th century:

1. The **problem of quadrature**: to determine the area within a curve. This is the fundamental problem of the integral calculus.

2. The **problem of tangents**: to determine the tangent line to a given curve. This is the fundamental problem of the differential calculus. This problem originated from the desire to determine the minimum and maximum of functions.

The great insight of Newton and Leibniz is to have clearly recognized the intimate connection between these two problems.

3. INTEGRATION

Finding the area within a polygon was no problem. But what about curves?
Archimedes could already find the area within special curves such as circles.
The merit of the integral calculus is its generality in being applicable to all
continuous curves.

To find the area under the curve y between $x = a$ and $x = b$, proceed as
follows (see Figure 2). Subdivide the part of the x axis between a and b into

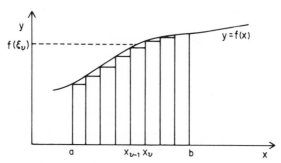

Figure 2.   Area as limit of a sum.

n strips, with boundaries $x_0 = a, x_1, x_2, \ldots, x_{n-1}, x_n = b$. Let ξ_1 be a point
in the first interval, ξ_2 a point in the second interval, and so on. Form the
sum

$$A_n = (x_1 - x_0)f(\xi_1) + (x_2 - x_1)f(\xi_2) + \cdots + (x_n - x_{n-1})f(\xi_n)$$
$$= \sum_{v=1}^{n} f(\xi_v)\Delta x_v,$$

where

$$\Delta x_v = x_v - x_{v-1}.$$

Let the number of points n increase without limit and let the length of the
longest subinterval tend to zero. Then the area A is the limit of the sum,

$$A = \lim_{n \to \infty} \sum_{v=1}^{n} f(\xi_v)\Delta x_v,$$

where the symbol in front of the summation sign means "the limit as n approaches infinity of." The limiting value A is called the **definite integral** of
the function $f(x)$, the **integrand**, between the limits a and b, and is written as

$$A = \int_a^b f(x)dx.$$

Some fundamental rules that follow from the above definition are given here.

$$\int_a^b f(x)dx = -\int_b^a f(x)dx.$$

$$\int_a^b f(x)dx + \int_b^c f(x)dx = \int_a^c f(x)dx.$$

$$\int_a^b cf(x)dx = c \int_a^b f(x)dx.$$

$$\int_a^b f(x)dx = \int_a^b \Phi(x)dx + \int_a^b \chi(x)dx \text{ if } f(x) = \Phi(x) + \chi(x).$$

NOTE. x is a "dummy" variable. We could equally well write for the area A $\int_a^b f(u)du$ or $\int_a^b f(t)dt$.

4. DIFFERENTIATION

Consider the tangent problem mentioned previously. Let a smooth curve be represented by $y = f(x)$ (see Figure 3). We defined the tangent to the curve

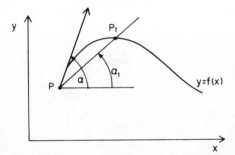

Figure 3. Tangent line at point P as limit of secant line.

at the point P by means of the following geometrical limiting process. In addition to P, we consider a second point P_1 on the curve. Draw the secant line going through P and P_1. Now let the point P_1 approach P along the curve $f(x)$. Then the tangent line is defined to be the line through P with slope $\tan \alpha$, where

$$\tan \alpha = \lim_{P_1 \to P} \tan \alpha_1 = \lim_{x_1 \to x} \frac{f(x_1) - f(x)}{x_1 - x} = \lim_{\Delta x \to 0} \frac{\Delta y}{\Delta x}.$$

The above limit is also written as $dy/dx = df(x)/dx = f'(x)$. $f(x)$ gives the height of the curve $y = f(x)$ for the value x. We can consider the slope as a new function of x denoted by $f'(x)$. The limiting process by which it is obtained from $f(x)$ is called **differentiation**, and $f'(x)$ is called the **derivative** of the function $f(x)$ at the point x. Because the slope is the gradient of the curve $f(x)$, the derivative $f'(x)$ is sometimes called the **gradient**. The intuitive idea of a function being **differentiable** is that it is continuous and smooth, that is, it has no sharp corners. Also, vertical slopes are excluded. More precisely, $\lim [f(x + h) - f(x)]/h$ must exist independently of the manner in which $h \to 0$.

The direction of the tangent to the curve $y = f(x)$ at a point P with coordinates x, y is that direction which makes an angle α with the positive x axis, determined by the equation $\tan \alpha = f'(x)$. Hence, as the curve $f(x)$ is traversed in the direction of increasing x,

$f'(x) > 0$ implies the curve is ascending.

$f'(x) < 0$ implies the curve is descending.

$f'(x) = 0$ means the curve is horizontal. A point x at which this occurs is called a **stationary point**. Such a point is either a local **maximum** or a local **minimum** of the curve, or a **point of inflection** (see Figure 4).

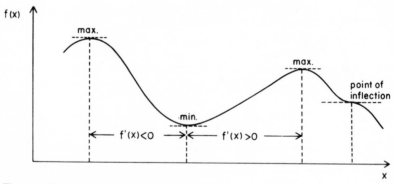

Figure 4. Stationary points of a curve.

It is to be emphasized that $f'(x) = 0$ is only a **necessary** condition for the existence of a minimum or maximum. For numerical computation as with a computer, finite sums are utilized to represent integrals, that is, the integral is approximated by the sum A_n with n finite. Similarly, a derivative is determined numerically by the expression $[f(x + h) - f(x)]/h$ with h finite (albeit small).

In a similar vein we may differentiate the function $f'(x)$ and obtain the **second derivative** of the function $f(x)$, or $f''(x)$. The geometrical meaning of it is as follows (see Figure 5).

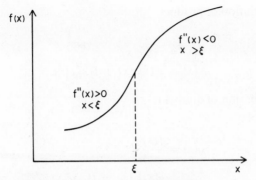

Figure 5.   A point of inflection ξ that is not a stationary point.

$f''(x) > 0$ means the value of the slope is constantly increasing and the curve is becoming steeper. We say it is **concave upward**.

$f''(x) < 0$ means the curve is **concave downward**.

$f''(x) = 0$ is a special case. If $f''(x)$ changes sign on going through such a point, it is called a **point of inflection** of the curve. In Figure 5, ξ is a point of inflection.

If ξ is a stationary point of the curve $f(x)$, that is,

$$f'(\xi) = 0,$$

then we can conclude, from the above considerations, that ξ is a maximum if $f''(\xi) < 0$, a minimum if $f''(\xi) > 0$, or a point of inflection if $f''(\xi) = 0$.

EXAMPLES

1. $y = f(x) = x^2$ is concave upward everywhere because $f''(x) = 2$ for all x.

2. $y = f(x) = x^3$.

$$y''(x) = 6x \begin{cases} + \text{ if } x > 0, \\ 0 \text{ if } x = 0, \\ - \text{ if } x < 0. \end{cases}$$

$x = 0$ is a point of inflection of the function $y = x^3$.

5.   SOME RULES FOR DIFFERENTIATION

1.   Differentiation of sum

If

$$k(x) = f(x) + g(x),$$
$$k'(x) = f'(x) + g'(x).$$

2. Differentiation of product

If

$$p(x) = f(x)g(x),$$
$$p'(x) = f'(x)g(x) + f(x)g'(x).$$

3. Differentiation of quotient

If

$$q(x) = \frac{f(x)}{g(x)},$$
$$q'(x) = \frac{g(x)f'(x) - f(x)g'(x)}{[g(x)]^2}.$$

When $f(x) = 1$, $q'(x) = -\dfrac{g'(x)}{g^2}$.

4. Compound functions

If

$$z = g(y),$$

and

$$y = y(x),$$

then

$$z = g[y(x)],$$

and

$$\frac{dz}{dx} = \frac{dz}{dy}\frac{dy}{dx}.$$

This relation is called the **chain rule**.

5. Inverse functions

If $y = f(x)$, and $x = g(y)$ is the inverse function,

then

$$1 = \frac{df}{dx}\frac{dx}{dy}$$

by the chain rule. But

$$\frac{df}{dx} = \frac{dy}{dx}.$$

Therefore,

$$1 = \frac{dy}{dx}\frac{dx}{dy},$$

or

$$\frac{dy}{dx} = \frac{1}{dx/dy}.$$

This result is intuitively obvious from the geometrical interpretation. The importance of this relation is that it is just *as if* the symbols dy and dx are quantities which can be treated like ordinary numbers.

6. FUNDAMENTAL THEOREM OF THE CALCULUS

The operations of differentiation and integration are intimately related. They are **inverses** of each other, like addition and subtraction.

Consider

$$F(x) = \int_a^x f(u)du,$$

where x is a variable upper limit. Graphically, $F(x)$ represents the diagonally shaded area of Figure 6. The fundamental theorem of the differential and

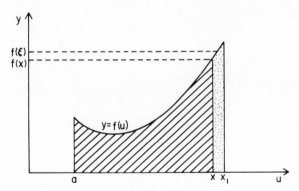

Figure 6.   Illustrates the fundamental theorem of the calculus.

integral calculus states that

$$F'(x) = f(x).$$

In words, the process of integration leading from $f(x)$ to $F(x)$ is undone or inverted by differentiation of $F(x)$. Clearly,

$$\frac{dF(x)}{dx} = \lim \frac{F(x_1) - F(x)}{x_1 - x} = \frac{\text{area of added rectangle}}{x_1 - x},$$

$$= \frac{(x_1 - x)f(\xi)}{x_1 - x}, \text{ where } \xi \text{ is an interior point between } x \text{ and } x_1,$$

$$= f(x) \text{ in the limit as } x_1 \to x.$$

An important subsidiary rule of differentiation is that

$$\frac{d}{dx} \int_{a(x)}^{b(x)} f(u)du = \frac{db}{dx} f(b) - \frac{da}{dx} f(a).$$

7.   EXPONENTIAL AND LOGARITHM FUNCTIONS

The exponential and logarithm functions are ubiquitous in biology and therefore it pays to make a special study of them and their properties. *Definition.*

$$y = \int_1^x \frac{du}{u}, \text{ for } x > 0,$$

is called the **logarithm** of x or **natural logarithm** of x. We write it as $y = \log x$. The variable of integration is denoted by u to avoid confusion with the upper limit. Geometrically, it represents the shaded area shown in Figure 7. The

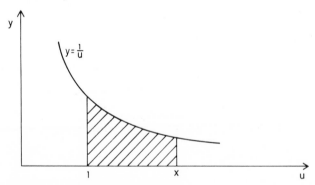

Figure 7.   Definition of $\log x$ as an area.

logarithm is negative if $x < 1$, and vanishes if $x = 1$. Therefore $\log 1 = 0$. From the definition and the fundamental theorem,

$$\frac{d}{dx} (\log x) = \frac{1}{x}.$$

From this we see that log x has a slope that is always positive, and therefore log x is a function, which is always increasing as x increases (a **monotonic increasing** function), and in fact log $x \to \infty$ as $x \to \infty$. Other important properties which follow from the definition of the logarithm:

$$\log ab = \log a + \log b.$$
$$\log a^r = r \log a,$$

where r is any number. From the latter we see that

$$\log\left(\frac{1}{2^n}\right) = -n \log 2.$$

Therefore, as $x \to 0$ through positive values, log x is negative and $\to -\infty$. Hence, for $0 < x < \infty$, log x goes from $-\infty$ to $+\infty$. From the graph of $y = \log x$ (see Figure 8), we see that there must be a number such that

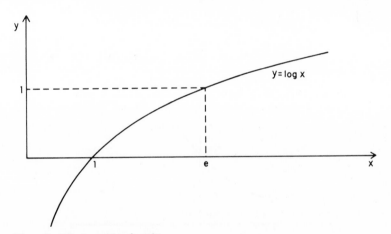

Figure 8. The logarithm function.

log $x = 1$. This number is called e. The inverse function we shall at first denote by

$$x = E(y).$$

PROPERTIES

$$E(0) = 1.$$
$$E(a + b) = E(a)E(b).$$
$$E(1) = e.$$

The last property follows from log $e = 1$ (See Figure 8).

$$E(r) = e^r,$$

where r is any number. In other words,

$$E(y) = e^y.$$

This is the common form of writing the exponential function. Since E and log are inverse functions,

$$E(\log x) = x$$

or

$$x = e^{\log x}.$$

Similarly,

$$\log(E(x)) = x$$

or

$$\log e^x = x.$$

Graphically, the exponential function is shown in Figure 9.

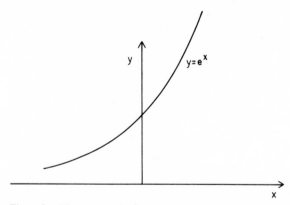

Figure 9.   The exponential function.

The exponential function for an arbitrary base a is defined as follows. Let a be a positive number, and define α as

$$\alpha = \log a.$$

Then

$$a \equiv e^{\log a} = e^\alpha.$$

Now define $y = a^x$ as

$$y = a^x = (e^\alpha)^x = e^{\alpha x} = e^{x \log a}$$

The last expression defines the exponential function for an arbitrary base in terms of the exponential function for the natural base e.

From this it follows that the formula for differentiating $y = a^x$ is

$$\frac{dy}{dx} = \log a e^{x \log a}$$

or

$$\frac{dy}{dx} = \log a \cdot a^x.$$

Before we saw that if $y = e^x$, then $x = \log y$ is the inverse function. Similarly, if

$$y = a^x,$$

the inverse function is

$$x = \log_a y,$$

and is called the **logarithm to the base a of y**. What is the relation between $\log_a$ and natural log or $\log_e$? As a particular example, let $y = 10^x$, so that, by forming the inverse relation,

$$\log_{10} y = x.$$

Also, by taking the natural logarithm of $y = 10^x$,

$$\log y = x \log_e 10$$

or

$$\log y = \log 10 \cdot \log_{10} y.$$

The last equation gives the relation between natural logarithms and the logarithm function to the base 10, which is called the **common logarithm**. Because common logarithms are sometimes denoted by log, the notation ln to denote the natural logarithm is frequently seen. We note that

$$\log 10 = 2.3026 \ldots.$$

Other important properties of e that are demonstrable are as follows:

$$e = \lim_{n \to \infty} \left(1 + \frac{1}{n}\right)^n,$$

$$e = \lim_{h \to 0} (1 + h)^{1/h},$$

$$e^x = \lim_{n \to \infty} \left(1 + \frac{x}{n}\right)^n.$$

8. THE CONCEPT OR ORDER

It is often useful to make comparisons between functions which become infinite as their argument increases without bound (for example, x^a, log x, e^x). Given two such functions, we may wish to know which grows "faster," so to speak. Mathematicians say, what is the **order** of their increase? Quite generally, we say that $f(x)$ **becomes infinite of a higher order than** $g(x)$ if

$$\frac{f(x)}{g(x)} \to \infty \text{ as } x \to \infty.$$

If instead, $f(x)/g(x) \to 0$ as $x \to \infty$, we say that $f(x)$ **is of a lower order than** $g(x)$. If $f(x)/g(x) \to$ constant as $x \to \infty$, then we say that $f(x)$ and $g(x)$ **have precisely the same order**.

EXAMPLE

Compared to $3x^2 + 6x + 5$, the functions $x^3 + 1, 25x^{3/2} + 10x + 12,$ and x^2 have, respectively, a higher, a lower, and the same order as $x \to \infty$.

Similar considerations apply to **infinitesimals**, which are variables whose limit is zero. If $f(x) \to 0$ and $g(x) \to 0$ as $x \to a$ (where a is finite or infinite), and $f/g \to 0$, f is said to be an **infinitesimal of higher order than** g. If $f/g \to \infty$, f is said to be an **infinitesimal of lower order than** g. If $f/g \to a$ finite limit, then f and g are said to be of **precisely the same order** as $x \to a$.

EXAMPLE

As $x \to 0$, x^3 is of higher, lower, and the same order as $(x^2 + x)$, x^4, and $(4x^3 + x^4)$, respectively.

If $f - f_1$ is an infinite of lower order than f, and $g - g_1$ is an infinite of lower order than g, then lim $f/g = f_1/g_1$. This is called the **replacement theorem** and permits us to ignore infinities of lower order. The replacement theorem also applies to infinitesimals: if $f - f_1$ is an infinitesimal of higher order than f, and $g - g_1$ is an infinitesimal of higher order than g, then lim $f/g = $ lim f_1/g_1.

EXAMPLES

$$\lim_{x \to \infty} \frac{ax^3 + bx^2 + cx + d}{a_1x^3 + b_1x^2 + c_1x + d_1} = \lim_{x \to \infty} \frac{ax^3}{a_1x^3} = \frac{a}{a_1},$$

provided $a_1 \neq 0$.

$$\lim_{x \to 0} \frac{ax^3 + bx^2 + cx + d}{a_1 x^3 + b_1 x^2 + c_1 x + d_1} = \lim_{x \to 0} \frac{d}{d_1} = \frac{d}{d_1},$$

provided $d_1 \neq 0$.

If $f/g \to +1$ as $x \to a$, we write $f \sim g$, read "f is **asymptotically equivalent to g**.

EXAMPLE

If

$$f(x) = x^2 + 2x^3,$$
$$f \sim x^2 \text{ as } x \to 0,$$
$$f \sim 2x^3 \text{ as } x \to \infty.$$

Can we use the exponent as a yardstick to measure the orders of all functions? The answer to this question is no. There are functions that become infinite of higher order than x^a no matter how large a is chosen. Such a function is the exponential function, for example. It is readily proved that

$$\frac{e^x}{x^a} \to \infty \text{ as } x \to \infty,$$

regardless of the value of a.

In addition, there are functions that become infinite of lower order than x^a, no matter how small the positive number a is chosen. For example, consider

$$\frac{\log x}{x^a}, a > 0.$$

Let $x = e^y$. Then

$$\frac{\log x}{x^a} = \frac{y}{e^{ay}} \to 0 \text{ as } y \to \infty.$$

Therefore, $\log x \to \infty$ grows more slowly than any positive power of x, no matter how small the exponent. Log x is of lower order than x^a for any a. In words, e^x grows faster and $\log x$ grows slower than any positive power of x.

9. TAYLOR SERIES

Suppose we have a polynomial

$$f(x) = c_0 + c_1 x + c_2 x^2 + c_3 x^3.$$

Can we express the coefficients c_i, $i = 0, 1, 2, 3$, in terms of $f(x)$ and its derivatives, evaluated at a particular point? The answer is yes. Consider the neighborhood of $x = 0$.

$$f(0) = c_0,$$
$$f'(x) = c_1 + 2c_2 x + 3c_3 x^2,$$
$$f'(0) = c_1.$$

Similarly,

$$f''(0) = 2!c_2,$$
$$f'''(0) = 3!c_3.$$

Consequently,

$$f(x) = f(0) + xf'(0) + \frac{x^2}{2!} f''(0) + \frac{x^3}{3!} f'''(0).$$

More generally, given any function which is infinitely "smooth," that is, it and its derivatives exist in the neighborhood of some point, then we can represent the function in the neighborhood of that point as a power series,

$$f(x) = \sum_{n=0}^{\infty} \frac{x^n f^{(n)}(0)}{n!},$$

where $f^{(n)}(0)$ denotes the nth derivative of $f(x)$ evaluated at $x = 0$. We say that we have **expanded the function in an infinite Taylor series**. Thus, for the generic point x_0, we can write the infinite Taylor series for the function $f(x)$ as

$$f(x) = \sum_{n=0}^{\infty} \frac{(x - x_0)^n f^{(n)}(x_0)}{n!}.$$

EXAMPLES OF INFINITE SERIES

1. $e^x = 1 + \dfrac{x}{1!} + \dfrac{x^2}{2!} + \dfrac{x^3}{3!} + \cdots = \displaystyle\sum_{n=0}^{\infty} \dfrac{x^n}{n!},$

where, by definition, $0! \equiv 1$. This infinite series is valid for all values of x. Incidentally, we can calculate the value of e from this formula by setting $x = 1$ in it. It is found that

$$e = 2.71828 \ldots .$$

2. $\log(1 + x) = x - \dfrac{x^2}{2} + \dfrac{x^3}{3} - \dfrac{x^4}{4} + \cdots = \displaystyle\sum_{n=1}^{\infty} (-)^{n-1} \dfrac{x^n}{n}.$

This infinite series is valid for $-1 < x \leqq 1$.

3. $(1 + x)^k = 1 + kx + \dfrac{k(k-1)}{2!} x^2 + \dfrac{k(k-1)(k-2)}{3!} x^3 + \cdots$

$$= \sum_{n=0}^{\infty} \frac{k!}{n!(k-n)!} x^n.$$

This infinite series is valid for $x^2 < 1$.

Suppose we break off the infinite Taylor series for the function $f(x)$ after $n + 1$ terms, and denoted the remaining terms as the error or **remainder** R_n. The resulting expression for $f(x)$, namely

$$f(x) = f(x_0) + (x - x_0)f'(x_0) + \frac{(x - x_0)^2}{2!} f''(x_0) + \cdots$$

$$+ \frac{(x - x_0)^n}{n!} f^{(n)}(x_0) + R_n,$$

is called the **Taylor series with remainder**. Its importance lies in the fact that, by discarding the remainder R_n, we obtain an approximate representation of the function $f(x)$. The term $f(x_0)$ in the above series is said to be of zero order in $(x - x_0)$, the second term $(x - x_0)f'(x_0)$ is said to be of first order, the next term is the term of second order, and so forth. Thus, the remainder term is of order $(x - x_0)^{n+1}$, and the remainder tends to zero with this order as $x \to x_0$. We say that we have **expanded the function $f(x)$ as far as the term of nth order**.

Of particular importance is the special case of the Taylor series for which the function $f(x)$ is expanded as far as the term of first order, and the remainder is neglected. This replacement is called the **linear approximation** to $f(x)$. The error term is of order $(x - x_0)^2$, which gets smaller, the closer the point x is to x_0. The approximation is shown graphically in Figure 10.

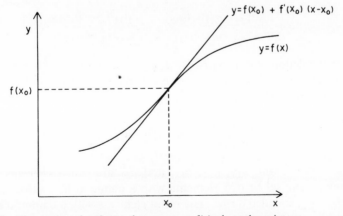

Figure 10. Linear approximation to the curve $y = f(x)$, about the point x_0.

APPENDIX B
DETERMINANTS, VECTORS, AND MATRICES

1. DETERMINANTS OF ORDER 2

Determinants have their origin in the solution of systems of linear equations with two or more unknowns. Consider the two equations

$$a_{11}x_1 + a_{12}x_2 = b_1,$$
$$a_{21}x_1 + a_{22}x_2 = b_2, \tag{1}$$

with unknowns x_1 and x_2. To solve this system, we can seek to eliminate either x_1 or x_2. To eliminate x_1, multiply the first equation by a_{21}, the second equation by $-a_{11}$, and add the two equations. A similar procedure permits the elimination of x_2. The result is

$$x_1 = \frac{b_1 a_{22} - a_{12} b_2}{a_{11} a_{22} - a_{12} a_{21}}, \, x_2 = \frac{a_{11} b_2 - b_1 a_{12}}{a_{11} a_{22} - a_{12} a_{21}}, \tag{2}$$

provided the denominator $a_{11}a_{22} - a_{12}a_{21}$, common to both expressions, is not equal to zero. This denominator is called the **determinant** of the system (1) and is denoted by

$$\begin{vmatrix} a_{11} a_{12} \\ a_{21} a_{22} \end{vmatrix} = a_{11} a_{22} - a_{12} a_{21}. \tag{3}$$

In this determinant, a_{11}, a_{22}, a_{12}, and a_{21} are called **elements** of the determinant. The set of two elements with a common first subscript, a_{11}, a_{12} or a_{21}, a_{22}, is called a **row**. The set of elements with a common second subscript, a_{11}, a_{21} or a_{12}, a_{22}, is called a **column**. With the definition (3), it is

364

readily seen that the solution (2) can be written as quotients of determinants,

$$x_1 = \frac{\begin{vmatrix} b_1 a_{12} \\ b_2 a_{22} \end{vmatrix}}{\begin{vmatrix} a_{11} a_{12} \\ a_{21} a_{22} \end{vmatrix}}, \quad x_2 = \frac{\begin{vmatrix} a_{11} b_1 \\ a_{21} b_2 \end{vmatrix}}{\begin{vmatrix} a_{11} a_{12} \\ a_{21} a_{22} \end{vmatrix}}. \tag{4}$$

2. DETERMINANTS OF ORDER 3

Consider the system of three equations in three unknowns,

$$\begin{aligned} a_{11}x_1 + a_{12}x_2 + a_{13}x_3 &= b_1, \\ a_{21}x_1 + a_{22}x_2 + a_{23}x_3 &= b_2, \\ a_{31}x_1 + a_{32}x_2 + a_{33}x_3 &= b_3. \end{aligned} \tag{5}$$

By elimination of x_3 from these equations, they can be reduced to a system of two equations with two unknowns x_1 and x_2, and solved by means of the previous formulas. Thus, it is found that

$$x_1 = \frac{b_1(a_{22}a_{33} - a_{23}a_{32}) + b_2(a_{13}a_{32} - a_{12}a_{33}) + b_3(a_{12}a_{23} - a_{22}a_{13})}{a_{11}a_{22}a_{33} + a_{12}a_{23}a_{31} + a_{13}a_{21}a_{32} - a_{11}a_{23}a_{32} - a_{12}a_{21}a_{33} - a_{13}a_{22}a_{31}}. \tag{6}$$

The denominator in the above expression is defined as the determinant of order 3 and written as

$$\Delta \equiv \begin{vmatrix} a_{11} a_{12} a_{13} \\ a_{21} a_{22} a_{23} \\ a_{31} a_{32} a_{33} \end{vmatrix} = a_{11}a_{22}a_{33} + a_{12}a_{23}a_{31} + a_{13}a_{21}a_{32} \\ - a_{11}a_{23}a_{32} - a_{12}a_{21}a_{33} - a_{13}a_{22}a_{31}. \tag{7}$$

A useful rule to remember how to obtain the six terms on the right from the determinant in (7) is to form the array

$$\begin{matrix} a_{11} a_{12} a_{13} a_{11} a_{12} \\ a_{21} a_{22} a_{23} a_{21} a_{22} \\ a_{31} a_{32} a_{33} a_{31} a_{32} \end{matrix} \tag{8}$$

in which the last two columns repeat the first two columns in their natural order. By taking the products of elements on descending diagonal lines with a + sign, and the products of elements on ascending diagonal lines with a − sign, we arrive at the right-hand side of (7).

An important property of the determinant is noted by observing the second subscripts in the six terms of the determinant. Directly from (7) we see that the sets of three second subscripts of the six terms are as follows,

$$123, 231, 312,$$
$$132, 213, 321. \tag{9}$$

These sets are in fact the $3! = 6$ possible permutations of the integers 1, 2, 3. Furthermore, the first three permutations are associated with a + sign in (7), while the last three permutations are associated with a minus sign. Think of these permutations arising from the natural sequence 1, 2, 3 by means of **transpositions** or interchanges of a given pair of integers. It is seen that no transpositions are required to obtain the first permutation in (9), the transposition (12) (meaning 1 becomes 2 and 2 becomes 1) following by (13) produces the second permutation, and (13) followed by (12) produces the third permutation. Thus, the permutations associated with a + sign require 0 or 2 interchanges. The permutations associated with a minus sign require only one transposition: (23) for the fourth permutation, (12) for the fifth permutation, and (13) for the sixth permutation. Actually, the number of transpositions required to transform one permutation to another is not unique, but it can be shown that the **parity** is unique: The number is either always odd or always even.

The solution to equations (5) by means of third order determinants is written as

$$x_1 = \Delta^{-1} \begin{vmatrix} b_1 a_{12} a_{13} \\ b_2 a_{22} a_{23} \\ b_3 a_{32} a_{33} \end{vmatrix},$$

$$x_2 = \Delta^{-1} \begin{vmatrix} a_{11} b_1 a_{13} \\ a_{12} b_2 a_{23} \\ a_{13} b_3 a_{33} \end{vmatrix}, \tag{10}$$

$$x_3 = \Delta^{-1} \begin{vmatrix} a_{11} a_{12} b_1 \\ a_{12} a_{22} b_2 \\ a_{13} a_{32} b_3 \end{vmatrix}.$$

3.   DETERMINANTS OF ORDER n

The generalization of the previous considerations to a system of n linear inhogeneous algebraic equations follows. Assume that the system of equations to be solved is

$$a_{11}x_1 + a_{12}x_2 + a_{13}x_3 + \cdots + a_{1n}x_n = b_1,$$
$$a_{21}x_1 + a_{22}x_2 + a_{23}x_3 + \cdots + a_{2n}x_n = b_2,$$
$$\cdots \cdots \cdots \cdots \cdots \cdots \cdots \cdots \cdots \cdots \cdots \cdots$$
$$a_{n1}x_1 + a_{n2}x_2 + a_{n3}x_3 + \cdots + a_{nn}x_n = b_n. \tag{11}$$

The unique solution to this system of equations is given by **Cramer's rule**:

$$x_i = \frac{\Delta_i}{\Delta},$$ (12)

provided Δ does not vanish, where Δ is the nth order determinant

$$\Delta = \begin{vmatrix} a_{11} & a_{12} & \cdots & a_{1n} \\ a_{21} & a_{22} & \cdots & a_{2n} \\ \cdots\cdots\cdots\cdots\cdots\cdots \\ a_{n1} & a_{n2} & \cdots & a_{nn} \end{vmatrix} = \sum_{(v_1 v_2 \cdots v_n)} \pm\, a_{1v_1} a_{2v_2} \cdots a_{nv_n}.$$ (13)

Here the summation extends over all possible permutations of the second indices. There are $n!$ such permutations, so that Δ contains $n!$ terms. Those terms containing permutations which are derivable from the natural ordered sequence or **basic permutation** $1, 2, 3, \ldots, n$ by means of an even number of transpositions have a $+$ sign before them. Those terms associated with an odd number of transpositions have a $-$ sign before them.

The determinant Δ_i is derived from Δ by replacing the ith column in Δ by the column of b's appearing on the right-hand side of (11).

4. BASIC PROPERTIES OF DETERMINANTS

1. It can be shown that it is equally possible to write Δ in terms of permutations on the first index,

$$\Delta = \sum_{(v_1 v_2 \cdots v_n)} \pm\, a_{v_1 1} a_{v_2 2} \cdots a_{v_n n}.$$ (14)

Here again, a $+$ or a $-$ sign is associated with those terms derivable from the basic permutation by means of an even or odd number of transpositions, respectively.

2. The value of a determinant changes sign when two columns or two rows are interchanged.

3. The value of a determinant does not change when its rows and columns are interchanged.

4. The number of terms in a determinant with a $+$ sign is equal to the number of terms with a $-$ sign, disregarding the intrinsic signs of the elements of the determinant themselves.

5. A determinant with two identical rows or two identical columns equals zero.

6. A determinant is a linear homogeneous function of the elements of any row, or of any column. A **linear homogeneous function** φ of the variables $x_1, x_2, \ldots, x_n$ has the form

$$\varphi = \varphi(x_1, x_2, \ldots, x_n) = a_1 x_1 + a_2 x_2 + \cdots + a_n x_n,$$

where the coefficients a_i do not depend on the x_i. The following three properties follow from this fact.

If all the elements of a row or column contain a common factor, that factor may be brought out in front of the determinant sign.

EXAMPLE

$$\begin{vmatrix} ka_{11} ka_{12} \\ a_{21} a_{22} \end{vmatrix} = k \begin{vmatrix} a_{11} a_{12} \\ a_{21} a_{22} \end{vmatrix}.$$

7.   If the elements of any row or column are sums of the same number of terms, then the determinant is equal to the sum of the determinants in which the elements of the row or column, respectively, are replaced by the separate terms.

EXAMPLE

$$\begin{vmatrix} a & b + b' \\ c & d + d' \end{vmatrix} = \begin{vmatrix} a & b \\ c & d \end{vmatrix} + \begin{vmatrix} a & b' \\ c & d' \end{vmatrix}.$$

8.   If all the elements of a row or column are equal to zero, then the determinant is equal to zero.

9.   If the ith row and jth column are deleted from a determinant, then the determinant of the remaining $n - 1$ rows and $n - 1$ columns is called the **minor** of the original determinant of order n belonging to the element a_{ij}, and denoted by Δ_{ij}. The product

$$A_{ij} = (-)^{i+j} \Delta_{ij} \tag{15}$$

is called the **cofactor** of the element a_{ij}. A determinant can be expressed in terms of cofactors of the elements of any row or any column, as follows:

$$\begin{aligned} \Delta &= A_{i1}a_{i1} + A_{i2}a_{i2} + \cdots + A_{in}a_{in}, i = 1, 2, \ldots, n, \\ \Delta &= A_{1j}a_{1j} + A_{2j}a_{2j} + \cdots + A_{nj}a_{nj}, j = 1, 2, \ldots, n. \end{aligned} \tag{16}$$

Either of these expressions is referred to as "expanding the determinant by minors."

EXAMPLE

$$\begin{vmatrix} a & b & c \\ d & e & f \\ g & h & i \end{vmatrix} = a \begin{vmatrix} e & f \\ h & i \end{vmatrix} - b \begin{vmatrix} d & f \\ g & i \end{vmatrix} + c \begin{vmatrix} d & e \\ g & h \end{vmatrix}.$$

10.   If the elements of a row or column, multiplied by the same factor,

are added to a different row or column, respectively, then the value of the determinant is unchanged.

EXAMPLE

$$\begin{vmatrix} 1 & 2 \\ 3 & 4 \end{vmatrix} = \begin{vmatrix} 1 & 2 - 2 \cdot 1 \\ 3 & 4 - 2 \cdot 3 \end{vmatrix} = \begin{vmatrix} 1 & 0 \\ 3 & -2 \end{vmatrix} = -2.$$

5. THE SECULAR DETERMINANT

An important determinant that is often encountered in the theory of systems of linear differential equations is the **secular determinant**, shown for definiteness when the order of the determinant is 3:

$$\Delta(\lambda) = \begin{vmatrix} a_{11} - \lambda & a_{12} & a_{13} \\ a_{21} & a_{22} - \lambda & a_{23} \\ a_{31} & a_{32} & a_{33} - \lambda \end{vmatrix}. \tag{17}$$

We wish to expand $\Delta(\lambda)$ in powers of λ. To this end, first rewrite it as

$$\Delta(\lambda) = \begin{vmatrix} a_{11} - \lambda & a_{12} + 0 & a_{13} + 0 \\ a_{21} + 0 & a_{22} - \lambda & a_{23} + 0 \\ a_{31} + 0 & a_{32} + 0 & a_{33} - \lambda \end{vmatrix}. \tag{18}$$

Each column is now the sum of two terms. We can expand it so that it consists of a sum of determinants whose columns no longer consist of sums. Thus,

$$\Delta(\lambda) = \begin{vmatrix} -\lambda & 0 & 0 \\ 0 & -\lambda & 0 \\ 0 & 0 & -\lambda \end{vmatrix} + \begin{vmatrix} -\lambda & 0 & a_{13} \\ 0 & -\lambda & a_{23} \\ 0 & 0 & a_{33} \end{vmatrix} + \begin{vmatrix} -\lambda & a_{12} & 0 \\ 0 & a_{22} & 0 \\ 0 & a_{32} & -\lambda \end{vmatrix}$$

$$+ \begin{vmatrix} -\lambda & a_{12} & a_{13} \\ 0 & a_{22} & a_{23} \\ 0 & a_{32} & a_{33} \end{vmatrix} + \begin{vmatrix} a_{11} & 0 & 0 \\ a_{12} & -\lambda & 0 \\ a_{13} & 0 & -\lambda \end{vmatrix} + \begin{vmatrix} a_{11} & 0 & a_{13} \\ a_{21} & -\lambda & a_{23} \\ a_{31} & 0 & a_{33} \end{vmatrix} \tag{19}$$

$$+ \begin{vmatrix} a_{11} & a_{12} & 0 \\ a_{21} & a_{22} & 0 \\ a_{31} & a_{32} & -\lambda \end{vmatrix} + \begin{vmatrix} a_{11} & a_{12} & a_{13} \\ a_{21} & a_{22} & a_{23} \\ a_{31} & a_{32} & a_{33} \end{vmatrix}.$$

The equality of equations (18) and (19) can be verified by adding up the determinants in (19) in successive pairwise fashion. By explicit evaluation, it is seen that $\Delta(\lambda)$ is a polynomial in λ of order 3,

$$\Delta(\lambda) = -\lambda^3 + S_1\lambda^2 - S_2\lambda + S_3, \tag{20}$$

where the coefficients S_j depend on the elements of the determinant $\Delta(0)$. Denote the determinant of order j obtained from the determinant $\Delta(0)$ by deleting $3 - j$ rows and their corresponding columns, as a **principal minor** of order j. Then S_j is the sum of all principal minors of order j of the determinant $\Delta(0)$, $j = 1, 2, 3$. In particular, $S_3 = \Delta(0)$.

In the above example,

$$S_1 = |a_{11}| + |a_{22}| + |a_{33}|,$$

$$S_2 = \begin{vmatrix} a_{22} & a_{23} \\ a_{32} & a_{33} \end{vmatrix} + \begin{vmatrix} a_{11} & a_{13} \\ a_{31} & a_{33} \end{vmatrix} + \begin{vmatrix} a_{11} & a_{12} \\ a_{21} & a_{22} \end{vmatrix},$$

$$S_3 = \begin{vmatrix} a_{11} & a_{12} & a_{13} \\ a_{21} & a_{22} & a_{23} \\ a_{31} & a_{32} & a_{33} \end{vmatrix}. \tag{21}$$

The relationship between S_i and the determinant $\Delta(0)$, and its principal minors is readily generalized when the order of the determinant is n. Thus, define the secular determinant of order n as

$$\Delta(\lambda) = \begin{vmatrix} a_{11} - \lambda & a_{12} & \cdots & a_{1n} \\ a_{21} & a_{22} - \lambda & \cdots & a_{2n} \\ \cdots\cdots\cdots\cdots\cdots\cdots\cdots\cdots \\ a_{n1} & a_{n2} & \cdots & a_{nn} - \lambda \end{vmatrix}. \tag{22}$$

Then $\Delta(\lambda)$ is an nth order polynomial,

$$\Delta(\lambda) = (-\lambda)^n + S_1(-\lambda)^{n-1} + S_2(-\lambda)^{n-2} + \cdots + S_{n-1}(-\lambda) + S_n, \tag{23}$$

S_j is the sum of all principal minors of the determinant $\Delta(0)$ that are of order j, and

$$S_n = \Delta(0) = \begin{vmatrix} a_{11} & a_{12} & \cdots & a_{1n} \\ a_{21} & a_{22} & \cdots & a_{2n} \\ \cdots\cdots\cdots\cdots\cdots\cdots \\ a_{n1} & a_{n2} & \cdots & a_{nn} \end{vmatrix}. \tag{24}$$

6.   HOMOGENEOUS EQUATIONS

Suppose the b_i appearing in equation (11) are all zero. Then the equations are said to be **homogeneous** equations. It is clear that the equations then possess the solution

$$x_1 = x_2 = \cdots = x_n = 0, \tag{25}$$

which is called the **trivial solution**. Do the equations possess a **nontrivial**

solution, one for which the x_i are not all zero? The answer is they do, if and only if the determinant of the equation system is zero.

A corollary of this theorem is that a system of linear homogeneous equations in which the number of equations is less than the number of unknowns has always a nontrivial solution.

7. VECTORS

An ordered set of n numbers a_1, a_2, a_3, ..., a_n is called an **n-dimensional vector**, or simply, **vector**, and denoted by the letter **a**. It is customary to write $\mathbf{a} = \{a_i\}$. The quantity a_i is called an **element** or **component** of the vector. If the elements are arranged in a column, then **a** is called a **column-vector**,

$$\mathbf{a} = \begin{bmatrix} a_1 \\ a_2 \\ . \\ . \\ . \\ a_n \end{bmatrix}. \tag{26}$$

If the elements are arranged in a row, then **a** is called a **row-vector**,

$$\mathbf{a} = (a_1 \ a_2 \cdots a_n). \tag{27}$$

Usually we shall be concerned with column vectors.

Two vectors **a** and **b** are said to be equal, $\mathbf{a} = \mathbf{b}$, if all their components are equal,

$$a_i = b_i, \qquad i = 1, 2, \ldots, n. \tag{28}$$

The sum of two n-dimensional vectors $\mathbf{a} = \{a_i\}$ and $\mathbf{b} = \{b_i\}$ is an n-dimensional vector **c**, where

$$\mathbf{c} = \{c_i\} = \{a_i + b_i\} = \begin{bmatrix} a_1 + b_1 \\ a_2 + b_2 \\ . \\ . \\ . \\ a_n + b_n \end{bmatrix} = \begin{bmatrix} c_1 \\ c_2 \\ . \\ . \\ . \\ c_n \end{bmatrix}. \tag{29}$$

Note that addition is commutative, $\mathbf{a} + \mathbf{b} = \mathbf{b} + \mathbf{a}$. Similarly, the difference of **a** and **b** is $\mathbf{c} = \mathbf{a} - \mathbf{b} = \{a_i - b_i\}$.

Multiplication of a vector **a** by a number α is defined as

$$\mathbf{b} = \alpha\mathbf{a} = \{b_i\} = \{\alpha a_i\}. \tag{30}$$

The **scalar product** of two n-dimensional vectors **a** and **b** is a number or **scalar** written as $\mathbf{a} \cdot \mathbf{b}$, where

$$\mathbf{a} \cdot \mathbf{b} = a_1 b_1 + a_2 b_2 + \cdots + a_n b_n. \tag{31}$$

Both forms of vector multiplication are **commutative**, $\alpha \mathbf{a} = \mathbf{a}\alpha$, and $\mathbf{a} \cdot \mathbf{b} = \mathbf{b} \cdot \mathbf{a}$.

Two vectors are said to be **orthogonal**, or perpendicular to each other, if their scalar product equals zero.

8. MATRICES

A rectangular array of elements consisting of n rows and m columns where n and m are integers, is called an $n \times m$ **matrix**, or simply a **matrix**. Let the **element** of the matrix A in the ith row and jth column be denoted by A_{ij}. Then it is usual to write

$$A = \{A_{ij}\} \tag{32}$$

to represent the matrix

$$A = \begin{bmatrix} A_{11} & A_{12} & A_{13} & \cdots & A_{1m} \\ A_{21} & A_{22} & A_{23} & \cdots & A_{2m} \\ A_{31} & A_{32} & A_{33} & \cdots & A_{3m} \\ \cdots\cdots\cdots\cdots\cdots\cdots\cdots \\ A_{n1} & A_{n2} & A_{n3} & \cdots & A_{nm} \end{bmatrix}. \tag{33}$$

The set of elements A_{ij} for i fixed and $j = 1, 2, \ldots, m$, is called the ith **row** of the matrix. The set of elements A_{ij} for j fixed and $i = 1, 2, \ldots, n$, is called the jth **column** of the matrix. An $n \times n$ matrix is called a **square matrix**, or a matrix of **order** n. The mathematical utility of matrices requires the definition of equality of matrices, and the definition of the operations of addition and multiplication of matrices.

9. EQUALITY, ADDITION, AND MULTIPLICATION OF MATRICES

If A and B are both $n \times m$ matrices, then we say A and B are equal, or

$$A = B, \tag{34}$$

provided that

$$A_{ij} = B_{ij} \text{ for every value of } i \text{ and } j. \tag{35}$$

If A and B are both $n \times m$ matrices, then addition or subtraction is defined for them as follows.

$$A \pm B = C, \tag{36}$$

where C is an $n \times m$ matrix, and

$$C_{ij} = A_{ij} \pm B_{ij}, \qquad \begin{aligned} i &= 1, 2, \ldots, n, \\ j &= 1, 2, \ldots, m. \end{aligned} \qquad (37)$$

Addition is commutative: $A + B = B + A$.

If α is a number or scalar quantity, then multiplication of the matrix by the scalar α is defined as

$$B = \alpha A, \qquad (38)$$

where A and B are $n \times m$ matrices, and

$$B_{ij} = \alpha A_{ij}, \text{ for all } i \text{ and } j. \qquad (39)$$

If A is an $n \times m$ matrix, and B is an $m \times 1$ matrix, then the matrix product AB is defined, and is based on the scalar product of two vectors. The result of the multiplication is an $n \times 1$ matrix C whose elements C_{ij} are given by the expression

$$C_{ij} = A_{i1}B_{1j} + A_{i2}B_{2j} + \cdots + A_{im}B_{mj} = \sum_{k=1}^{m} A_{ik}B_{kj}, \, i = 1, 2, \ldots, n, \quad (40)$$
$$j = 1, 2, \ldots, 1.$$

Thus, C_{ij} results from the scalar product of the vector which is the ith row of A, and the vector which is the jth column of B. This rule of multiplication is sometimes called the row-by-column rule. Obviously, matrix multiplication is **not commutative** because BA is not even defined unless $n = 1$. Usually, only square matrices are encountered. Even so, if the two matrices A and B are both of order n, it is generally the case that $AB \neq BA$. Consequently, because matrix multiplication is not commutative, its order must be observed carefully. Matrix multiplication is **associative**: $ABC = (AB)C = A(BC)$. The **distributive** laws also hold true: $(A + B)C = AC + BC$, and $C(A + B) = CA + CB$.

If $\mathbf{a}$ is an n-dimensional vector, and A is a matrix of order n, then the product $A\mathbf{a}$ is again an n-dimensional vector $\mathbf{b}$, where the components of $\mathbf{b}$ are given by the expression

$$b_i = \sum_{j=1}^{n} A_{ij}a_j, \, i = 1, 2, \ldots, n. \qquad (41)$$

If, in the matrix A, the diagonal elements A_{ii} are nonzero, $A_{ii} \neq 0$, and all the off-diagonal elements are zero, $A_{ij} = 0, \, i \neq j$, then the matrix A is said to be a **diagonal** matrix. A very convenient property of diagonal matrices is that multiplication between them is commutative. Thus, if A and B are diagonal, and of the same order, then $AB = BA = C$, where C is also a diagonal matrix whose diagonal elements satisfy the relation $C_{ii} = A_{ii}B_{ii}$.

An important special diagonal matrix is the **unit matrix** I, the diagonal matrix all of whose diagonal elements are unity. It is written symbolically as

$$I = \begin{bmatrix} 1 & & & & 0 \\ & 1 & & & \\ & & 1 & & \\ & & & \ddots & \\ 0 & & & & 1 \end{bmatrix}, \tag{42}$$

where the large zeros denote the fact that all the off-diagonal elements are zero. The matrix elements of I are denoted by δ_{ij}, $I = \{\delta_{ij}\}$, where δ_{ij} is the **Kronecker delta**, defined as follows:

$$\delta_{ij} = \begin{cases} 1 \text{ if } i = j, \\ 0 \text{ if } i \neq j. \end{cases} \tag{43}$$

An important property of the unit matrix is that $IA = AI = A$, where A is any matrix. I is the only matrix with this property. Note that $I^2 = I$.

Consider a diagonal matrix A, all of whose diagonal elements are equal to the same number α. Then the result of multiplying a matrix B by A is to form a new matrix C, all of whose elements are α times the corresponding element of B: $C_{ij} = \alpha B_{ij}$. Consequently, multiplication by the diagonal matrix A is equivalent to multiplication by the scalar α, $AB = (\alpha I)B = \alpha(IB) = \alpha B$.

With the square matrix $A = \{A_{ij}\}$ is associated the determinant called the determinant of A, and written alternatively as

$$\begin{vmatrix} A_{11} & A_{12} & \cdots & A_{1n} \\ A_{21} & A_{22} & \cdots & A_{2n} \\ \cdot & \cdot & & \cdot \\ \cdot & \cdot & & \cdot \\ \cdot & \cdot & & \cdot \\ A_{n1} & A_{n2} & \cdots & A_{nn} \end{vmatrix} = |A| = \det A. \tag{44}$$

If $C = AB$, then $|C| = |A|\,|B|$, or the determinant of a product is equal to the product of the determinants.

The **inverse matrix** A^{-1} of the square matrix A is defined by the relation

$$AA^{-1} = I. \tag{45}$$

A necessary and sufficient condition for the matrix A to have an inverse is that its determinant not vanish, $|A| \neq 0$. If $|A|$ vanishes, the matrix A is said to be **singular**. If $|A| \neq 0$, A is said to be **nonsingular**. A commutes with its inverse A^{-1} so that $AA^{-1} = A^{-1}A = I$.

AUTHOR INDEX

375

SUBJECT INDEX